P9-DBL-496

RECLAIMING
OUR
FOOD

RECLAIMING
OUR
FOOD

How the Grassroots Food Movement Is Changing the Way We Eat

Tanya Denckla Cobb

FOREWORD BY GARY PAUL NABHAN

PHOTO ESSAYS BY JASON HOUSTON

Storey Publishing

For Cecil, who sacrificed, supported, and cheered.

Dedicated to all the stewards of our food.

The mission of Storey Publishing is to serve our customers by
publishing practical information that encourages
personal independence in harmony with the environment.

Edited by Nancy W. Ringer and Gwen Steege
Art direction and book design by Dan O. Williams
Text production by Gary Rosenberg
Indexed by Christine R. Lindemer, Boston Road Communications

Text © 2011 by Tanya Denckla Cobb
Photography © 2011 by Jason Houston

All rights reserved. No part of this book may be reproduced without written permission from the publisher, except by a reviewer who may quote brief passages or reproduce illustrations in a review with appropriate credits; nor may any part of this book be reproduced, stored in a retrieval system, or transmitted in any form or by any means – electronic, mechanical, photocopying, recording, or other – without written permission from the publisher.

The information in this book is true and complete to the best of our knowledge. All recommendations are made without guarantee on the part of the author or Storey Publishing. The author and publisher disclaim any liability in connection with the use of this information.

Storey books are available for special premium and promotional uses and for customized editions. For further information, please call 1-800-793-9396.

Storey Publishing
210 MASS MoCA Way
North Adams, MA 01247
www.storey.com

Printed in the United States by Versa Press
10 9 8 7 6 5 4 3 2 1

Library of Congress Cataloging-in-Publication Data

Cobb, Tanya Denckla, 1956–
  Reclaiming our food / by Tanya Denckla Cobb.
    p. cm.
  Includes bibliographical references and index.
  ISBN 978-1-60342-799-9 (pbk. : alk. paper)
  1. Vegetable gardening–United States. 2. Organic gardening–United States.
  3. Community gardens–United States. 4. Local foods–United States.
  I. Title.
SB324.3.C63 2011
641–dc23
                                        2011020050

Contents

Foreword

BY GARY PAUL NABHAN

Reclaiming Our Food comes to us at a critical moment — a watershed — in the history of American farming, food processing, and distribution. Never have so many Americans used food banks, food stamps, food pantries, and soup kitchens. Never has so much food-producing land been converted to residential or industrial uses, or its water and soil degraded to the point of no return. Never have so many farmers and ranchers thrown in the towel due to rising production costs and diminishing savings accounts that might otherwise have allowed them to invest in better land stewardship and water conservation. With fossil fuel and groundwater reserves depleted, global climate uncertainty rising, and the economic downturn continuing, it is no wonder that many of us are convinced that we will never again be able to grow, harvest, distribute, and access food as inefficiently and uncaringly as we have over the last century.

While three out of every four meals in America still come through national grocery chains, big-box stores, drive-ins, and industrial-strength cafeterias, the other quarter of America's food is now coming to us through the kinds of community-based, independently owned and operated innovations that Tanya Denckla Cobb has so elegantly described in these pages. She has taken the pulse of a fledging alternative food system that is barely out of the nursery but growing by leaps and bounds. The threefold growth of farmers' markets in the United States since 1995 — with an increase of 16 percent more markets per year and a 114 percent increase over the last decade — is but one of many indications that the grassroots food movement is truly changing the way we eat. So is the growth of community-supported agriculture projects, from 60 CSAs in 1990 to well over 3,500 today. With 850 CSAs forming during the two years immediately following the global financial crisis of 2009, perhaps these grassroots efforts were the only portion of the food economy that did see much growth.

But what Cobb so deftly demonstrates is that communities are not turning to these innovative strategies for economic reasons alone. We are motivated out of concern for the health of our children and our elders, and out of concern for the health of the land. We are not necessarily opting for "local" food to reduce food miles, fuel costs, and greenhouse gas emissions; what matters to us is that our food can be traced to human hands, hearts, and minds that brought it to us through intelligence, patience, and hard work. The gift of this book is that we see the faces and hear the voices of the many "agricultural creatives" — the innovative land stewards, food distributors, chefs, and early adopters

of "foodshed consciousness" — who are remaking our food systems at this very moment. Not only are they hoping to bring health and even wealth to rural communities and urban food dead zones, where toxic industrial foodstuffs have accumulated, but they are doing so for moral, religious, ecological, and social reasons that the commodity food giants have not yet fully embraced. Rather than trying to turn around the *Titanic* as it heads toward the perils of giant obstacles in its path, these agricultural creatives have hand-crafted lifeboats that can raft our most precious foods to safety during these tumultuous times.

More than anyone else before her, Tanya Denckla Cobb has documented just how diverse the grass-roots food movements are in terms of the participation of cultures and faiths, races and classes, and folks who live at every spot along the rural to urban continuum. This is not merely about wine and cheese tastings or sit-ins against junk food in schools. Republican, Democrat, Tea Party, and Green Party members are all engaged, sometimes bringing in different ideological perspectives, but often because they converge around the notion that we all have a right to access and eat food that truly nourishes our bodies and communities. Rather than only fighting bad food practices and policies, these people model, exemplify, and support good food practices and policies. Moreover, this book is not laden with generalities; it provides you with enough detail that you can take a strategy found to be successful in one community and adapt it so that it may be applied in your own hometown or foodshed.

Best of all, Cobb's eloquent prose and Jason Houston's compelling photo essays can help us reimagine our food system, so that each of us is inspired to be part of some "foodshed design team" that changes the way we access food so that it is more efficient and equitable, resilient and diverse, nourishing and healing. Looking at the many smiles and expressions of wonder in Houston's photos, it is also abundantly clear that this alternative food system is far more fun and rewarding than pushing a shopping cart through a supermarket and standing in line to buy tasteless, rootless, fructose-laden crud. Read this book, taste its wonders, and a fresh new world will open up to you every day you sit down at a table to break bread.

Food and Community:
Growing a Grassroots Movement

This book tells a fresh — and refreshing — story about America. While economic hardship and political strife may dominate headlines, throughout our country many of us are choosing to make a positive difference in our own backyards.

We are moving away from the impersonal, shielded, and inaccessible, toward the personal, transparent, and accessible. When it comes to food, this movement seems to have an especially powerful force. We are moving away from anywhere-grown food, anonymous farmers, and opaque production and processing methods, toward more local and regional foods, farmers we can meet, farms we can visit, gardens we can grow. We are moving away from highly processed, nutrient-poor foods with staggeringly long lists of unpronounceable ingredients, toward minimally processed, nutrient-rich foods with simple, known, trusted ingredients. In all types and sizes of communities — urban, suburban, and rural; rich, poor, and middle class; homogeneous and diverse; old and young — we are rebuilding local food systems. We are reclaiming our food.

The grassroots food movement is a broad tent encompassing a multitude of initiatives, as you'll learn from the people and projects featured in this book. Some focus on food, some on the environment, and some on community. Some concentrate on growing food, while others focus on its equitable distribution. Some are creating markets, processing centers, business incubators, and distribution networks. Others are creating community support systems through jobs training programs, community food gardens, and gleaning networks.

Why are so many people involved in the grassroots food movement? Their motivations are diverse, from something as simple as wanting to sell fresh tomatoes to local restaurants to a deeply spiritual yearning to heal the land and our relationship with it. Some people

want to encourage children to eat healthier foods, or to teach youth responsibility, job skills, and leadership. Some people are altruistic or faith-driven, inspired from within to provide healthy, fresh foods to the often forgotten or overlooked members of our society — the low-income, incarcerated, homeless, homebound, disabled, and hungry. Some people want to reduce the amount of money they spend on food, while others just want enough food to feed their family. Some are striving to prevent the loss of cultural heritage. Others want to put into action a philosophical desire to reduce isolation in our fast-paced modern life and to nurture the growth of meaningful community.

No matter the starting point, no matter how richly diverse the motivations or approaches, the stories in this book demonstrate that, over time, successful grassroots food projects ultimately converge around two central points: local food and community. A community can grow a more sustainable and resilient economy by growing its local food system, and a healthy local food system will nurture and grow community spirit. In other words, it's both food and community, stupid.

Those who brand the local food movement as elitist are more myopic than wrong, as elitists surely can be found in most any movement. But this book demonstrates that the desire to reclaim our food resonates at all levels of society and is manifesting in real, ground-level community projects that serve all levels of society.

The people portrayed in this book are their own brand of pragmatic leaders. Some are social innovators, some are shrewd businesspeople, some are technical inventors, some are brilliant systems thinkers, and some are lucky enough to be all of the above. Some have struck out on their own — as urban or rural farmers, or as restaurant owners or chefs — while others have catalyzed people or neighborhoods to come together for common cause. And all are striving to do it economically and sustainably, whether for a restored environment, for improved health, for empowerment, or for community.

As I traveled and talked with people in writing this book, I struggled to understand what might connect all of these seemingly disparate grassroots projects. Along the way, I visited farmers and ranchers on the Navajo reservation in Arizona. Here I heard about the Beauty Way, the Navajo belief in the way of harmony, balance, and relationship, and I listened to these farmers, ranchers, and elders speak passionately about wanting to restore traditional relationships of harmony and balance with land and food. Suddenly the connecting thread became apparent. Across America, the grassroots food movement seems to be arising from a common feeling that we have lost our center. Across our nation, we see spiritual restlessness, children disconnected from nature, people disconnected from each other, a proliferation of foods that fail to nourish either body or spirit, and a lack of community, neighborliness, and relationship. This book tells the story of people who are seeking to find a new center, to create meaning and purpose in their lives, to restore harmony and balance in their relationships with the land, food, and each other.

Food, as the sustenance of body, spirit, and culture, is a powerful agent of change — for better or worse. At its worst, we've learned that daily fare that is highly processed, rich in calories but poor in the broad range of nutrients needed to maintain health, benefits the corporations that produce it, not the people who eat it. Decades of this unmindful daily fare, and the resulting epidemics of obesity, diabetes, and heart disease, are now stressing our health-care system (not to mention our collective pocketbook) and creating the dismal prospect that our children may lead shorter lives than their parents.

At its best, however, our daily fare can be a powerfully positive force for individual and community healing and health. This is a core value of the grassroots food movement. The stories in this book trace the many ways in which grassroots food leaders are influencing the future sustainability of all our communities. By sharing their stories, I hope to inspire others to action, to facilitate networking, and — by offering lessons from these pioneers — make it easier for tomorrow's grassroots leaders to reclaim our food.

I: Food from Home

Supporting Backyard Gardeners

If you are a food gardener, you are a member of an increasingly popular and widespread activity in America. More than one-third of America's households are growing and serving food from their very own gardens. Why are so many people gardening? The National Gardening Association (NGA) estimates that a well-maintained food garden yields on average a $500 return, based on average market prices of produce. Though an economist might wonder whether the effort is worth the return, an NGA survey indicates that most gardeners are motivated to grow their own vegetables because they believe they will have better quality, taste, and nutrition.[1]

What isn't commonly discussed in gardening surveys are all the other possible motivations for people to grow their own vegetables, raise their own honey, pick their own berries, or gather their own eggs. The reasons offered by people I have met are many and varied — teaching children where their food comes from, de-stressing, leading a more balanced life between being indoors and outdoors, becoming more healthy by getting outside and working in the garden, even connecting with a larger spirit and meaning in life.

The stories that follow are rich in lessons. Perhaps one of the most important lessons is that, despite our growing romance with and dependence on technology, twenty-first-century communities *can* support home food gardening, and more are choosing to do so. Sometimes people just need a helping hand. Perhaps they have the space for a garden but no resources and no idea how to start a garden. Or perhaps they have the resources and desire but no personal time to start a garden. Here you'll learn how communities can provide that helping hand.

And let us not forget *fun*. It's true that gardening can be hard work, causing long-unused muscles to groan in legs, arms, back, and hands. But like basketball, bicycling, jogging, soccer, and other sports that often require people to break a sweat, gardening today has become for many people more of a hobby sport than the necessity it might have been during the World War II victory garden era. As such, it should be no surprise that food gardening is becoming more of a community social activity — with fun activities to celebrate different seasonal and garden events, as well as activities intended to educate people about where our food comes from. In this and later chapters, you'll learn how people are using food gardening as a tool for whimsical, educational, and artistic fun, while also renewing our sense of community.

Gardening Is Elitist?

Of all the rationales against growing your own food, the notion that gardening is elitist is the most fascinating. The arguments seem simple enough: lower-income people don't have the time or money to invest in gardens and animals, while those who do are simply widening the gap between the haves and have-nots. If you add up the costs of infrastructure and feed, and the time in daily chores and maintenance — presumably we can substitute any home gardening chore, whether it's raising produce, chickens, goats, or bees — the hourly wage for gardening will be 50 cents or less. One university professor of planning writes, "I would hope that we could all use our time more productively. This is even more important for low- to moderate-income families. Granted, some engage in these activities for pleasure, but again this gets us away from the purpose."[2]

The professor's argument is hard to dissect, precisely because it reflects deeply entrenched thinking about what is right or wrong, good or bad, productive or not productive. It is good to use our time productively. It is bad to be unproductive. Low wages are not productive. Pleasure may be good, but it is not productive. It's okay to raise vegetables for pleasure, as long as you don't call it a productive use of your time.

Yet it is precisely this kind of thinking — the belief that productivity can and should be measured in units of time and output — that led us to the current state of affairs that produces food for the masses in the most productive, inexpensive manner possible. When we set aside as unimportant a concern for quality, for relationships, for "meaning" in our daily food and daily activities — values considered by some to be ridiculously silly and romantic — then there is no argument. Yes, when we set these values aside, there would be little reason to

ever grow our own food. Let others do it faster, cheaper, more efficiently. And, most especially, if you rank low on the income scale, don't get caught up in the hype of thinking that you could improve the food on your table or better your life, for you would simply be wasting your energy, time, and money on nonproductive activities that will do nothing to move you out of poverty.

Still, in books such as Michael Pollan's *The Omnivore's Dilemma*, Barbara Kingsolver's *Animal, Vegetable, Miracle*, and Richard Manning's *Food's Frontier*, to name a few, others have effectively countered this argument by simply pointing to the results: a food system that is double-edged testimony to the amazing power of efficiencies — successfully feeding billions of people but at an incalculable long-term cost to our land, waters, habitat, health, communities, and spirit. These thoughts are heavy and depressive, not a pleasant way to view our abundance of food. Voltaire's Candide, when despairing over the human condition, decided in the end to cultivate his own garden. Today this is the way out and up for many: cultivating our own gardens, physically and metaphorically, in our backyards, in our neighborhoods, in our communities, in our region.

To me it seems that people who are interested in growing vegetables — or raising chickens or tending beehives — are more interested in the reality of the possible than the theory of the desirable. Instead of despairing over a system that is out of their control, people are finding respite and renewal in seeking out what they can control. And if we are to believe what people report in the many stories in this book, this pragmatic spirit is sometimes equaled, if not even exceeded, by another kind of spirit — an internal spirit, or a joy in connecting with the soil, plants, animals, world, and people around us.

Giving Gardens to People in Need
The Home Gardening Project Foundation

The Home Gardening Project Foundation, based in **southern Oregon**, got its start by providing free raised-bed vegetable gardens to those in need, from the disabled to the elderly, single-parent mothers, and large families. The HGPF now also assists community garden groups across the nation that would like to do the same.

Occasionally, you have the luck and privilege of meeting someone whose depth of passion and commitment is a point of light almost too blinding to see. The world of community food attracts people who are naturally caring, passionate, and authentically committed; after all, there's no glitz or glitter in a garden, only black gold. In only a few minutes of talking with Dan Barker, you figure he is either a crank or the real deal. And as the conversation progresses, it all comes clear: He is refreshingly direct, yes, and deservedly so, as a Vietnam veteran still suffering from post-traumatic stress disorder. But far more, Barker is a philosopher, poet, writer, artist, student of the Tao, wacky humorist, and most definitely the real deal. He is our generation's Johnny Appleseed of home gardens, responsible for thousands of free raised-bed home gardens, giving the poor, the disadvantaged, and the disenfranchised a place to grow flowers and food.

"They ask me why I do this, and I say it needs to be done," Barker writes in *Queen Jane*, a winding story of his experience with giving away home gardens. With a clarity so rare in these days of careful political speak, Barker readily discusses the inherent spirituality of his work.

> The work is religious, there are three sides to it: one involves arduous physical labor – building soil frames, wheel-barrowing soil, four gardens a day, four days a week, until the goal is reached or surpassed. A second side absorbs the incessant tales of suffering, sees the misery and despair, the rotted teeth, the heart problems, the amputations, the disfigurements of body and soul. The third side demonstrates the capacity to run a business dedicated to the allevia-

tion of that suffering. The books and the reports must be straight, so I can circle in with a garden where there wasn't one before, and with a bit of work a new occupation of time and spirit is opened.[3]

Barker started small, driven by an encounter with a gun against his head during a convenience-store robbery. The robbery — followed by stern advice from a friend to quit feeling sorry for himself and, if he wanted to change the world, just go ahead and find a way — propelled Barker to action. He had fallen in love with the language of seed catalogs. And though community gardens had taken off in the 1970s, he didn't think they catered to all people who might benefit, such as the elderly, single mothers, and the disabled. In 1983 he obtained a small $5,000 grant from the City of Portland to build 20 gardens for free for people in need.

Installing Free Vegetable Gardens

Barker's formula works. He builds two or three 5-foot-by-8-foot raised beds for each family, making them "double-highs" for people who would have trouble squatting or bending. He fills the beds with top-quality soil, builds a trellis, gives the family packets of seeds, and instructs them in succession and intensive gardening. He returns in a few weeks with plant starts, and through the summer he stops by now and then to see how they're doing and to offer mentoring. He also sends them a newsletter with tips.

"The gardens are a particular benefit to older people and for people in wheelchairs," says Barker. "These folks are often sorely impoverished, as they are paying so much for medical care. Older people are between a rock and a hard place." He says that in Portland

90 percent of the participants in the Home Gardening Project were women — usually older women who were either widowed or caring for their sick husbands, or young single mothers.

In his first year, Barker says, 95 percent of the participants were able to produce food for their own table. He says success is high with his program because participants are self-selecting and want to succeed. When he started, people were suspicious of the idea of a giveaway garden. He distributed two hundred flyers and waited. No calls came. Finally, after convincing a friend with muscular dystrophy to host the first garden, his phone started ringing. And it hasn't stopped ringing since.

The main expenses in giving people home gardens are materials for raised-bed wood frames and paying a living wage to an "executive builder," or the person who picks up the shovel. If someone wants to start a Home Gardening Project, Barker estimates it requires a minimum start-up fund of $25,000. While that might not be enough to actually pay yourself, it is at least enough to get started and pay the executive builder, who, he says, should be paid at least three times the minimum wage. With $25,000 you can put in about 25 gardens, says Barker. And it also enables you to rent a tractor for loading, buy insurance — a must, he says — and file for nonprofit status, as well as buy the lumber and soil. He says that second-year seeds can be scavenged easily from hardware store leftovers.

Barker's method of advertising by flyers and local media is part of his formula for success. Communicating to his intended clients that this will be *their* garden, once installed, and that they will need to go out and work it — water, weed, fertilize — turns around their perception of the project. "Nobody believed that anybody would be interested in this," says Barker. But instead of seeing themselves as recipients, people begin to see themselves as producers. "Now there is a physical object in their yard," he explains. "So if they are going to walk outside at all, they are going to run into it! We call it a 'significant occupation,' as they're contributing to it themselves."

How to Start Giving Away Gardens

In his booklet *Building Free Vegetable Gardens for People in Need*, which is available as a free download from the project's website (see the resources), Dan Barker writes:

A good way to find recipients is to go to the neighborhood most feared in your community, and take a look around. If the houses are small and in a state of deterioration, if the yards are barely maintained, dead cars littering the streets, old women hiding inside with their TVs, old men sitting on their porches hungry for socializing, children running unattended, young men swaggering for trouble – that's a good place to start.

Or, contact your local senior citizen organizations, Multiple Sclerosis Society chapter, housing authority, mental health services, food bank or food stamp distribution agency, or churches.

Your city government [may] have a map of disenfranchised neighborhoods. Pick one or two, visit them, and determine which neighborhood would most benefit from the service of distributing gardens. Selecting from that map will likely make your Project eligible for government sponsored self-help funding.

Announcing your intention to give away gardens can be accomplished through public service television announcements, leaflets, [and] neighborhood association newsletters. It is important to let the public know what you're up to. Gaining public support and recognition lifts the spirits of everyone in the community and further ensures the potential of your Project becoming an ongoing enterprise. The more public notice, the more money, the more money, the more gardens, the more gardens, the happier and healthier the people.

When working with neighborhood associations it is best to start in mid-January. Present your Project to their meeting and request that they include the announcement in their newsletter in time for you to receive calls beginning in late February or early March.

In his first year Barker built 21 gardens. "People were amazed by the gardens and started talking with each other," he says. In the second year Barker provided all the first-year gardeners with a start-up kit of seeds, manure, and composters. He says that 85 percent continued gardening and became better gardeners in their second year.

A Good Idea Takes Hold

Every year afterward requests for gardens increased until, in 1989, with additional unexpected funding, Barker was able to install as many as 117 gardens. Word of his work was spreading, and people in other cities began calling to seek help in starting their own similar projects. The media, including *Smithsonian* magazine, began covering his work.[4] He estimates that, thanks to hundreds of organizations in the United States and elsewhere that have taken his idea and run with it, as many as 50,000 gardens have been built.[5] Barker points to a project in Flint, Michigan, where the Flint Urban Gardening and Land Use Corporation transformed abandoned and unkempt yards into block gardens, as an example of how a community can adapt his model to its own needs for social and physical sustenance.

Barker is particularly proud to point to the successful offshoot effort of Richard Doss in Olympia, Washington. While still a student at Evergreen College, Doss wanted to interview Barker for a paper. Barker wrote back and said, "I don't write people's papers for them." But Doss ventured down anyway. When he returned to Evergreen College, Doss did write the paper, but he also was inspired to start the Kitchen Garden Project, a college-based approach to giving away gardens.

Barker explains how Doss transformed the idea into a course for college credit, with students building the gardens. Since 1993 the Kitchen Garden Project has built over 2,200 gardens for low-income people, and it continues to give away between one hundred and two hundred gardens each year. The Kitchen Garden Project follows Barker's original model by giving recipients three raised beds, a trellis, fertile soil, seeds, starts, a gardening guide, and the opportunity to work with a garden mentor. "They have a real working board, a

rarity for nonprofits, and a huge number of contributors," says Barker. "It's like the whole town is behind the project." Under Doss's direction the project has grown into a successful nonprofit organization called Garden-Raised Bounty, or GRuB, which has garnered national acclaim.

Barker is a firm believer in the therapeutic value of a garden for individuals and for the community. "A lot of people in the scarier neighborhoods are agoraphobic. They don't go outside. Giving them a garden gives them a reason to go outside, and it also helps to quiet down the neighborhood, as the bad guys don't like to be seen."

Eventually, however, the stress of working in distressed neighborhoods took its toll on Barker. "People get edgy and assume you're there for other reasons, not building gardens." He had been threatened, shoved, shot at, and insulted until in 1996 he decided that Portland was the wrong place for someone with PTSD. He still may not be able to sleep without nightmares, but the rural life in his new home in southern Oregon is a little slower and safer. After he passed the torch of the Home Gardening Project to a new director, the organization evolved into the Portland nonprofit now known as Growing Gardens, which continues his vision of installing home gardens for free. (For more on Growing Gardens, see page 17.)

Seeding Free Garden Projects

Barker is nothing if not peskily persistent. After 15 years of driving Portland's streets with a truck full of wood and soil, bags of seeds, and starts of broccoli, tomatoes, and eggplant and receiving hundreds of calls from around the nation from people inspired to start their own free garden projects, Barker realized he could multiply the providence of free home gardens by providing start-up funds to other projects. So today Barker is at work raising money and giving it away to free garden projects in places of need. With funding from the Wallace Genetics Foundation in 2003, Barker supported projects in Boston; Portland, Oregon; and Flint, Michigan.

Barker will never grow a big nonprofit organization. He is all about action at the grassroots level. He just wants to build gardens. So if someone were to drop $5 million in his lap, he told *Biography Magazine* in 2003, he knows exactly what he would do: he would distribute checks for $25,000 in two hundred cities.[6]

"We're going to be stuck in this deflationary pickle a long time," says Barker. "And we're also changing our general attitude toward what constitutes well-being in this country. Hopefully, this change will be disseminated throughout the rest of the democratic world, as well as those places where tyranny reigns." And Barker believes giving away home gardens to people in need is a form of community preventive medicine. "Gardens are a way to raise the quality of life in a way that does no harm." Nearly 30 years after he started in 1983, Barker is still trying to change the world, one garden at a time.

A Home Garden Project Checklist

- Have you identified and located a consistent soil source?
- There are five phases of work in the activity of building gardens for people in need: fund-raising, building, teaching/monitoring, fall manure distribution, and spring start-up. Who will do the actual labor, and how will that labor be organized?
- How will you identify and recruit participants, and what income criteria will you use to qualify them as recipients?
- Liability insurance is necessary to do this work, as well as unemployment insurance, worker's compensation, and truck insurance. Do you have that coverage arranged?
- What materials will you use to construct the raised beds, and what will be their dimensions?
- Compost is a low-cost material for enriching garden soil. Will you manufacture composters and distribute them, with information on how to compost? If so, what materials will you use?
- What planting and gardening techniques will you teach? How will you teach – through gardening classes, a gardening club, a newsletter, or an instructional booklet?
- If some recipients are too frail to plant for themselves, will you provide them with someone who will plant for them?
- How frequently will you monitor the gardeners? And who will do it?
- How will you evaluate the success of the gardeners?
- How will you notify the public of your accomplishment?
- Do you have ambition to continue this work beyond the first season of garden building? If so, how will you go about securing funding? Do you have a plan to become self-sustaining?
- The ethos of spending public money demands strict accounting and prudent expenditures. How will you pursue this ethos, and how will you document your expenditures?
- Do you have a board of directors? If so, what tasks are assigned to which members?

Adapted with permission from a publication of the Home Gardening Project Foundation

Cultivating Gardeners
at Home and at School
Growing Gardens

with research by Robin Proebsting

Growing Gardens' mission is to promote home-scale organic food gardening to improve nutrition, health, and self-reliance while enhancing the quality of life and the environment for individuals and communities in **Portland, Oregon**. The group works primarily with low-income populations and schools, assisting them in cultivating gardens, increasing awareness of and interest in fresh local produce, integrating gardening into classroom curricula, and offering practical courses in cooking, preserving, and other aspects of garden-related living.

Some people think of gardening as an upper-class hobby, enticing only to those who have the time and money for it. Some think that lower-income people simply can't be expected to be interested in growing their own food because they lack the land and resources or, as they often work two or more jobs, because they must also lack the time and energy to tend a garden. And if they have kids? Well, they must be too busy running between work and child-care duties. But in northeast Portland the nonprofit organization Growing Gardens, which promotes organic home food gardening, is blowing these stereotypes away. Since 1996 it has installed more than seven hundred home food gardens, and it is unable to meet the demand for home food gardens among the low-income population it serves.

"Growing Gardens is all about making sure that people have access to good, fresh food — healthy fruits and vegetables," says Debra Lippoldt, executive director. "The neat thing is that people are actually growing for themselves, and we're just helping them get the resources — whether it's learning about something, or just getting the materials."[7]

"This is the biggest stress reliever in the world," says Monique, a single mother of seven children, who lives in a small apartment in north Portland. When she started, she had no gardening experience and was interested only in growing cucumbers. Now she grows much more in her garden and has even progressed to starting her own plants indoors — a process that enchants her children. "It was instant joy," she says, "because it made me realize they weren't just going through the motions with me."[8]

"My garden is like my kids," says Isabel, a young immigrant from Mexico City who grows food for family and to share with neighbors. Growing Gardens helped Isabel and her neighbors plant a container garden, but Isabel soon got permission from her apartment building's manager to expand and plant a larger garden. She is proud that her tomatillos are not grown with chemicals like the ones she used to buy in Mexico City. She talks to her plants and says they can understand Spanish or English. When she's sad, Isabel says, the garden cheers her.[9]

"If it wasn't for Growing Gardens, this beautiful life of our flowers and plants and raspberries . . . I would give up, really and truly," says Violet, an elderly woman whose backyard is now brimming with plants. She says that she and her husband rarely shop for groceries because their garden is so productive.[10]

This is just a glimpse into the lives that Growing Gardens is changing — all by the simple act of providing home gardens. "Our vision is to inspire as many people as possible to grow their own food," says Caitlin Blethen, manager of the Youth Grow program. "Our focus is to work with low-income populations."

Through years of learning by trial and error, Growing Gardens knows that installing the garden is just the gardener's first step in a long process of learning. To increase the first-time gardener's success, and to increase the likelihood that the gardener will actually keep planting seeds for years to come, Growing Gardens has innovated a unique and deliberate safety net for its home gardeners. It is a program worthy of emulation, as it reflects an understanding that people thrive with different kinds of support and interaction.

From the moment that participants enter the Growing Gardens program, they are guided at every step of the way. First, they must make a three-year commitment to stay in the program. In return a volunteer team installs the home garden, and Growing Gardens provides a start-up kit of seeds and tools, free workshops, a monthly newsletter, and, perhaps most important, a personal mentor who will visit at least four times in the season. "I ended up with a wonderful mentor," says Monique. "We communicate all the time, and she shows up randomly, not just during the growing season."

At the end of each season, Growing Gardens surveys its home gardeners. The majority, by far, have saved a significant amount of money by growing their own fruits and vegetables, says Rodney Bender, the garden programs manager. And what's more, they also are eating better — not just because they're eating fresh produce from their own gardens, but also because they're doing more thinking about the kinds of food they eat, and what they buy when they go food shopping.

Growing Lifelong Vegetable Eaters

In addition to growing new home gardeners, Growing Gardens is also working with elementary school children in after-school garden clubs and summer garden camps through its Youth Grow program. "Connecting kids to their food sources is really valuable," says Caitlin Blethen, "because . . . they're more likely to eat what they've grown. A lot of times, they'll say, 'Grooooss, broccoli!' But when they see it on the plant, and pick it, they'll eat it. That's what we're hoping for — to create good connections between kids and fresh vegetables, and where they come from, so they can continue to grow their own food in the future."

In 2009 Growing Gardens worked with four Portland elementary schools. "We work with schools where at least 50 percent of the student body is eligible for free and reduced-price lunch," explains Debra Lippoldt. All four schools participate in the city's Schools Uniting Neighborhoods program, a community partnership program that provides resources for schools to coordinate and offer extracurricular educational, recreational, social, and health services to their students and surrounding neighborhood. Those services in part encourage the formation of active parent programs and after-school clubs.

Schools are stressed and busy places, especially resource-poor schools that serve low-income populations. Given this reality, Growing Gardens realized it would be inappropriate to try to create new demands or new structures for the gardening activities it hoped to implement. Instead, Growing Gardens piggybacked onto existing school structures: The active parent program allows Growing Gardens opportunities to meet with parents to explore their needs, answer questions, and hear their suggestions. And the after-school clubs are a perfect platform for the garden club program.

Students can sign up for different clubs three times a year, with each club lasting about eight weeks and meeting once a week for two hours after school. Begun in 2000, Growing Gardens' after-school garden clubs always include hands-on physical activities. Children plant seeds, transplant them, compost, tend worm bins, harvest and prepare vegetables, identify insects, and work as a team.

When Growing Gardens works with a school, it makes a commitment to offer the garden clubs for at least three years. The clubs are free, with parental sign-off, and each can handle up to 15 children. Some schools allow children to enroll in the garden club as many times as the child might wish, while others impose term limits. Growing Gardens' experience is that children benefit from ongoing participation in the garden club, learning more about their food, gardening, and environment, and some children have even participated over several years.

"Kids really love the garden clubs," says Lippoldt. Their success with the garden clubs has led Growing Gardens to develop other school-based programs such as summer garden camps and school gardens. More recently, they have decided to focus their summer efforts on a series of parent-child workshops — on planting seeds, planning a dinner garden, harvesting and eating vegetables, and worms.

Because many children in Youth Grow want to have a garden at home, Growing Gardens decided to synergize the connection between youth in after-school garden clubs and their parents. In 2009 Growing Gardens reserved 10 spaces in its home gardening program for parents at the schools where it was conducting after-school programs, and all 10 slots were filled. Not all parents who signed up had children in the after-school clubs, but some did. This is a creative way to connect the dots between learning in schools and learning at home, as well as between gardening for fun and gardening for fresh food. By connecting these dots Growing Gardens is weaving a complex community web to support healthy food and eating.

Training School Garden Coordinators

By early 2010 the word was out about Growing Gardens' success in working with schools, and the demand for its services from parents and teachers at other Portland schools had exploded. "We've had about 40 schools in the last year and a half come to us for help," says Lippoldt.

Unfazed, Growing Gardens turned its attention to figuring out how to meet this overwhelming demand in a way that would foster sustainable change throughout the school system. Necessity bred the invention of a unique five-day, 35-hour certificate training program for school garden coordination. People who want to become a school garden coordinator learn the ins and outs and best practices for creating quality sustainable garden-based education programs.

Lippoldt says the certificate training attracts individuals who have a relationship with the school — parents, individuals who want to work with school gardens, and teachers. So far the training has attracted folks from Portland, but it's easy to see how the certificate training may soon attract individuals from other cities and states who wish to bring the wisdom back to their own school systems.

Though the certificate training is still in its early years, Growing Gardens has a clear vision of what it will accomplish. Lippoldt hopes that graduates of the training will help schools grow and sustain their own garden-based education programs. And that, she hopes, will lead the schools to pool resources to hire part-time coordinators, who can network and support the garden-based programs throughout the entire school system.

"This is why we started the school garden certificate training," says Lippoldt. "We are a small organization, and there is more need than we can fill." In Portland the schools requesting help have a wide range of conditions — some with on-site gardens that were abandoned, some located near a community garden, some more well endowed. Growing Gardens focuses

School Garden Coordinator Certificate Training

Growing Gardens' training program focuses on the best practices for creating a sustainable garden-based education program, including:
- Organizing the school community
- Resource development
- Garden planning and implementation
- Working with youth
- Garden-based activities
- Serving garden-grown produce in the cafeteria
- Program evaluation

Bright Idea: Annual Tour de Coops

Growing Gardens premiered its first "Tour de Coops" in 2003 and has found it to be a fun and educational fund-raiser. This self-guided tour takes participants to see backyard chicken coops all over east Portland. For $15 each person buys a booklet listing the stops along the tour, and families and friends can go at their own pace. In 2009 eight hundred people participated, raising $12,000 for Growing Gardens. At the same time, the tour is a community-building event where people can meet their neighbors while talking about chickens (fore more about backyard chickens, see page 23).

its efforts on those schools serving low-income communities, but it hopes the school garden coordinator certificate training will ultimately assist all Portland schools.

Like the training offered by the American Community Gardening Association in starting sustainable community gardens (see page 69), and like the Food Project's training in starting youth-based community farms (see page 136), this certificate training emphasizes the importance of beginning with the community itself. Without community support, experts say the effort is wasted. Years of experience — successes and failures — have produced at least one consensus "best practice": without the understanding and support of the community — whether a school, neighborhood, village, or city — a community garden can't survive.

"Food really binds people together," says Blethen. "All humans need to eat and need healthy food. Gardening is a great way to provide healthy food." Growing Gardens sees itself as one piece of the bigger food-security puzzle. Its vision is to inspire anyone who wants to grow more food. With more than seven hundred home gardens, at least four school garden programs, and umpteen garden workshops to its credit, Blethen says the group has still barely made a dent. Others could argue that Growing Gardens has created a kind of tipping point, casting enough seed on fertile ground that it has already begun to multiply and naturalize on its own.

Lessons Learned
Installing Home Gardens for Low-Income Populations

Build a volunteer program to install the gardens.

In Portland, Oregon, the nonprofit Growing Gardens (page 17) builds and installs an average of 60 or more home gardens each year with a volunteer program drawing on six hundred to seven hundred community members. Garden installation is the most popular volunteer activity, says Debra Lippoldt, executive director. When Growing Gardens sends out a request for volunteers for garden installation, the slots are usually filled within three days.

This remarkable feat is testimony to the power of tapping into people's natural interest in growing and eating good food and coming together as a community, according to Lippoldt. Growing Gardens trains the volunteer crew leaders, who lead teams of 8 to 10 when installing the new gardens. On a typical spring or fall Saturday, Growing Gardens is able to install six new household gardens by sending out three crews that are each able to complete two new gardens. Volunteers also teach workshops and serve as mentors for new home gardeners.

Create double-dug beds, not raised beds.

Conventional wisdom might suggest that the easiest way to install a new home garden is with raised beds. However, Growing Gardens found that after several years gardeners would request more soil to keep the beds "raised." So instead, Growing Gardens creates gardens by double-digging in the spring and sheet-mulching and cover-cropping in the fall. Each household garden consists of two 4-foot-by-8-foot beds.

Test the soil.

Soils can vary significantly in quality and contaminants, so it's important to establish a protocol for testing the soil prior to installing a home garden. Local governments are a good source of assistance, as they often have federal funding to test for lead in the soil. Portland's Water Bureau funds Growing Gardens to test its garden sites. If soil tests suggest a potential problem (in Portland lead contamination is common), Growing Gardens will recommend gardens of raised beds lined with plastic and filled with uncontaminated soil brought in from an outside source.

Establish basic requirements for participation.

Over the years Growing Gardens has developed key requirements for participation in its home gardener program, to make sure its program continues to assist its target population. Participants:

- Must meet the same income-level requirements as SNAP (Supplemental Nutrition Assistance Program) or the federal Women, Infants and Children (WIC) program, or 185 percent of the federal poverty level. Applicants must fill out a form on which they self-report the number of people in the household and total household income, which determines their eligibility.
- Must attend the orientation, where Growing Gardens makes sure they understand that they will be fully responsible for their own garden.
- Must commit to participate for three years.
- Are encouraged to attend at least two educational workshops; in their first year they are able to attend all workshops for free and receive priority entrance.
- Are encouraged to volunteer during the growing season at Growing Gardens' events, such as its "seed sorting party."

Provide ongoing support for home gardeners.

Home gardeners may be thrilled with their newly installed gardens, but some may have no idea how to manage them. "People come at all different levels of expertise,"
says Lippoldt. Some people are already experienced gardeners and just need access to a garden, she says. Others know nothing about how to manage a garden.

To deal with these different levels of expertise, Growing Gardens has developed several methods of gardening support. Their new home gardeners:

- Receive a mentor who has committed to working with the new home gardener for at least one full season. The mentor will make four visits to the home garden during the growing season and also ideally will attend the garden installation. Growing Gardens is trying to create networks of home gardeners in the same neighborhood, so they can share seeds, recipes, and ideas and spread the community spirit. In addition to offering mentorship from program staff, they also try to link experienced gardeners with first-year gardeners.
- Receive seeds on a monthly basis for planting the next month. In the second and third year, participants receive seeds for the entire season at once.
- Receive a monthly newsletter with tips and advice.
- Are invited to a "plant distribution day," where plants grown by volunteers are distributed. First-year gardeners receive basic supplies, such as bamboo poles, tomato cages, plant starts, hoses, and even raspberries and blueberries. Second- and third-year gardeners are also welcome to take plants, after the first-year gardeners have finished.

Most unusual, Growing Gardens offers a series of "Learn and Grow" workshops, some of which are offered in Spanish. Originally the workshops were intended for people in Growing Gardens' home gardener program, but soon the program learned that others in the community wanted to attend as well. So Growing Gardens has opened the workshops to the broader community, and to ensure that money is not a barrier, they use a sliding-scale "donation" of $5 to $20. Also, participants in workshops can work in others' backyard home gardens in lieu of paying a workshop fee.[12]

Another key feature is that Growing Gardens grows its own volunteer program by encouraging home gardeners who have graduated out of the three years to volunteer as mentors, help install other gardens, grow starts, and more.

A New Breed of Farmer: Selling Gardens to City Folk

Seattle Urban Farm Company

with research by Robin Proebsting

The **Seattle** Urban Farm Company, a for-profit enterprise, provides services to home-owners and businesses in designing new gardens, developing multiyear gardening strategies, installing ready-to-go vegetable gardens, weekly garden maintenance, pest management, and the design and installation of backyard chicken coops.

As college students, Colin McCrate and Brad Halm lived at Denison University's "Homestead," what may be the nation's only student-run intentional community with a focus on ecological sustainability. Here they learned to tend chickens, goats, and gardens and to live "off the grid."

Today, as pioneers of an entirely new kind of business, McCrate and Halm are still off the grid in how they think about approaching their life and work. When McCrate founded Seattle Urban Farm Company in 2007, it was among the first in the nation to fill a new but increasingly common desire: people want to grow their own food. And though people may want to garden, and people may like the idea of gathering fresh vegetables and herbs from their gardens for home-cooked dinners with the family, who has the time for any of this?

McCrate and Halm thought they had an answer. Perhaps people could have their carrots and eat them, too, alongside their fast-paced lifestyles, if someone else designed, installed, and even maintained the garden for them. Perhaps this little bit of a helping hand would tip the scales, so that a garden could move from a desired impossibility to an easy reality.

Each had been working in various garden and farming jobs after college, and each shared a love of growing food. "Our strength was definitely knowing a lot about gardening and growing food," says Halm. Neither had any business training, however, and others suggested they go slow, creating a business plan to figure everything out before taking the plunge. McCrate wasn't having any of that. He told them, "People want gardens right now. This is the time to do it." So with a little bit of savings and some loans from family, McCrate launched the Seattle Urban Farm Company.

Building a Business around Backyard Gardens

Soon McCrate and Halm were leading the way in making dreams of a home food garden come true. Hundreds of gardens later, they are able to reflect on some of their choices and lessons. A business plan would have helped, they now agree, as they went through very difficult times, but neither regrets their choice. As the first business of its kind in Seattle, the company gained important name recognition. Besides, they say, struggling through a business plan would probably have robbed them of their enthusiasm and initiative.

The Seattle Urban Farm Company didn't need to do much marketing to be successful, says McCrate. "There wasn't evidence that people wanted it, then all of a sudden there were newspaper articles written about us every week. We were there at the right time. People in Seattle wanted it to happen."

Every urban garden project is somewhat different. Sometimes people know exactly what size garden they want and where they want it. Sometimes what they want is not realistic, given the conditions of their yard. McCrate says they may need to talk people out of setting up gardens in undesirable locations, such as very shady spots or areas that don't drain properly. People often have strong ideas about how they want their property to look and function, so Seattle Urban Farm Company helps them design a garden within their desired budget. This is not always an easy task. McCrate explains that a client may want a 3-foot-high raised bed, so she doesn't have to bend over to work the garden, but the lumber cost alone might exceed her budget. Every garden requires a conversation and careful consideration. "There is always something that can be done; it just depends on how much people want to do, and how much they want to spend," says McCrate.

The Seattle Urban Farm Company offers a very flexible menu of services, primarily in garden planning, installation, and maintenance, but the company is also there to help when unexpected issues arise. Its simplest service might be to just design the vegetable garden for a family that wants to dig and build the garden on its own. For this the company customizes the crops to the family's desires and provides a plan for what should be planted when and where. Or the company might also install the garden, including drip irrigation, plant the seeds and seedlings, and return weekly to weed and check on the watering system. Sometimes they even help the family decide how to use the harvest. At times McCrate and Halm are called in the spring to do just the spring garden cleanup and planting. Other times a family who had intended to maintain the garden themselves have forgotten what they are supposed to do and will call for help.

Supplementing with Coops

When they began, McCrate and Halm thought they could create complementary off-season work by building backyard chicken coops to supplement their urban garden installation service. Often clients wanted the company to build a customized coop from salvaged materials. This soon proved impractical, as it required significant time in salvaging the materials, designing a coop around those materials, then actually building the coop. McCrate said they learned they would need to charge a lot, and it became too expensive. Now, whether for ready-made designs or a customized design, they use a mix of environmentally responsible store-bought materials as well as some materials from local salvage companies. The company also offers a monthly cleanout of the coop, but McCrate says nobody has ever requested this service.

"Raising chickens is really easy," says McCrate. "They really change the atmosphere of your yard. They're awesome!" But he also warns that chickens are more complicated than gardens. People need to store the food properly; spilled feed can attract rats. And if the coop is not properly secured, ground and aerial predators can be a problem. McCrate and Halm believe that raising chickens can be rewarding and a good experience for children, and they plan to keep this as part of their business. Overall, however, the interest in chicken coops is small compared with the interest in gardens.

A New Breed of Farmer

Perhaps the business of installing urban gardens is ideal for a new breed of farmers. This business is for farmers who don't want to own their own farmland and who may prefer living in an urban or suburban community rather than the rural countryside. It's a business with very low start-up costs — a minimal amount of equipment and a small backyard greenhouse for starting seedlings. Also, given McCrate's description of his company's work, another departure from the mind-set of the classic rural farmer — "leave me alone so I can farm" — is that the urban farmer needs to be entrepreneurial and enjoy constantly meeting and working with new people. Becoming an urban farmer is arguably an ideal profession for someone who wants to facilitate connecting people with the land and food.

Neither Halm nor McCrate envisions staying in the city forever, and they talk about trying to find a model for small-scale farming that could be profitable and sustainable. But for now Seattle Urban Farm Company is proving good for them. "It was important for me to have my job be something I love," says Halm. "I see so many people get into jobs they don't enjoy. I think that's a poor way to live. I want a business where I can be happy and spiritually satisfied with my day-to-day work."

McCrate agrees. "I'll only move on if I find something else that feels right and that inspires me to keep working hard every day."

Lessons Learned
Running a Garden Installation Business

Use quality materials.

Purchase the best-quality materials with the best reputation, says Colin McCrate of Seattle Urban Farm Company. His company always imports soil for the garden beds it builds, and though it's important to balance cost, quality, and access to your material sources, he has learned that it's better to err on the side of quality. When his company was getting started, it bought soil from sources that were easy to access and offered a good price. But the soils were unreliable over time and sometimes resulted in customers calling them back with concerns. Though even the best sources can have inconsistencies over time, he says, they now buy soil only from the highest-quality sources.

Implement a record-keeping system.

Unlike an average landscaping job, which McCrate says may take a month, the typical backyard garden installation takes one or two days. This has significant business implications, as it means the company must be constantly doing new projects to stay busy (and solvent). "You're always scrambling to get a new project and get it done efficiently," says McCrate. Though each project is different, he says that eventually you can apply the same lessons, becoming better and more efficient.

To manage this constant scramble, Halm says, the most important thing is to establish a system. You need to have a seeding schedule, a way to keep track of tools, a system for billing, and a system for tracking the hours on different jobs. "For a while our record keeping was sporadic and haphazard," he admits, adding that,

like most other farmers he's met, he personally hates keeping records. But he's learned that all of these systems are crucial. "I've found on all the farms I've worked on, and this one, record keeping and systematizing has been crucial to keep things successful and reliable."

Diversify your customer base to include commercial businesses.

The Seattle Urban Farm Company began with a focus on installing home gardens, but after only one year they learned that it would be important to balance the short-term home garden projects, which last one or two days, with longer-term commercial projects, to reduce the constant scramble from one job to the next.

To be successful over the long term, McCrate says it will be important for his company to expand its work with restaurants and developments, because they are bigger-ticket and longer projects, they advertise their new garden, giving his company free marketing, and they also send him new customers. Chefs are an eager clientele, he says, and there is particular interest in rooftop gardens. The gardens can be a "big selling point" for the restaurants, McCrate suggests.

For future expansion McCrate wants to connect with architects and builders. He envisions a time when Seattle will have a new status quo, where every new development will automatically have a garden. If prospective projects come through — garden installations for newly constructed homes and a large-scale production farm for a high-tech company — his vision may begin to take hold.

The Freshest Eggs You've Ever Had . . .
Backyard Chickens

Small flocks of backyard chickens can be found in an ever increasing number of cities, suburbs, and rural homesteads. Communities and nonprofits across the nation are working with local governments to make backyard chickens — and the fresh, nutritious eggs they provide — a possibility for all people.

A gentle clucking may soon join your community symphony of sounds — the tweets, chirps, and coos of birds, the hum of cicadas and tree frogs, the growls and barks of dogs, the snap of doors opening and closing, the whine and whir of string trimmers and lawn mowers, the rumble of cars and trucks. And if you haven't yet heard hens clucking in your neighborhood as they scratch, scoot, and preen, then you're sure to hear people clucking plenty about their absence.

Chickens are coming back home to roost — to lay eggs, to be butchered for Sunday dinner, and to round out the family menagerie for educating children about the natural world. More and more communities throughout the United States are changing their laws to permit backyard chickens — specifically hens, not roosters. Cities that are saying "yes" to backyard poultry are of all sizes and shapes, including Boston, New York City, Los Angeles, San Francisco, Chicago, Madison (Wisconsin), Ann Arbor, Fort Collins, South Portland (Maine), Seattle, Portland (Oregon), and Everett (Oregon).

Backyard chickens have catalyzed jokes, heated arguments, and citizen campaigns, as well as concerns for social justice, public health, and sustainability. Mayor Dave Cieslewicz doesn't mind ruffling feathers with his blogs in Madison, Wisconsin, which reversed its poultry ban in 2004. "Chickens are really bringing us together as a community. For too long they've been cooped up," he wrote that year. "It's a serious issue — it's no yolk."[13] And five years later, in 2009, he blogged, "It's not just a shell game. Four chickens (the maximum allowed under the Madison ordinance) can produce two or three eggs a day. Over time that can save some real scratch. So it's good eggonomics in hard-boiled times, which is something to crow about (though not in Madison, where roosters are not allowed)."[14]

On a more serious note, backyard chickens did trigger a prolonged discussion about community food sustainability in Madison. Mayor Cieslewicz argues that local food brings the promise of improving the tables of the needy, not just those who can afford it. "I'd like to see a comprehensive city strategy that gets good quality, healthy and affordable locally produced food into the hands and the kitchens of people who really need it," he writes. "Set in the heart of some of the richest agricultural land in the world, Madison is well positioned to be a leader in urban agriculture. But our progressive tradition should also make us a leader in using those advantages to improve the lives and the health of everyone."[15]

Madison's desire for food justice, to make fresh, healthy, and, when possible, locally produced food affordable and accessible for all its residents, is not unique. These same issues and concerns are being taken up by other communities, small and large, as the pendulum that swung away from backyard fowl in the twentieth century is now swinging back. While critics may snicker and find fault with virtually every argument for backyard chickens, proponents are successfully removing legal barriers and leading communities to appreciate their previously unarticulated social, educational, and health benefits. For those who are eyeing the possibility of a local yolk, here is a quick guide to the pros and cons of backyard chickens. Decide for yourself which side of the coop you're on.

Benefits of Backyard Chickens

Mounting evidence shows that backyard chickens have multifaceted benefits for their owners.[16]

BETTER EGGS. Different studies cite different numbers showing that backyard hens lay more nutritious eggs. In general chickens allowed to live on grass and feed on insects lay eggs with less cholesterol, less saturated fat, more vitamin A, more omega-3 fatty acids, more vitamin E, and more beta-carotene than industrial-production eggs. They also just plain taste better — if you haven't had a truly farm-fresh (or backyard-fresh) egg, you haven't really had an *egg* — and they have a firmer white and more brightly colored yolk. Besides that, a good hen lays about 180 eggs a year.

HEALTHIER YARDS. If you invest in a mobile mini coop on wheels to "pasture" your chickens by moving the coop around the yard, your chickens will scratch the soil, building tilth while also cleaning your yard of ticks, grubs, earwigs, and other bugs, which add protein to their diet (and produce that better egg!).

ENRICHED COMPOST. Three small hens are said to generate less waste than one average dog. Chicken litter is an excellent source of nitrogen and can be composted, creating a rich soil amendment for the garden.

IMPROVED SOCIAL NETWORKING. Backyard chickens are a never-ending source of conversations, according to chicken enthusiast Robin Ripley, and she vows that her chickens have improved her social network, as people are always interested in and curious about them.

QUIET HOME ENTERTAINMENT. When "properly socialized," says Ripley, chickens can even do tricks. She writes, "My chickens will come running from the other side of our property when they hear me call 'Where are my chickens?' My hens will jump on my lap and let me pet them."

EDUCATION. Some people argue that chickens offer an ideal opportunity for teaching children about food as well as responsibility, as children can easily assist with daily feeding and watering, cleaning the coop, inspecting the chickens to make sure they're healthy, and, of course, gathering eggs.

Arguments against Backyard Chickens

Most of the reasons people — and town and city administrators — argue *against* backyard chickens cluster around a few specific fears.[17]

SMELL. Chickens themselves do not smell, but chicken manure in large quantities can smell, which is why the industrial-size poultry houses have gained a reputation for pungent ammonia-like odors. Backyard chickens, however, do not produce quantities of manure. Three hens will produce less manure than an average dog.

NOISE. Hens do not crow; only roosters do. Hens sometimes cluck when they lay an egg. At night they move inside the coop and perch in complete silence until morning.

PUBLIC HEALTH THREAT. Unlike dogs and cats, chickens cannot spread parasites or pathogens to humans. A major concern has been whether backyard chickens could be a vector for transmitting avian influenza to humans. There is general consensus that the kind of avian influenza that is contagious to humans has not been found in North America. Bird flu is spread by contact with the contaminated feces of wild birds, primarily migratory waterfowl. Unlike rural farm birds, which might commingle with migratory birds or drink from a shared pond, backyard chickens are kept in an enclosed pen where contact with migratory waterfowl is unlikely. As the sustainable agriculture advocacy group GRAIN concluded in 2006, "When it comes to bird flu, diverse small-scale poultry farming is the solution, not the problem."[18] *Consumer Reports* echoed this same conclusion when it wrote that salmonella infection is more likely to result from "cramped confines of factory farms than in free range, backyard chicken runs."[19]

PEST ANIMALS (COYOTES, HAWKS, RACCOONS, AND RODENTS). If chicken feed is stored properly, it will not attract wildlife. If chickens are raised in an enclosed pen, they are not susceptible to predators. And chickens will actually eat nuisance insects such as mosquitoes, ticks, slugs, and fly larvae.

Basic Requirements for Backyard Chickens

Humane treatment of farm animals is commonly defined by the "five freedoms," as developed by the Farm Animal Welfare Council, an advisory body to the UK government. These include:

- Freedom from thirst, hunger, and malnutrition – by ready access to fresh water and a diet to maintain full health and vigor
- Freedom from discomfort – by providing a suitable environment including shelter and a comfortable resting area
- Freedom from pain, injury, and disease – by prevention or rapid diagnosis and treatment
- Freedom to express normal behavior – by providing sufficient space, proper facilities, and company of the animals' own kind
- Freedom from fear and distress – by ensuring conditions that avoid mental suffering

To enjoy the Five Freedoms, hens need shelter, food, water, adequate space, environmental conditions (such as adequate ventilation and light) conducive to good health, and the opportunity to socialize and engage in fundamental behaviors, which for them include:

- Scratching (foraging by scraping the ground with their claws)

- Roosting (resting on a stick or branch)
- Dustbathing (thrashing around in the dirt to clean feathers and remove parasites)

For shelter and protection from predators, hens need:

- An enclosed house, with a locking door, which is known as a coop. Coops should contain a nest box, in which hens will lay their eggs, and one or more perches per bird.
- Access to the outdoors, either by free ranging or by use of an enclosed outdoor space that allows them ground on which to scratch and peck. For hens without access to bare earth, a dust bath, made of any combination of sand, soil, ash, food-grade diatomaceous earth (to control parasites), or other similar material, should be provided.

Another recommendation with humane implications is a restriction on hens younger than four months. As well as reducing the number of unexpected roosters, this provision is intended to reduce impulse purchasing of chicks and subsequent abandonment of no-longer cute-and-fuzzy hens.

Adapted from "Guidelines for Keeping of Backyard Hens" (March 2010), a study by the City of Vancouver to establish guidelines for backyard hen owners

Legalizing Livestock in Urban Environments

The Goat Justice League

with research by Robin Proebsting

The Goat Justice League was founded by homeowner Jennie Grant in an effort to legalize backyard goats in **Seattle, Washington**. Having won that battle, the league continues to advocate for goats in city yards, offering education and support for homeowners in Seattle and beyond.

Snowflake and Maple can thank a little sibling rivalry, as well as a progressive city council, for their cushy backyard pen with a spectacular view overlooking Seattle's Lake Washington. Several years after Seattle passed an ordinance allowing goats in the city, Snowflake is the reigning founding member of an elite herd of urban miniature goats that call Seattle home. It all started because Jennie Grant, sometimes called the godmother of urban goats, and founder of Seattle's Goat Justice League, felt just a wee bit of competition with her sister, who lived in California and, like Jennie, was raising backyard chickens. How could she outdo her sister?

Years later, with two goats that provide her family with milk, chèvre, and mozzarella cheese, and six chickens that put eggs on the table, Grant has no regrets. Sometimes she yearns for a break from the daily routine of milking for half an hour every morning and evening and the cleaning and feeding chores. "But there's something about it that's appealing," she says. "Some people might not want to make this commitment. Raising goats will probably never be a widespread, popular thing — because most people don't have time. You have to do it because it's more important than the milk."

Campaigning for City Goats

Goats in the city? Grant can recite just about every objection that people can dream up. Goats smell, make too much noise, and are an escape risk — imagine a goat on the loose! She is ready with answers. It is the male buck goat that smells, not the female dairy goat and neutered male goats. Goat urine is less odorous than cat urine, and goat manure is small, compact pellets, easy to remove and compost. In terms of noise, Grant says goats normally don't bleat much, and their bleats are usually not loud. In fact, when they feel threatened, instead of bleating more, goats generally become very still and quiet. As for the escape risk, as long as the goats have a secure pen, there's no fear of them getting out, like a cat slithering through the fence or a dog digging out under.

Still, Grant did the most important thing of all: when she decided to raise a couple of miniature goats, she asked her neighbors if they might have objections. Thankfully, none did. But it wasn't too long before an objection quite unforeseen was raised. A distant neighbor pointed to Snowflake and Brownie as the perpetrators of a serious illness contracted by her daughter. Grant was horrified and immediately took in samples of her goats' blood for testing. Though the lab reports were clean, the damage had been done: Grant learned that her goats were illegal. Classified as farm animals, the goats required a minimum of 20,000 square feet — over four times the size of Grant's small city lot.

Grant went into action, launching a campaign to have the city legalize miniature goats. She recruited nearly a thousand allies who signed a petition, along with Richard Conlin, a neighbor and city councilor sympathetic to her cause. Conlin's office "checked with the Health Department and Seattle Animal Control, and found that there are no significant issues associated with keeping miniature goats in the City."[20] Seattle had already earned a reputation for being a pioneer in matters of local food. So it wasn't a surprise when the council vote for the new ordinance was unanimous. The law classifies miniature goats as small animals rather than farm animals, and the new licensing requirement treats them like dogs, cats, and potbellied pigs.

The city now permits three goats per city lot (with no minimum site size), and some city departments have hired goats to clear blackberry brambles. "This is part of our idea that sustainability involves both the large and the small acts," Councilman Conlin told the *Seattle Post-Intelligencer*. "[The goat ordinance] doesn't apply to a whole lot of people, but there are a significant number of people who are interested in it."[21]

Legalization: A Three-Part Strategy

Grant says that some cities, such as Oregon's Portland and Everett, never got around to making goats illegal. But most cities do have ordinances prohibiting farm animals. Grant says she gets a steady stream of letters from people elsewhere in the nation who want to legalize goats. She suggests three strategies.[22] The first, and most powerful, is to make the personal liberty argument, avoiding issues of environmental or animal rights, which she says will cause people to tune

you out. In Grant's view it is simply about making sure that city laws don't prevent people from doing what they want, so long as it doesn't harm anyone else. And there is very little anyone could object to, as dairy goats do not smell, will keep themselves clean if given half a chance, and even have interesting personalities, according to Grant.

The next strategy is to argue that making classification distinctions between pets and farm animals is culturally biased. "Because they are raised to be eaten, never named, never loved, and always treated as a commodity and not as a fellow creature, we deem eating these 'farm animals' as okay. We also make sure, as does the factory farming industry, that we never come face to face with these animals in any form until they are shrink-wrapped beyond recognition at the grocery store," Grant writes. So which is worse, she goes on to ask — to treat an animal with kindness and then eat it, or to let others treat an animal badly and then eat it? Citing cultural studies of relationships between humans and their animals, she concludes that the industrialization of meat production has shaped the hard — and culturally based — distinction between a pet and a farm animal.

The third strategy for legalizing urban goats is to argue that the reasons "farm animals" were banned from cities in the first place had less to do with sanitation than with development priorities. As cities became built up, they took over the vacant lots that had once been used as pasture and passed laws

Key Facts about Goats

- A miniature dairy goat weighs less than 100 pounds.
- Goats need companionship, so they are always raised in pairs.
- One pair requires little space, about 20 by 20 feet.
- One pair of goats, each bred in alternate years to ensure continual milk production, will produce on average between 1 quart (in winter) and 1 gallon (in summer) of milk per day.
- Goats like to climb and do best with some vertical spaces or stairs.

- Goats require shelter from the weather, because they despise getting wet.
- Goats are browsers, meaning they will eat mostly nongrassy species such as brambles, shrubs, and small trees. They will eat grass if that is all that is offered. Goats love nettles, thistles, dandelions, chickweed, leafy spurge, greasewood, and poison ivy. They enjoy bitter-flavored plants that other ruminants purposely avoid.

banishing livestock, which were said to detract from their "sophisticated" city appeal. But such laws, Grant argues, were directed at cows and other large grazing animals, rather than goats (which will do quite well in a small yard), and they were simply part of a long chain of events that took us farther away from our food.

These are heady arguments that might appeal most to political historians and philosophers. I suspect Grant was successful less for these arguments than for the supporting constituent passion carrying her forward. If there is one thing that will get an elected official's attention, it is constituent passion. Humor also played a role, says Grant, as the idea of legalizing goats and the name of her organization, the Goat Justice League, made people smile.

Now Seattle boasts more than 36 backyard goat owners, bringing the urban herd to more than 100 strong. Interest in raising backyard goats continues to be strong, and Grant's "City Goats 101" workshop for Seattle Tilth, a local gardening and conservation organization, usually fills up. She warns that attending the workshop can be risky. "My friend Didi came to it, sure that it would convince her to forget the idea of getting goats, but the reverse happened."[23]

In Seattle, where blackberries are so prevalent that they are almost invasive, Grant forages as much wild blackberry as possible for her goats. Still, there are significant costs to raising goats; in 2009 Grant spent about $85 a month to purchase feed for her two goats and six chickens. But she was able to sell three young goats for $250 each, offsetting the cost of the feed.

Grant sees a pastoral future for Seattle, populated with small animals. "We would be a really charming city if we were a place people could keep minifarms with chickens, goats, a vegetable garden, and fruit trees," she told the *Seattle Times*.[24]

Enjoying Goat Milk

With just two dairy goats, Grant gets all the milk she needs for her family, plus enough to make chèvre and mozzarella, and more to give away. "It's good on cereal," says Grant. And perhaps the best testimonial to its taste is that her preteen son drinks four glasses of the goat milk each day.

Though Grant says goats can usually be safely bred every year, she breeds just one of her goats each year. To find a stud buck, Grant suggests contacting your local Cooperative Extension office or networking at a county fair. Some goat breeds will lactate for up to two years after giving birth, but Grant tends to play it by ear, sometimes letting a goat go dry after a year, sometimes milking for the full two years, depending on each goat's situation. For her a key factor is not being overrun with goat milk during the high-lactation summer months. "I don't want to spend my entire summer making cheese," she says.

A goat's lactation reaches peak production eight weeks after she gives birth, before her kids are weaned. After this, the goat produces on average 1 gallon of milk per day, though Grant says that production can vary a *lot*, depending on the goat and the season. In the deepest part of winter, for example, a goat's production will decline to an average of just 1 quart per day. Summer, when milk yield is up, is Grant's season to make chèvre and mozzarella, with 1 gallon yielding about 3 cups of chèvre, and she makes enough to stock up her freezer and share with friends and neighbors. "I think about the cost of a small packet of chèvre at the grocery store and wince," she says. For her chèvre has become an everyday commodity that she can slather on toast or use in quiche and other baked dishes with impunity. That small luxury alone would likely convince some people to add a couple of goats to their family menagerie.

Goat milk offers multiple nutritional benefits: It contains high levels of CLA (conjugated linoleic acid), which is an anticarcinogen, as well as beta-carotene and vitamins A and E. It provides the ideal balance of omega-3 and omega-6 fatty acids.[25] Compared with cow milk, goat milk has smaller fat molecules, which makes it easier to digest, and it contains 13 percent more calcium, 25 percent more vitamin B_6, 47 percent more vitamin A, 134 percent more potassium, 27 percent more selenium, three times more niacin, and four times more copper. One drawback, however, is that goat milk contains less than 10 percent of the amount of folic acid found in cow milk, so it must be supplemented with folic acid when used in formula or as a milk substitute for infants and toddlers.[26]

Brush-and-Bramble Goats?

Goats are also becoming more common partners in urban and rural development, with rent-a-goat operations popping up across the country. Goats are said to have a low impact on the environment due to their cloven hooves. Because they are natural climbers and love steep slopes, uneven terrain, and brush, they may be a cost-effective and more sustainable replacement for machinery and herbicides in controlling brush. Goat Patrol in North Carolina specifically advertises its goats for eradicating invasive species of poison ivy, honeysuckle, kudzu, and out-of-control privet hedges. "Goats are becoming very popular with construction companies," writes Richard Graham for the TreeHugger website. "They are finding that it's a win/win for everyone. Many companies use the goats to clear brush so surveys can be done. They have also found it can be very cost-effective when considering how much it is to clear the land, then haul the debris and dispose of it. Plus, permits are not needed for the goats to clear which saves time and money. And it also improves communities by bringing people together to see the goats, not to mention the lack of noise that heavy machinery generates."[27]

In an interview with "Living the Country Life Radio," Tammy Dunakin, owner of Rent-A-Ruminant in Vashon, Washington, says, "It generally takes 60 goats about three to five days to clear about a quarter of an acre — which is 10,000 square feet — and that's relatively dense vegetation. Sometimes it can take a little longer, sometimes it takes a little less, it just depends. If it's really steep, that can slow it down, or if it's really dense and there's a lot of debris that they have to get around."[28]

But goats are not always the boon they are made out to be, argues Dr. Jon Gelbard of Conservation Initiative, an international nonprofit working to preserve biodiversity and increase global sustainability. Goats might be helpful in clearing brush and weeds, he

---

Lessons Learned
Raising Livestock in the City

Get your neighbors' permission.

When Jennie Grant, founder of the Goat Justice League, decided that she wanted to raise two goats at her Seattle home, the first thing she did was to ask for her neighbors' support. This is important for maintaining neighborly relations, and it also provides an opportunity to educate people about goats by answering any questions they may have. Of course, this does not mean there won't be objections from some quarter, but reaching out in goodwill is an important step to easing neighbors' concerns.

Have a backup.

It's helpful to recruit a nearby friend or neighbor who is willing to share milking responsibilities as well as the abundance of milk, says Grant. There are times when you will want or need to leave town, and it's important to train someone as a backup who can step in — and enjoy the rewards of milk — while you're gone.

Seek assistance and advice from community leaders.

If you're not sure what your community allows in terms of livestock, check with your local department of planning. Equally important, meet with your locally elected representatives to share your interest and intention and to seek their support. When you meet with them, be prepared to answer their questions and to educate them about livestock. If you are seeking to change your local community laws, seek their advice for appropriate strategies to change the law and ask for their support. Grant also says that if you want to raise goats, it doesn't hurt to bring them freshly made chèvre, so they can understand one of the benefits of backyard goats.

writes, or their grazing might actually favor weed invasion. The impacts of clearing by goats may depend on the native plant species, which are able to tolerate animal grazing east of the Rockies and less able to tolerate heavy grazing west of the Rockies. The author cites an unofficial study from his own front yard: "My landlord outside of Portland once stopped mowing our nice green lawn and brought in sheep and goats to take care of it. She divided the lawn (two adjacent 1-acre home plots) into three pastures. Within a month or two, after she rotated the animals out of each 'pasture,' what had been a nice green lawn turned into a mess of thistles and other weeds that had never been there."[29]

Gelbard's suggestion that goats are poor lawn mowers may be correct, says Grant, but unfair. She cites a Greek expression, "Goats look up, sheep look down." Goats are browsers of trees and brush, while sheep graze on grass. Lawn mowing is not their forte.

What does all of this have to do with food? As we find more uses for goats — clearing debris, brush, overgrown sites, and invasive species like kudzu — goats also become more economically attractive to raise in and near communities, and it is only a short step from there to goat becoming a more mainstream part of our diet. According to an agricultural specialist at the Cornell Cooperative Extension, "U.S. goat numbers are increasing . . . with meat-goat production being the fastest growing livestock industry in the country today. Ethnic consumers are the backbone of the meat-goat industry in the U.S. Demand for goat meat continues to increase as the population in the U.S. becomes more ethnically diverse."[30]

So I'm placing my bet now: by 2020 brush-and-bramble-raised goat meat will be like the grass-fed beef of today. Only far better. Goat meat is an anti-obesity nutritionist's dream come true, with the same amount of protein but less fat than even chicken. For a 3-ounce portion, the USDA reports that goat has only 2.6 grams of fat, compared to 6.3 grams in chicken and 7.9 grams in beef. For saturated fat goat wins again, with only 0.79 gram, compared with 1.7 in chicken and 3.0 in beef.[31] Goat is prominent in the cooking of the Caribbean, Southeast Asia, and other tropical countries, and as American palates expand into these cuisines, goat is sure to follow.

Why not raise brush-and-bramble meat goats in rotations through urban landscapes that need clearing? Grant warns that one obstacle to this idea is that many ornamental plants are poisonous to goats, and laurel is especially deadly. But she likes the idea. And if Grant likes the idea, then maybe the idea has a future. I'm no expert, but if there aren't other major, unforeseen obstacles to incorporating goats into different aspects of community life, then community goats may be here to stay.

Building a Buzz across the Nation
Community and Home Beekeeping

coauthored by Coogan Brennan

The grassroots beekeeping movement is building momentum as an ever increasing number of communities begin to see the benefits of keeping bees, not just for their honey but also for their essential pollination services. Home beekeepers, too, are growing in number, in rural, suburban, and even urban locales.

When the rooftop of Chicago's city hall started buzzing with beehives in 2003, urban beekeeping had reached new heights — literally. The impetus for the project came from Mayor Richard Daley, who was inspired by a trip in 2000 to Germany, where rooftop gardening and apiculture have been a more common sight for years. The two rooftop beehives were installed by local beekeepers Stephanie Averill and Michael Thompson. Thompson went on in 2004 to found the Chicago Honey Co-op, a beekeeping operation that, with 60-plus hives tucked around the city, strives for sustainability, runs a honey CSA, and gives jobs to underemployed people. Thompson and his co-op crew collect honey from the city hall rooftop hives twice a year, in early September and again in October, harvesting approximately 200 pounds of honey per hive. All proceeds from the sale of this honey benefit Chicago Cultural Center projects.[32]

Chicago is hardly alone in its beekeeping fervor. While its city hall may be one of the first of government-sponsored urban apiaries, beekeeping is becoming more common throughout North American cities. In 2010, as part of the drive to become the "greenest city in the world," the city of Vancouver installed a demonstration hive on its city hall roof. Its official beekeeper, Allen Garr, in addition to tending hives all over the city, also looks after hives at the Vancouver Convention Centre, where a nearly 6-acre green roof, the largest in Canada, is home to four hives and about 60,000 bees.[33]

Another effort in Vancouver by a volunteer organization named Environmental Youth Alliance (EYA) launched a $90,000 project to increase the city's bee population. By 2010 the EYA had distributed 100 small bee "condos" — essentially stand-alone bee frames housing small colonies of 36 bees each — to homeowners willing to garden organically, ensure an abundance of flowering plants, and report on the status of their colonies. And in parks and public areas, the EYA placed 53 larger condos — some housing 72 bees each — and a few megacondos housing 720 bees.[34]

The grassroots beekeeping movement is finding creative new ways to expand its ranks. In Pittsburgh a group named Burgh Bees, founded in 2006 by an apiarist who previously worked with the Chicago Honey Co-op, offers beekeeping classes to city residents, complete with urban apiaries for hands-on lessons. The group is also working to create a cooperative urban apiary where, like a community garden, beekeepers can tend their own hives as they would an individual plot.

And 2010 brought a landmark victory for urban beekeepers when New York City finally removed honeybees from its health code's register of "venomous insects." Now, instead of slinking around to tend beehives at unnamed, secret rooftops and gardens, New Yorkers can tend their bees legally. And once beekeeping became legalized, participation in the New York City Beekeepers Association's course on urban beekeeping more than tripled in the space of just a few months.[35]

Do Bees Belong in the City?

While beekeeping might seem more fitting for rural locations, there are reasons that apiaries are a growing movement in cities. Will Allen of Growing Power (see page 74), among others, believes that more honey can be produced by an urban than a rural apiary.[36] One reason this might be possible is that pavement and other impervious surfaces in urban areas create a "heat island" effect of warmer temperatures, which provides bees a longer season for production. Even in Chicago, renowned for its chilly winters, Lake Michigan creates different microclimates across the city, so the same types of trees and plants bloom at different times, thereby extending the growing season three weeks longer than outside the city. Beekeeper Stephanie Averill says that this longer season is great for bees: "They can get a lot of work done in those extra three weeks."[37]

Some suggest that cities may be a good home for bees because pesticide use is less frequent than in the country or suburbs. Others point out that city honey is best for city residents, since it contains local pollen, which is thought to be a natural inoculant for people with seasonal allergies.

Urban beekeeping also has invigorated a different demographic than the traditional beekeeper, attracting professionals who see beekeeping as an exciting new hobby. They set up hives atop brownstones and apartment buildings and in vacant lots — essentially, wherever space is available. Their enthusiasm can be sensed in the number of blogs and websites devoted to urban beekeeping that have emerged.

Commercial Beekeeping Is Big Business

Without honeybees our food supply would be in dire straits. The American Beekeeping Federation says that about one-third of all our food — most vegetables, fruits, and berries — is directly or indirectly derived from honeybee pollination.[38] The USDA estimates that honeybees are responsible for the annual production of food crops worth $14.6 billion; in California, it reports, "the almond crop alone uses 1.3 million colonies of bees, approximately one half of all honey bees in the United States, and this need is projected to grow to 1.5 million colonies by 2010."[39] And in terms of honey alone, the USDA reports that producers with five or more colonies were managing in 2008 about 2.3 million colonies that yielded about 164 million pounds of honey. But production is declining. Just one year later, in 2009, a greater number of colonies (about 2.5 million) yielded *less* honey (only 144 million pounds).[40]

Much of this honey production is accomplished by large commercial migratory beekeepers who travel with their hives, exposing their bees to many different environments to maximize their efficiency during the year, rather than allowing their hives to go dormant in the off-season. Migratory beekeepers are often paid for their pollination services, with large agricultural enterprises arranging for the hives to arrive at their fields when the plants are flowering, to ensure pollination and fruit development.

Bees Colonies Are Threatened

Beekeeping is just one more part of our food system that has consolidated over the last few decades. And now, just as large-scale meat and vegetable production is facing outbreaks of bacterial contamination, large-scale commercial beekeeping is facing its own threat: colony collapse disorder (CCD), in which a hive loses all or most of its adult bees, for as yet unknown reasons. The hive often still has a live queen, immature bees, and honey. The missing bees aren't dead in the hive; they're simply gone.

Beginning in 2006, colony collapse disorder began to devastate honeybee colonies around the world, with commercial beekeepers reporting losses of 30 to 90 percent of their hives.[41] In 2009 the British Columbia Honey Producers Association reported that approximately nine hundred commercial migratory beekeeping operations were left in the United States. And as the CCD epidemic entered a third year, the association says, "The worry is that these businesses may too collapse. Peoples' livelihoods are at risk, as is the unique skill set of these specialized farmers. Honey bee pollination is imperative for sustaining the world's food supply."[42] In the same vein the Mid-Atlantic

Apiculture Research and Extension Consortium reports that, as of February 2007, "many of the beekeepers reporting heavy losses associated with CCD are large commercial migratory beekeepers, some of whom have lost 50–90% of their colonies. Surviving colonies are often so weak that they are not viable pollinating or honey producing units."[43] While the cause of CCD is still unknown, numerous contributors have been suggested, including increased use of pesticides, a pathogen, infection by the parasitic varroa mite, poor nutrition, and migratory stress.

Grassroots Beekeeping Can Help

Perhaps one of the best arguments in support of the expanding grassroots apiary movement is the following from the USDA Agricultural Research Service: "The number of managed honey bee colonies has dropped from 5 million in the 1940s to only 2.5 million today. At the same time, the call for hives to supply pollination service has continued to climb. This means honey bee colonies are trucked farther and more often than ever before."[44]

Pollination is an unsung but critical cornerstone of a successful community food system. As managed colonies decline, communities seeking to foster a resilient food system would be wise not to take for granted the honeybee's natural pollination "service." Indeed, a community-supported, thriving grassroots beekeeper network is an ideal foundation for every community food system, providing pollination for local farmers, community gardens, orchards, and urban farms, as well as a local source of liquid gold.

Yard Farming: The New Victory Garden

Food is now on the planning agenda, and on the minds, of Americans as never before. More and more are interested in organic foods, and interest in buying local food has blossomed into a true movement. And there is much on the horizon to suggest the need to think even more about how and where our food is grown and a real need to rethink our food systems – we are increasingly concerned about its cost, the security of its supply, and the ability to grow sufficient amounts for a world population poised to grow to more than 9 billion.

What is afoot is no less than a revolution in food production. What we need is to replace our model of a cheap, intensive, globalized food production system with one where much more is grown at home, locally, and regionally; with fewer fossil fuel inputs; with environmental stewardship and carbon sequestration as central goals. What we need is a profoundly more resilient community-based food system.

What are the elements of a new model of community food resilience? Nurturing a healthy community of farmers is a key step, and of course conserving and protecting the productive farmland in and around cities is also key. Finding new ways to connect local farmers with local consumers is part of the task, too: CSAs, metropolitan buying clubs, farmers' markets, and stocking local food in local markets and grocery stores, restaurants, and other venues. This shift can result in renewed appreciation for the places in which we live. The Rain City Grill, a restaurant in Vancouver, British Columbia, now offers something it calls the 100-mile tasting menu, which features a story of the food offered – where and how and by whom the food (and wine) was produced – encouraging customers to savor the place and landscape as much as the food itself. An outgrowth of the 100-mile diet begun by Alisa Smith and J. B. Mackinnon (eating only food sourced from within a 100-mile radius of their home, as described in their entertaining book, *Plenty*), it has helped to trigger a new sensibility about local and regional food.

But perhaps our first line of attack ought to be not the 100-mile diet but the 100-foot diet, the promise of growing much of what we need in our neighborhoods and in and around our homes. Perhaps we should return to the notion of the victory garden and work to incorporate much more production of food in and around our neighborhoods and communities. After all, in the early part of the last century victory gardens were extremely common; at the height of their production in 1944 they were estimated to provide some 40 percent of the vegetables and fruit consumed by Americans. To be sure, the idea of edible landscaping has been around a long time and has been steadily applied in urban design and landscape architecture, but what perhaps is needed is a profound rethinking of the American home. Could we reimagine our conventional urban and suburban neighborhoods and living environments as opportunities to grow food?

One of the more interesting recent examples of the idea of yard farming can be found in a business called Community Roots, the brainchild of Kipp Nash, which has taken root in Boulder, Colorado. Kipp has assembled a collection of yards in and around his neighborhood and is intensively cultivating these suburban spaces. The homeowner sometimes helps, but for the most part Kipp and his volunteers do the actual farming. In return, the homeowner gets to pick and eat some of the produce. Most of the production, however, is sold through Kipp's CSA and the Boulder farmers' market. The amount of food produced on these small spaces is impressive indeed, and Kipp gets multiple harvests per bed.

The potential fruit production of a typical suburban lot is tremendous, and the new victory garden is capable of producing much more than most imagine. This is dramatically demonstrated by Greg Peterson, who operates what he calls the Urban Farm on a third of an acre in Phoenix, Arizona. Greg has now planted more than 70 fruit trees on the Urban Farm, which doubles as the site for fruit cultivation classes, fruit tree sales and distribution, and a host of other urban agricultural endeavors (there are chickens and vegetables being raised on the site as well). His secret to fitting in so many fruit trees: he plants smaller varieties, and he creatively utilizes the edges of the lot (planting fruit trees becomes a point of conversation and friendship building with his adjoining neighbors).

What the Phoenix and Boulder stories suggest is a profoundly different attitude about the conventional American home — that this vision, this quintessential American Dream, needs revising and expanding to match the new challenges Americans face. New priorities could emerge in the design, building, and marketing of homes. Perhaps homeowners could be invited by builders to think about food production in the process of building a house. In addition to the customary things that new homeowners are offered — the usual choices about paint colors and countertops and interior layout — homeowners could be offered choices among food-production packages: Do you want the blueberries/apple tree/peach tree combo, or are your food tastes more exotic (you'd rather go for the currants/figs/paw-paws combo)? And where would you prefer to do your vegetable gardening: in front-yard row crops, in backyard raised-bed planters, or perhaps even in pots on the balcony or rooftop? And will small greenhouses, root cellars, and built-in canning closets become standard home features? As the cost and availability of food continue to rise in importance, these sorts of questions will be increasingly common.

As with many changes in lifestyles, we may need some helpful coaching about how and when and where food could be grown around our home. There are already commercial companies that will plant, tend, and harvest food in and around your home, and there are various degrees to which residents could be involved, perhaps viewing as a kind of apprenticeship learning how to plant, tend, and harvest the urban or suburban garden. Community classes on freezing and canning, organic gardening, and vermiculture (worm farming) could function as neighborhood socials or block parties.

There are obstacles, to be sure. Some are legal (typical limits placed on chickens and farm animals), while others are aesthetic and perceptual (lawns are for growing grass, not food), historical (not many of us know how to grow food anymore), or posed by a harried American lifestyle. The downturn of the economy and the growing interest in local food, though, might suggest that never has there been a better time to rediscover the home victory garden.

– by Tim Beatley

Supporting Backyard Foragers
ForageSF

ForageSF, based in **San Francisco**, is a wild-foods advocacy group whose mission is "to connect Bay Area dwellers with the wild foods all around them." The group offers wild-foods harvesting tours and classes, a community-supported foraging (CSF) monthly basket, and The Underground Market, a private, members-only club where foragers and at-home food producers can sell their wares.

Foraging is a largely forgotten art, particularly in urban and suburban environments. In 2008 entrepreneur Iso Rabins founded ForageSF to reestablish foraging and wild foods as staples of the San Francisco Bay Area food system.

Foraging for food may seem like a fringe activity, but ForageSF maintains that wild foods are abundant in the world around us, even in cities, and available to all. The group is working to establish foraging as a viable commercial enterprise, able to both support foragers and to provide consumers and retailers with reliable supplies of wild foods. One of its endeavors is a community-supported foraging (CSF) program, which is based on the community-supported agriculture (CSA) model. Subscribers pre-pay for a share in the season's harvest. Each month ForageSF purchases a bulk amount of wild foods from independent foragers, divvies it up, and delivers it to its subscribers. The shares might contain anything from acorn flour to mushrooms, ramps, curly dock, miner's lettuce, watercress, nettle, plantain, nuts, and even wild turkey.

Beyond supporting entrepreneurial foragers, ForageSF is working to establish foraging as a "backyard" endeavor as well, and education is a cornerstone of this effort. The group offers wild food walks in city parks, among other places, with knowledgeable guides advising participants on everything from identification and preparation of wild foods to ethical harvesting and ecological stewardship of wild places. Seasonal mushroom walks (or, more appropriately, hikes) are popular, with expert mycologists taking groups to the coastal mountains to participate in mushroom gathering. Given the city's maritime location, seafood is considered a staple of the local food system, and ForageSF offers hands-on classes on fishing and harvesting native eels, crabs, seaweed, and more.

Kevin Feinstein leads foraging walks in parks and public lands around the Bay Area and as far north as Mendocino County. These outings focus on safely identifying everything from annual greens to medicinal bark to wild mushrooms, but he also gives all types of tips for cooking and otherwise using the bounty found right under our feet.

Mushrooms in Salt Point State Park, north of San Francisco, vary widely in appearance and abundance. Being able to identify all mushrooms — edible and nonedible — is an essential component of mushroom foraging, according to Feinstein.

2: Community
Coming Together around Food

Today's children, collectively, more than any previous generation, are isolated from the daily reminders of how food is produced and its seasonal cycles. Baby boomers, though they may not have had direct contact with farmers, might remember the days when fruits were available only seasonally, when milk could be delivered to the door, eggs could be eaten raw in eggnogs, and steak tartare was a rare delicacy, not a risky venture into foodborne illness.

The "greatest generation" might remember having even fewer degrees of separation from their food production, when meat was obtained from a local butcher (and the butcher knew the farmer who had raised the animal) and vegetables came from backyard gardens or nearby farmers. Today more people than ever before in our history may have so many degrees of separation from their food production that they couldn't distinguish a chicken from a rooster, a cow from a buffalo, a carrot from a parsnip, or a turnip from a potato.

And so what? Who cares? Why does this matter?

The "who cares," it turns out, is a broad, cross-cutting spectrum of our population. And the "why" is no single, simple reason, but a complex intersection of values and economic incentives. As we begin to understand the unintended consequences of this distancing from our food production and we measure the cost in terms of damage to our environment and to our physical health, more and more people do care. Diverse professions — public health, medicine, nutrition, child psychology, agriculture, environmental protection, education, and more — are seeking to change our relationship to food.

As part of this new movement, community farms and gardens are powerful tools. They offer hands-on learning about soil, water, ecosystems, environmental health, animals, and, of course, food and nutrition. Children and adults alike can see, hear, smell, touch, and even taste the food that is produced. These farms and gardens are a form of up-close, living education, and they are influencing individual choices about what gets put on the table, as well as local, state, and federal food policies.

Community farms and gardens may be the first gateway for creating a local food system, and they are often an important stepping-stone leading to broader community food action.

The projects featured here represent just the tip of the iceberg of what is happening in our country and other nations. I hope you'll find them as inspiring as I do and perhaps even be motivated to initiate or join an effort in your own community. Food, it turns out, does more than nourish the body. When integrated into the community fabric, a strong local food system can build community relationships, promote social and physical health, and foster community resilience.

A Collaboration to Connect Community with the Land
Community GroundWorks

coauthored by Regine Kennedy

Community GroundWorks is a nonprofit organization managing Troy Gardens, a 31-acre mixed-use parcel combining cohousing units for low-income families, community gardens, a CSA farm, and prairie and woodland conservation areas, all within the city limits of **Madison, Wisconsin**.

Troy Gardens is an unusual story of a community coming together to restore a more natural relationship between people, nature, and food. The site that Troy Gardens occupies, situated in Madison, Wisconsin, not far from Lake Mendota, has always held special significance for whoever has been fortunate enough to live nearby. Before Troy Gardens became the unique place it is today — a quilt of community gardens, mixed-income "green" housing, an organic farm, and restored prairie and woodlands — the site had been many things to many people, including farmland that provided food to a nearby mental health facility and, before the city or state even existed, sacred ground to local Native Americans.

The process of conceiving Troy Gardens began when neighbors decided they wanted to save the 31-acre site from becoming another residential subdivision. Legally, the land was considered vacant, ripe for development, but in reality the land was far from empty. In fact, for 15 years the community had been cultivating part of it for community gardens and enjoying the rest as an informal nature preserve: a place for bird-watching, walking dogs, and restoring spirits. The community valued the land for its intrinsic character and natural beauty.

Shortly after the state announced its decision to sell the land in 1995, the community rallied by forming a remarkable coalition of individuals and nonprofits — a community land trust, a conservation land trust, a community garden coalition, the local university, an antipoverty agency, and an award-winning coalition of neighborhood organizations. These unlikely partners began working together when they realized they shared a common interest: they were not going to sit idly by while the land they cherished was plundered by bulldozers to build yet another development. Without the dedication of these friends, Troy Gardens would never have been realized.

Working together, walking the land together, the coalition came to the table to discuss, sometimes heatedly, their visions for conserving, preserving, and using the land. One bedrock principle, according to Marcia Caton Campbell, at the time with the Department of Urban and Regional Planning at the University of Wisconsin in Madison and now the Milwaukee director of the Wisconsin-based Center for Resilient Cities, was an agreement that the site had to be used in ways that would benefit the neighborhood. The unusual diversity of skills, knowledge, and funding opportunities brought to the table by the various organizations is what eventually shaped the unique future of the land. The coalition agreed on five distinct functions: farm, gardens, prairie, wilderness, and affordable housing.

When approached by the coalition with its collaborative plan, the state took the property off the market and gave the coalition a long-term lease on the land, with the provision that it could purchase the property with a conservation easement. Several years later, after exploring numerous funding avenues, the partners were finally successful in obtaining a long-term, low-interest loan from Madison's community development block grants and bought the land.

Of all the pieces desired by the coalition partners, Caton Campbell remembers, housing was the most disputed. But if the entire site was to be saved from its subdivision destiny, it required support from all the coalition members, some of whom insisted that affordable housing needed to be part of the picture. And the coalition needed to agree on where and how much land to dedicate to each use. They decided to allocate 5 acres for housing, grouping all the housing on only a small part of the land, so that the rest could be left for woodlands, prairie, farm, and community gardens.

Beginning with a grant in 2000 to assist the community gardens and the farm, Troy Gardens has grown over a decade to a budget of about $350,000, according to Christie Ralston, who is now associate director and natural areas coordinator at Community GroundWorks, the organization that stewards the land and facilitates programs at Troy Gardens. The land is a vision of diversity, with hundreds of volunteers providing countless hours of service. Ralston credits key volunteers for managing the community gardens, writing the monthly newsletter, and working as stewards in the natural areas of Troy Gardens. On work days both volunteers and community gardeners work to keep the common areas clean. On fun days the entire community is invited to enjoy the land with such nature walks as "Exploration in Edibility" or a summer festival featuring music, hayrides, and food.

Integrating Gardens with Housing

Including affordable housing in the plan for Troy Gardens was critical to holding the original coalition together, and the group determined that a cohousing development would work best. Unlike many other cohousing developments, the coalition designed and built the units before recruiting homeowners. Two-thirds, or 20 of the 30 units, are designated as permanently affordable, but all units have identical base configurations. Within the development, homeowners selected where they wanted to live and were able to purchase upgrade packages, resulting, essentially, in 30 custom homes, according to Greg Rosenberg, formerly the executive director of the Madison Area Community Land Trust, and now academy director of the National Community Land Trust Network.

Unlike most cohousing developments, Troy Gardens does not yet offer a common house for community gatherings, but residents do share a variety of activities, such as child care, a monthly community potluck, site maintenance, and gardening. The development has many unique features. For instance, Rosenberg and the site design team strove to keep pavement out of greenspace areas and pushed parking to the eastern edge of the housing site, allowing homes on the other side to look out on the farm and community gardens. Rosenberg says that the designers were criticized for not offering parking next to each unit, but when homeowners selected their preferred units, the units farthest from the parking area were selected first. The design also used green building techniques as much as possible within the budgetary limits. Balancing green building techniques with

Bright Idea: Selling Gardens to the Community

As its gardening program evolved, Community GroundWorks found itself fielding many requests for gardening advice from across Madison. As part of its core mission, the group wanted to build awareness of and engagement with agriculture and local food around the city. At the same time, the group needed funding for its efforts. The result? Madison FarmWorks, an innovative business that will design a productive and aesthetically pleasing vegetable garden for $250 and install the garden for $450 (2010 prices). FarmWorks vegetable gardens are now growing throughout the city; you can even find one near the entrance to the state capitol. If it takes off, FarmWorks could become an important income generator to support the nonprofit work of Community GroundWorks.

affordability was challenging, according to Rosenberg, but having packages available that included upgrades such as solar photovoltaic and thermal panels or tankless water heating made these elements possible for many homeowners (including for the affordable units). As testimony to its success, Troy Gardens cohousing has received the AARP and National Association of Homebuilders "2007 Livable Communities Award" and the 2008 Home Depot Foundation "Award for Affordable Housing Built Responsibly."

For those interested primarily in community gardens, Troy Gardens offers an example of how broadening the scope of a project can enhance community benefits. By integrating gardening with housing, the project serves multiple community goals, bringing greater benefit to all. Rosenberg suggests that a critical lesson was thinking through how to use the natural layout of the land to link the housing with the gardens. Food production is integrated into the design of the housing site, and residents interested in growing their own food can do so just across their large "yard."

Managing the Community Gardens

The Troy community garden is huge, consisting of more than three hundred plots, some of which are reserved for Troy Gardens' cohousing homeowners. Gardeners pay between $10 and $65 each year, depending on family size and income level, for a standard 20-foot-by-20-foot plot. The fees go toward providing water throughout the site and shared community garden tools for everyday use.

More than half of the Troy community gardeners are low-income — a mix of Hmong, Spanish-speaking, and English-speaking people — and their families rely heavily on what they produce. Ralston noted that communication can be a challenge, but Community GroundWorks is able to keep everyone informed through a general meeting at the beginning and end of each season, providing translators at work days, and translating sections of the newsletters.

One unusual feature is that, while Community GroundWorks encourages its gardeners to use organic gardening methods and offers an organic, no-till section, it also offers a separate tilled section where it allows the use of synthetic fertilizers. Another unusual feature is that gardeners are required to volunteer three hours for every plot they manage (though they may choose to pay $10 per hour instead of volunteering). Gardeners may work multiple plots but are limited to no more than four plots per person, and they are restricted from selling the produce they grow because the gardens are supported by an antipoverty program that envisions the gardens as providing food security for participating families.

As part of its mission to foster healthy communities, Community GroundWorks also works actively with children through various programs. One program is its Kids' Garden, which occupies about 10 plots within the community garden. There, kids can learn where their food comes from, connect with nature, express themselves artistically, and develop leadership skills. Children plant everything in their garden themselves, learning along the way what the food is, what it tastes like, and what to do with it. In fact, says Nathan Larson, the Community GroundWorks education director, growing food themselves makes some kids much more adventurous — they'll actually try food they've never tasted before!

Cooking is part of the curriculum as well. Kids prepare the food they've grown themselves — the use of a solar oven and hand-cranked blender brings energy conservation into the conversation — and take the food home to their families. Adding the knowledge they've gained to the sense of accomplishment they've earned, says Larson, creates a powerful combination for transforming young lives.

With an operating budget of about $100,000 per year, funding, says Larson, is the biggest challenge to maintaining the youth programs. Partnerships help ease the burden, although programs may change from year to year, depending on funding sources and community needs. The Oakhill Correctional Institution donates seedling plants for the Kids' Garden through its horticulture vocational program. And a partnership with obesity prevention specialists at a pediatric clinic at the University of Wisconsin School of Medicine and Public Health led to a pilot program called "Garden Fit." This program aims to help children avoid the troubling and paradoxical trend of becoming less fit during the summer by engaging them in gardening and other physical outdoor activities.

Managing the Community Farm

Claire Strader, manager of Troy Community Farm, was the first person hired at Troy Gardens. She runs the 5-acre farm with the help of an assistant farm manager and various interns. The farm is set up as a community-supported agriculture (CSA) project, with 120 shareholders dropping by weekly during the season to pick up their share of the harvest. The farm also sells produce to a variety of local stores, local restaurants, and a farmers' market. In addition, Strader teaches classes about organic gardening at the local high school and the University of Wisconsin at Madison.

Since its start in 2001, the farm has been certified organic, and because machinery is limited by both budget and a farming philosophy that prefers intensive hand-scale techniques rather than mechanical inputs, much of the work is done by hand. Strader says the farm faces a number of challenges — no barn, irrigation, or on-site cooler. Despite these challenges — or perhaps because the farm has not yet made major capital investments — the farm is running in the black.

In a bold effort to grow its CSA and provide more local produce to the community, the farm plans to expand its acreage by using neighborhood yards, becoming Madison's first organic urban-yard farm. Strader believes this expansion will connect the farm more closely with the community and also provide the farm with its first irrigated land. Troy Community Farm is showing that an urban farm can be innovative, self-sustaining, and a viable way to put more fresh food on community plates, while bringing community together in common purpose. The farm is an important pathway for Community GroundWorks to accomplish its mission of fostering a meaningful relationship between people and the land.

Lessons Learned
Building Coalitions to Support a Community Food Project

Build a broad coalition.

Two large community organizations working together, on their own, can easily end up in conflict and call it quits. A broader coalition can help mediate the tensions, providing the pressure and support needed for core partners to stay focused on the larger goal of meeting the community's needs and interests. At Troy Gardens in Madison, Wisconsin (page 41), the creation and implementation of a joint vision depended completely on the ability of individual players to "de-emphasize individual organizational goals for the overall good of the project."[45]

Find common ground for unusual partners.

Partnership between two different land trusts with very different goals and expertise in land conservation – affordable housing versus open space and natural area restoration – manifested in Troy Gardens a unique outcome better than anything either partner could have achieved alone. This partnership enabled the early Troy Gardens coalition to weave together an unusual plan for buying and conserving the land in perpetuity.

Each land trust offered a different mechanism for land purchase and ownership. The Madison Area Community Land Trust (MACLT) is able to acquire and hold land for the benefit of the community, including providing permanently affordable housing to first-time homebuyers whose annual income is at or below 80 percent of the county median income. The Center for Resilient Cities (formerly the Urban Open Space Foundation), on the other hand, is able to acquire and hold conservation easements on land for natural, cultural, and recreational uses.[46]

Instead of working at odds, in this case competing for the land, as can often happen among community nonprofits, the two groups combined their experience in different domains of community development, finance, housing, and natural areas preservation. MACLT was able to obtain a long-term, low-interest loan from the City of Madison's Community Development Block Grant program to purchase the site from the state. It owns the site and developed the housing. MACLT provides a long-term lease for the 26 acres of greenspace for $1 per month in exchange for Community GroundWorks' commitment to manage the land. The Center for Resilient Cities holds a conservation easement on the 26 acres of greenspace to provide an additional layer of protection to safeguard the land.

Create guiding principles or a code of cooperation.

In 1999 the Troy Gardens Advisory Council created a list of "Guiding Principles" for its coalition. These seven principles help the coalition stay focused on the original goals, even when founding members move on and others join the effort, which is often a point where nonprofits can flounder and lose or shift focus.

Follow key principles for working with partners.

Communities today are often supported by intersecting networks of citizen organizations, many of which are vying for dollars and community support. To survive, particularly in times of economic hardship, these networks often find strength in working together. Virginia's Lynchburg Grows (page 116) could not have accomplished as much as it did in its first five years without the support of its community partners – 25,000 hours of work from volunteers from a variety of community organizations, plus another 8,000 hours from five particularly determined individuals. Lynchburg Grows attributes its successful relationships with partners to the following key principles, articulated by its founders.

DON'T DUPLICATE. Make sure you're not duplicating what is already being done in your community.

FILL A NEED. Find out where community resources and energy are already invested, and learn what needs are unfilled. Do not try to undertake a project for which there is no community interest.

DON'T TAKE CREDIT. You can accomplish miracles as long as you don't need to take the credit; it's not about padding résumés.

MAKE IT THE COMMUNITY'S PROJECT. Do not try to "own" the project: The project has to be given to the community, grown with the community, and built on relationships and partnerships within the community. If you do these things, the community will be there for you, helping you at every step.

MAKE NEW PARTNERS. Don't stay in your own silo. Reach out to the usual partners, such as civic clubs, schools, YMCAs, churches, hunger relief organizations, Boys and Girls Clubs, local government. But also reach out to unexpected or unusual partners – youth detention centers, rehabilitative services, senior centers, assisted living residences, and social justice organizations.

CONNECT WITH LOCAL GOVERNMENT. Local government is a powerful resource and ally. Find out which department is most appropriate for your project, whether it's community planning or development. Sometimes it just takes knocking on the city manager's door! It's critical to find the person in local government who can help you weave your way through the organization.

CONNECT WITH LOCAL ECONOMIC DEVELOPMENT. Local economic development organizations may have resources to assist your project. Check with the local chamber of commerce, economic development commission, and even individual developers.

DON'T TAKE ON AN 800-POUND GORILLA. Start small! Understand the resources you have on any given day, and use them strategically.

TELL THE STORY. Getting the word out about what you do is just as important as what you do. Use every possible community outlet for your story, such as radio interviews, press releases and newspaper articles, appearances on local TV stations, and presentations to local civic organizations. Most important, be ready with a "story to tell" and your successes. Everybody loves a good story.

APPROACH PEOPLE WHERE THEY ARE. Different people and organizations have different needs and ways of understanding. For potential funders, be ready to provide them with the strategic plan. For the compassionate heart-centered groups, be ready to tell the human story.

Involve potential partners early in the project concept.

The case of Troy Gardens suggests that it is better to seek out partners early in a collaborative process. The earlier partners come together, the more they will create a jointly shared investment and sense of ownership in the project. And, the more the partners share an invested ownership in the outcome of the project, the more chances the project will actually come to fruition and enjoy a long life.

The experience of Janus Youth in Portland, Oregon (page 105), underscores this point. When Janus Youth was forced to sell the property where it had hosted a garden program for runaway and homeless youths, the supervising staffperson, Tera Wick, didn't go off into a corner and dream up her next step alone. Instead she noticed an opportunity – the USDA Community Food Projects grant program – and organized a community meeting of Portland residents concerned with food access. At this meeting she asked, "How can we respond to this USDA request for proposals?" and facilitated a productive brainstorming that led to the idea of a garden in the St. Johns Woods public housing community.

From my own work in the field of environmental conflict resolution, I can attest that early collaboration is an important step in creating community interest and buy-in, and to avoid stepping on the toes of potential future community partners. For building trust, earlier engagement is always better than later. And experience suggests that it is better to err on the side of inviting more potential partners than less, as inclusivity is appreciated and fosters broader community support.

Build relationships with community members and groups.

The point of entry for any community project is always the community and your existing relationships, says Dennis Morrow, executive director of Janus Youth. It doesn't matter whether you are a youth program, a community service organization, or a YMCA, says Morrow; the key always is to develop or build on a relationship with the community. Then you have a place to start.

Tera Wick, the "mother" of the Village Gardens program at Janus Youth, says that another key for success is a willingness to form new relationships. In fact, she says, at the very first community meeting when she was asking how Portland should respond to the USDA grant program, one of the points that emerged was to develop partnerships and build relationships. The act of naming "relationship building" as a task and deliverable has important consequences. In fact, Wick points out, well-established social science theories of social capital and networks[47] and recent work on the social determinants of health by the World Health Organization[48] suggest that reconnecting people and community organizations is key for improving quality of life and fighting the factors that lead to poverty, food insecurity, and inequities in health. So, when one of the groups that came to Wick's first meeting was a public housing advocacy group that suggested the need for something positive in the community that people could plug into over the long term, Wick listened. And, together, they began the St. Johns Woods program.

Janus Youth has taken the need for relationship to heart, believing that a collaborative approach is important not only within a program, but with all community groups and partners. It actively pursues partnerships with just about everybody possible – local governments, universities, public health institutions, grocery stores, farms, and other gardening programs.

Rachel Reinhart, program director at the Jones Valley Urban Farm (JVUF) in Birmingham, Alabama (page 158), puts it bluntly. "Don't expect people to eat your food," she says, "if you haven't invited them to the table." Reinhart very much wanted to launch JVUF's Seed 2 Plate program, providing hands-on on-farm workshops for kids, but lacked funding. JVUF invited a "person of influence" to a meeting about the program, and Reinhart believes that making this invitation and taking the time to build a relationship is what finally led to funding for the program.

Before starting, learn what the community really needs.

"Don't do something that people don't care about," advises Julia Rivera, organizing director for Nuestras Raíces in Holyoke, Massachusetts (page 123). Rivera emphasizes the important of meeting and connecting with community people, finding out what they need, to build a strong base of support for a project – a community coalition. "This is what has made us unique," she says. "We didn't just start." She explains that the group's very first garden, La Finquita, was started when immigrant farmers from Puerto Rico, whose families were struggling with poverty and hunger, said that they needed to grow food for their families. They looked out their windows and saw abandoned lots, and they wondered why they couldn't begin to grow food there.

Similarly, Rivera says that Raíces Latinas, an educational and training program for women, began because the women wanted and needed it. They wanted to take jobs and start businesses, and they needed help getting there. The group listened and responded. "You don't work for the community and do things for a community," explains Rivera. "You have to be a part of the community."

The Recipe for Troy Gardens

Troy Gardens has been, and still is, far from a perfect project. But there are many lessons embedded within our experience at Troy Gardens that might be instructive to other communities wanting to do a sustainable development project. . . .

First, we start with *community organizing*, because that is the only way to pull together a community to figure out what vision it might have for itself. We were fortunate to have great organizing by Tim Carlisle of the Northside Planning Council. If not for his diligence and skill, these 31 acres might have been turned into just another vinyl-clad subdivision.

The second ingredient in our recipe is *tenacity*. Projects like this are very challenging. You must push through more than one brick wall to make them happen.

Third is *patience*. Projects, particularly outside-the-box projects, take a long, long time. Our system of municipal zoning and development approval does not lend itself easily or swiftly to multifaceted projects that combine housing, open space, urban agriculture, and the many sustainability features that are found at Troy Gardens.

Fourth is putting *agriculture* at the heart of your project. The threatened loss of the community gardens, which had existed on the site for many years, sparked the organizing effort that led to everything else. Community gardens saved Troy Gardens and remain the heart and soul of the place. Community gardens are the best engine of community building that I have ever seen, crossing every conceivable boundary of class, race, and culture.

Fifth is *partnership*. I don't know of any organization that can carry out all the elements required to do a project like this alone. Partnering expands your horsepower. It also requires you to give away control. But it really isn't giving away control if you couldn't pull it off by yourself in the first place.

Sixth is dense *clustering of housing* to preserve open spaces. By grouping all the housing in a 5-acre portion of the site, 26 acres were freed up for a farm, community gardens, and a restored prairie. If every home at Troy Gardens had been designed and developed with its own backyard, preservation and restoration of so much open space would never have been possible.

Seventh is the *community land trust model*. I put it seventh because I talk a lot about building homes for seven generations of homeowners, which is what the CLT model is all about. Because the community was so committed to the permanent protection of the green space at Troy Gardens, people understood the value that the stewardship function of the community land trust model could bring to Troy Gardens. When we talked about our 99-year renewable ground lease, people understood the benefit this type of long-term orientation would bring to Troy Gardens. Nowadays, people talk about Troy Gardens being around forever. It is the CLT model that allows people to have confidence that "forever" might be attainable.

Eight is *accessibility*. This has several dimensions: physical accessibility of housing for everyone, regardless of disability/ability; accessibility to home ownership for people of modest incomes; accessibility to green spaces for the whole community; access to fresh, healthy food through urban agriculture; and access to public transportation, including the ability to walk or bike to places important to you.

Ninth is *government support*. Federal, state, county, and city support were essential to Troy Gardens. This project would never have happened without it. . . .

Tenth is *belief*. You can't do a project like Troy Gardens unless you believe in the power and beauty of the dream you are trying to achieve. Belief helps you find a way through interpersonal conflicts that can grind you down unless all the players and partners hold on to a shared vision for what they are trying to accomplish. Sol Levin [one of the original community leaders in the development of Troy Gardens] had faith that things would work out for Troy Gardens long before there was any logical reason to think so. His faith was infectious and helped to carry a whole lot of people over the finish line, through difficult times and countless obstacles.

— by Greg Rosenberg, executive director, Community GroundWorks

Excerpted from "Troy Gardens: The Accidental Ecovillage" in The Community Land Trust Reader, edited by John Emmeus Davis (Lincoln Land Institute, 2010)

A Community Comes Together to Save a Farm

Indian Line Farm

Indian Line Farm in **South Egremont, Massachusetts**, was one of the founding CSAs in North America. Although at one point it seemed like the farm might go under, with the help of the local community, it now operates under a unique ownership model, with a land trust owning the land, the farmers owning the buildings, and The Nature Conservancy owning conservation rights. The farm is now a thriving business, focused primarily on its CSA program.

Community-supported agriculture (CSA) in North America got its start in 1986, when two CSA programs, independent of each other, were founded: Indian Line Farm in South Egremont, Massachusetts, and Temple-Wilton Community Farm in Wilton, New Hampshire. The group behind Indian Line Farm, influenced by the teachings of Rudolf Steiner, hoped to put into practice the idea of producing locally what is consumed locally. They formed an association to lease land from one of the founding members, Robyn Van En, and began farming as a CSA, selling shares in the farm's produce to local community members.

In 1997 Van En died, at the age of 49, and her son inherited the farm and homestead. Unable to manage the upkeep, he was forced to sell the farm. The farmers who had been working the land, Elizabeth Keen and Al Thorp, wanted to buy it, but they couldn't afford to. The local community, concerned with preserving the working landscape of the region and supporting the local farming economy, stepped in to help. Keen and Thorp partnered with a local land trust organization, called the Community Land Trust in the Southern Berkshires, and The Nature Conservancy to craft an unusual purchase-and-sale agreement.

The nonprofit land trust purchased the property, with funds raised from the local community. It sold conservation restriction rights to The Nature Conservancy, which had been working in the area to preserve a stretch of unique wetlands. It sold the buildings on the property to Keen and Thorp, and it gave them a 99-year lease on the land. Having ownership of the buildings gives the farmers a financial stake in the farm, along with the eventual equity that comes from owning property. The land lease is inheritable and renewable, giving the operation a sense of a true family farm, and it is accompanied by certain land-use agreements, such as the land being used for a working farm, with sustainable farming methods, and certain limits set to keep operations within the land's ecological carrying capacity.

One of the biggest issues facing new farmers today is affording land. At Indian Line Farm, the local community has relieved Keen and Thorp of the burden of land debt, allowing them to be economically viable and to focus on sustainable production. And by cooperating with and supporting the farmers in this way, the community has been able to preserve the agrarian culture and landscape that historically have characterized life in the southern Berkshires, and to honor the legacy of one of the founding CSAs in North America.

Top: Buying into community-supported agriculture and picking up your weekly share is the first level of engagement. *Bottom left and right:* CSA members also join in to help with maintenance and other seasonal events, like fall garlic planting, providing much-needed extra hands throughout the year.

The CSA model provides capital up front for the farm and standardizes the weekly harvest with a set number of pickups each week. Such planning allows the farmers stability and efficiency throughout the year.

Hoophouses are set adjacent to the fields. Their adjustable "walls" can be raised or lowered, protecting plants in colder weather and allowing air to circulate in warmer weather.

Top: Midseason crops come in both variety and abundance. *Bottom:* Lettuce is planted at staggered intervals, so that mature heads can be harvested for weeks on end.

Pick-your-own beans, peas, and tomatoes are a hit with the kids and also help teach everyone about where their food comes from.

Elizabeth Keen, with baby Colin on her back, plants garlic in the fall cover crops.

Lessons Learned
Running a Successful CSA

Effective CSAs build relationships with customers and suppliers.

People often join a CSA for more than food, says Mary Seton Corboy of Philadelphia's Greensgrow Farm (page 96). They are also looking for a relationship. The CSA staff must have the people and communication skills to cultivate those relationships. They must also be able to establish and maintain good relationships with any suppliers who become partners in the CSA. For suppliers, reliable and consistent outlets for their produce are important, and it's the job of the CSA staff to give them accurate information about when and how much of their products the CSA will need.

When a crop fails, transparency supports trust from shareholders.

The premise behind a CSA is that shareholders, who purchase their share at the beginning of the season, lower the farmer's up-front direct costs and risk by spreading the cost and risk among many people. This sounds good – in theory. People who join a CSA are told about the risks in farming and usually sign a contract acknowledging that as shareholders in the farm they are not buying a hard and fast guarantee. They say they understand weather and pests are factors.

But in reality, people are forgiving only up to a point. Greensgrow's Corboy learned this the hard way when a severe drought ruined the seed germination for her hydroponic operation. She learned that, despite what people say, they do expect the farm to do well, and they do not truly expect to share the farm's loss.

In a situation like this, Corboy says you must be completely transparent: explain how the situation came to be, and the challenges, in a way that acknowledges the shareholders as a vital part of your operation; this enables them to be more forgiving. "You need to become a Ph.D. of the human condition," Corboy advises. And, too, when a failure occurs and you can't fully deliver what you promised, people need to be given the choice of opting out. Customer service means the customer is always right – and it is an essential attitude, even for farmers. Customer service is the foundation for the two keys of long-term success: relationships and trust.

Develop cooperative relationships with other producers to fill CSA shares.

Because of the vagaries of weather and pests, the ideal CSA will build in a certain amount of insurance by creating relationships with other farms that will provide either a steady diversity of produce for the basket or supplemental produce on demand.

Corboy adds that a cooperative is also *more* than just insurance. Farming, she says, "is by nature a cooperative enterprise. So why not *admit it*, and *go with it*. People are not going to drive down the road to buy a pint of peaches and then drive another 10 miles to buy a pint of blueberries." She says that more and more CSAs are recognizing the benefits of a cooperative approach. Not all farms are going to have the same problems at the same time, so a cooperative approach benefits all.

The experience of Lynchburg Grows in Virginia (page 116) shows that forming a partnership with an established producer is key for those who are just starting out. It's often hard to grow a CSA, as you may also be just starting the farm and learning the ropes of what works or doesn't work so well on your farm. You may lose an entire crop, or several crops, to animal destruction or disease or harsh weather. Lynchburg Grows found a partner in Appalachian Sustainable Development (ASD; page 253), a distributor of organic produce with a good track record. Though it was nearly 190 miles away in southwest Virginia, ASD was already trucking its produce throughout Virginia and was happy to find another outlet for it.

A partnership like this is mutually beneficial – it helps move the other organization's produce and provides you with a reliable source of weekly produce for your CSA. If Lynchburg Grows doesn't need all of the ASD produce it has contracted to purchase to fill its weekly CSA baskets, it sells the remainder at a farmers' market. It marks up ASD produce enough to cover its packaging and selling costs (about 30 percent), which is far less than the average 80 percent retail markup, so their farmers' market customers still receive a bargain price.

Create variations on a CSA to support low-income shareholders.

Low-income populations may have trouble coming up with the hefty up-front investment that most CSAs require at the beginning of the season. To give low-income families affordable access to the fresh food available in its CSA, Milwaukee's Growing Power (page 74) offers the Market Basket program, in which people can order a basket at the beginning of the week

and pick it up at the end of the week, paying for the basket at pickup. Moreover, because Growing Power can tap into the wholesale prices offered through its Rainbow Farmer's Cooperative (box, page 76), its baskets are priced nearly 25 to 30 percent below supermarket cost.

The Jones Valley Urban Farm in Knoxville, Tennessee (page 158), has offered a low-income CSA share for one-quarter the price of a normal share – only $5 for a weekly food basket, versus the normal $20. But people didn't sign up. Even $5 was viewed as prohibitive. So the farm is trying another tactic – trading a CSA share for work. In 2009 it offered a free CSA share in exchange for a half day of work each week, and this system had more success.

Elizabeth Keen of Indian Line Farm in South Egremont, Massachusetts (page 49), one of the founding CSAs in the United States, works with a staff member in the fields.

Putting Agriculture Front and Center in Planned Developments
South Village

by Tim Beatley

South Village, a new housing development in **South Burlington, Vermont**, is pioneering an innovative approach to development design: the idea that farming is not a nuisance but a planned community amenity. Residents will be able to grow their own food or join the on-site CSA. The project's clustered village design makes it possible to conserve 75 percent of the site, creating space for not only agriculture but also the preservation and restoration of wetlands and natural landscapes.

Andres Duany, leading proponent and practitioner of the so-called New Urbanism — a design and planning trend that seeks to shift car-dependent suburbanization to more compact, mixed-use, walkable towns and neighborhoods — has lately declared: "Agriculture is the new golf." By that he means that there is an interesting and hopeful trend emerging of including farms and gardens in housing development designs from the beginning, so that these agricultural features are essential, indeed primary amenities to future residents, in the same way that golf courses have been in the past.

And the numbers suggest that this trend is catching on in a major way. Ed McMahon of the Urban Land Institute estimates the number of developments in the country that incorporate agriculture as a "key community component" in their planning as at least two hundred.[49]

One especially interesting example is South Village, near Burlington, Vermont. Here a former dairy farm being developed into a new residential community is not retreating from its agricultural past but highlighting and building onto it. With the help of Will Raap, cofounder of the Intervale (see page 92, South Village is being designed to include a working farm, which will operate a CSA, as well as extensive gardens for residents who want to grow food themselves. Proximity to this actively farmed land is seen

not as a liability but as a positive to new homebuyers. And the farm is just one piece of a larger effort to heal the land here. The plan will protect some 75 percent of the 220-acre site as undeveloped land, with work under way to restore extensive wetlands and wildlife habitat.

South Village has many of the other typical qualities of a New Urbanist community: clustered homes, narrow streets, small front and back yards. These features allow for the conservation of open land and facilitate slower traffic, walking, social interactions, and neighborliness. When the village is fully built, there will be more than three hundred homes, ranging from condos to townhouses to detached single-family units. The project also has a number of "green" features, including on-site storm-water management and homes certified by the Leadership in Energy and Environmental Design (LEED and Energy Star programs).

The South Village developers are crafting some unusual governance structures to help promote the ecological goals of the community. One is the establishment of a 0.5 percent real estate transfer tax that will generate funds for future conservation efforts; whenever a home in the development is sold, the transfer tax is collected and deposited in a stewardship fund. (This tax mechanism has been a key funding element in other conservation-oriented communities, such as Prairie Crossing in northern Illinois and Hidden

Springs in Boise, Idaho.) Also, taking advantage of a unique provision of Vermont law, South Village is organizing itself as a *low-profit* limited liability corporation (LLLC) to make it possible for foundations to support it financially. (Foundations with endowments are required to spend a certain amount each year on program-related investments, and support for the South Village farm should qualify.) The South Village LLLC will lease the land from the developers and will likely take over as the homeowners association for the community, with responsibility for maintaining the common land and for administering the stewardship fund.

Another important goal at South Village is to integrate the farmers who work the development's commercial farm into the community, allowing them the opportunity to buy homes in South Village. As Raap notes, "Our goal is actually to embed the farmers in the community." But will they be able to afford to buy a home here? "Ultimately, that should be one of the tests," Raap acknowledges. The developers were able to negotiate a density bonus from the City of South Burlington, in exchange for keeping some of the homes more affordable. Nevertheless, the lowest-priced home here will still be quite expensive when compared with what a typical farming income might support.

There are many other examples of development projects that include agriculture and food production not as an afterthought, but as a primary and central amenity. These include notable conservation communities such as Prairie Crossing in northern Illinois and Serenbe in Georgia, just outside Atlanta. But in some ways this is not an especially new idea. One of the very best examples, for instance, remains Village Homes, a 1970s neighborhood in Davis, California, and an iconic green development, completed many years before the topic of sustainability became fashionable. Its agricultural features include a community garden, a vineyard, and an orchard, as well as a block of three hundred almond trees that are commercially harvested, the profits from which help pay for the neighborhood's gardening staff. The extent to which fruit trees are scattered everywhere, seemingly in every imaginable variety, is one of the unique aspects of Village Homes. It is about as close as you can get to living immersed in a garden, which can lead to different views about acceptable methods for tending the garden — in Village Homes, for example, pesticides are forbidden.

Together these developments demonstrate the immense value of including food production in any new housing or neighborhood design. Increasingly agriculture is seen not as a liability or a nuisance but as a positive amenity and a life-enhancing community feature. As the desire to know where our food comes from, and increasingly to participate directly in its production, grows, more and more housing consumers will look for ways to vote with their feet (that is, where they plant themselves) as well as their palates.

City-Sponsored Community Gardening
P-Patch Community Gardening Program

with research by Robin Proebsting

Seattle's community gardening program, known locally as the "P-Patch Program," provides and manages garden plots in more than 70 locations throughout the city for more than two thousand people, most of whom wouldn't otherwise have physical space or financial means to create a garden. In the process of managing these gardens, the city cultivates volunteer leaders, thereby building its long-term civic capacity.

If you're wondering whether community gardens would be a hot commodity in your area, consider this: In 2010 Seattle's "P-Patch" community garden network had more than two thousand plots at more than 70 sites.[50] And despite this wealth of gardens, nearly two thousand would-be gardeners are on the city's waiting list.

And if your community is considering creating community gardens in low-income neighborhoods and housing projects, but naysayers believe that low-income people have neither the time for nor an interest in growing food, consider this: Fifty-five percent of the P-Patch gardeners in Seattle are low income. Fifteen percent are under the federal poverty level, 48 percent live in multifamily dwellings, and 77 percent have no garden space where they live.[51]

The Benefits of City Support

Seattle's P-Patch program — founded in 1973, when the city bought the Picardo Farm (from whence comes the name "P"-Patch) — could rightfully be considered our nation's queen of community garden programs. And it offers sustained proof through nearly 40 years that community dwellers at all income levels do want to garden. Indeed, Seattle can't keep pace with the demand, for as the number of gardens increases, so does the demand.

P-Patch gardeners represent only 1 percent of Seattle's population, yet the city clearly believes the program is an important service for the larger community, as it not only financially supports the program

but continues when possible to expand the gardens. Jim Diers, former director of the Seattle Department of Neighborhoods, makes the case that Seattle's community gardens provide a host of benefits by:

- Creating attractive open space that doesn't require city maintenance
- Attracting wildlife to support the city's varied ecosystem
- Educating the public about the environment
- Building a sense of community with events and by connecting people who wouldn't otherwise interact
- Serving as community centers with play structures, picnic benches, and other features that help draw people into the gardens
- Promoting social equality, as they are available to everyone, even those who cannot pay the annual fee[52]

Gardeners pay a minimal annual fee of $25 for a plot, plus $12 for each 100 square feet of garden, though the city may waive or lower the fees for those who can't afford them. Each gardener is also required to donate eight hours to maintain the garden public areas, and while it isn't required, the P-Patch gardeners donate more than 10 tons of food annually to the city's food banks, and 30 gardens have active food banking programs.[53]

More than anything else, city support is critical for any kind of community gardening effort, says

Richard MacDonald, supervisor of the P-Patch program. He cites Laura Lawson's *City Bountiful* as evidence: During World Wars I and II, when community gardens were supported by federal and local governments, they thrived. And when this support was withdrawn, community gardens fell by the wayside. Of course, if a local government is not able to support community gardening, other options do exist. Some communities have proved that strong community partnerships can equally successfully fill this role; for example, Southside Community Land Trust in Rhode Island (page 65), South End/Lower Roxbury Open Space Land Trust in Boston (page 62), Community GroundWorks in Madison, Wisconsin (page 41), and Indian Line Farm in South Egremont, Massachusetts (page 49).

Citizen Leadership Is Essential

One of the biggest challenges, says MacDonald, is simply managing a program that involves so many community residents. One way the five city staff accomplish this task is by cultivating and working with volunteer citizen leaders. In exchange for the invaluable management provided by the volunteer leaders, city staff develop new resources for them, such as mapping, and help them solve the problems that inevitably arise when some four thousand citizens are regularly rubbing elbows around their bok choy and beans.

The volunteer leaders' main job is to ensure that plots are maintained properly. When gardeners aren't able to weed and water, they're given notice, making room for others on the wait list to garden. Each year city staff recruit new volunteer leaders and spend individual time walking them through the garden and learning their skill sets. "In an ideal world volunteer leaders recruit and train the new leaders," says MacDonald. City staff also help gardens organize periodic get-togethers and work with garden leaders to identify the various responsibilities and tasks. In gardens that host official community events with art or food, the volunteer leaders may need more intensive help.

City strategies for assisting gardens reflect the nature of the neighborhood, says MacDonald. If the garden is in a very upscale neighborhood, people may already have resources and be savvy about how to get things done. If the garden is in a neighborhood where people have language and income challenges, then it's important to figure out how garden leaders will access the basic resources. MacDonald says they are often contacted by school or church youth groups who want to start a new garden or have been assigned garden plots but who don't have gardening skills or resources; in these instances, the city has used AmeriCorps volunteers to assist these low-resource groups.

Preserving Community Gardens through a Land Trust
South End/Lower Roxbury Open Space Land Trust

with research by Benjamin Chrisinger

The South End/Lower Roxbury Open Space Land Trust is a nonprofit working to, as its mission states, "acquire, own, improve, and maintain open space for community gardening and pocket parks in the **South End** and **Lower Roxbury** neighborhoods of **Boston** for the public benefit in perpetuity."

In the very heart of Boston, the South End/Lower Roxbury neighborhood is home to one of our nation's first experiments in using a land trust to preserve neighborhood gardens. The neighborhood's history gives meaning to the term "melting pot" — Puritan settlers, the nation's first recorded interracial marriage in 1690, waves of Germans, Irish, Jews, Africans, Puerto Ricans, Caribbean islanders, Latin Americans, and Asians. Here the South End/Lower Roxbury Open Space Land Trust (SELROSLT), founded in 1991, protects 16 gardens and pocket parks, little jewels of cool green tucked here and there. And the trust continues to seek opportunities for protecting or creating new gardens in the neighborhood.

Each of the SELROSLT community gardens has a distinct look and feel, reflecting the character of its gardeners and surrounding neighborhood. Some are tended by low-income residents, others by residents with greater means. Most have compost bins, some have drip irrigation, some have stern instructions posted for conserving water, and all have gardeners' agreements that detail clear "dos" and "don'ts" for gardening with others in close quarters.

The Worcester Street Community Garden is tucked behind a decorative iron fence, between apartment buildings and homes, and covers nearly an entire city block. Its 110 gardens, each 10 feet by 10 feet, are neatly laid out with newly rebuilt raised vegetable and flower beds. At its core is a highly abstract garden pavilion design selected by the gardeners themselves

in conjunction with a design-build class at the Boston Architectural College. The West Springfield Community Garden is an explosion of vegetables and flowers surrounded by brick buildings, and its growth seems to creep upward into neighboring apartment balconies that are crowded with container gardens. This garden was founded by a neighborhood community development corporation for tenants living in its buildings. The Lenox/Kendall Community Garden is a large open space polka-dotted by large white buckets hanging upside down on wooden stakes and sacks of plastic bags flapping in the breeze, which the mostly African-American gardeners use for watering and gathering produce. Here, too, is a mix of vegetables and flowers — but collards and kale predominate.

The Bessie Barnes Community Garden is a working landscape with barbecue grills, tools, and pots, while the nearby Bessie Barnes Memorial Park garden offers a minihaven of Oriental design for quiet meditation. The Frederick Douglass Peace Garden is a mix of cobbled performance space and a button oasis of groomed grass and ornamental plantings inside a short, white picket fence, where small events and weddings can be held amid the city bustle. In another neighborhood the Rutland/Washington Garden has a more manicured look, boasting a gazebo suitable for afternoon tea.

Then there's the queen of all 16 gardens, the Berkeley Street Community Garden, one of the largest community gardens in Boston, with more

Bright Idea: Create a Signature Fund-Raising Event

The South End/Lower Roxbury Open Space Land Trust is famous for its annual South End Garden Tour, which takes participants through private and public gardens around Boston's South End neighborhood, from patio and rooftop gardens to parks and community gardens. The tour, which in 2011 is in its eighteenth year, is the nonprofit's signature fund-raising event. Much like the annual garden tours that garden clubs all over the nation host, this event serves the dual purpose of educating participants about gardening and inspiring them to take it up. And what an opportunity it is! Weather affects the outcome, of course, but in recent years the tour has netted SELROSLT more than half of its $50,000 annual budget.

than 150 plots. Begun in 1971, this garden is close to Chinatown, and many of its plots are tended by Chinatown's residents. It feels as though you've entered another world. I'm reminded of a Mekong floating boat market, with its three-dimensional quality of movement and highly organized chaos. Many plots are like large cages, with fencing on all sides and overhead, giving the impression of large box art that one must peer into for a long time to understand. Inside, the gardeners have used scraps of wood, bamboo, metal, cardboard, buckets, cloth, plastic, rope, and string to create three-dimensional layered growing systems that shade, support, hang, dangle, and lean. Beans and other vining vegetables snake up side fencing onto netted ceilings, shading greens during the hotter months of the growing season. These gardens are producing food for entire families as intensively as possible. Interspersed with these wild but highly functional structures are pockets of orderly and serene English gardens, casual American plots, and completely eccentric gardens.

Preservation through Land Trusts

What binds all these gardens together is a community passion for preserving them from development — forever. In South End/Lower Roxbury's case, the community found an unusual path for achieving its goal: forming a neighborhood-based nonprofit land trust.

A land trust protects land by purchasing the land outright, by contracting ultra-long-term (such as 99-year) leases, or, if the private landowner retains ownership, by holding and managing conservation easements that govern how the land can be used by future generations. National land trusts — such as The Nature Conservancy, the American Farmland Trust,

the Wilderness Land Trust, and the Wildlife Land Trust — usually have specific narrow goals: conserving land with rare and endangered species, rare habitat, important farmland, or open space. Regional and local land trusts are often more rooted in goals specific to their community, such as preserving rural character, dwindling woodlands, migratory routes, water resources, or scenic viewsheds.

SELROSLT is among the few land trusts in the nation that preserve neighborhood gardens. It's hard to estimate the number of community gardens across our nation, says Betsy Johnson, cofounder and director of SELROSLT and former president of the American Community Gardening Association. Though there isn't a formal method for counting, Johnson thinks there may be as many as 20,000 community gardens across the country, not including schools, where new gardens are now being installed in increasing numbers. Still, of those 20,000 community gardens, Johnson estimates that community garden land trusts are less than 1 percent, or a small club of about 50 members.

While land trusts are not for everyone, they are an important tool that can preserve community gardens for future generations. It's likely that land trusts aren't used more frequently because grassroots leaders simply may not be aware of their potential or see them as too complicated, or government ownership and involvement makes them unnecessary. Where land trusts have been used, they reflect the goals and desires of the community. Some have stayed incredibly lean, such as SELROSLT, which operates entirely with volunteers, on a budget of only $50,000, with a storage locker for an "office." Others have grown into large nonprofits with large staff, budgets, and complex programs.

In addition to SELROSLT, some of the most prominent community garden land trusts are found in Rhode Island (see the box on page 65), Philadelphia, Chicago, New York City, and Los Angeles. Before SEL-ROSLT was formed, the Philadelphia Horticultural Society joined forces with Penn State in 1986 to create the Neighborhood Gardens Association, an incorporated Philadelphia land trust, to protect long-standing community gardens from development. This trust operates entirely on private dollars and now owns 24 community gardens.[54]

Similar development pressures led Chicago to create NeighborSpace in 1996. Unlike its Philadelphia and Boston counterparts, this land trust was initially funded entirely by public dollars, an unusual situation that gave it easier access to government but also made it susceptible to changes in government goals. Since then, NeighborSpace has broadened its financial base, and it now owns or leases 57 community gardens throughout Chicago.[55]

In New York City in 1999, when the city threatened to destroy 114 of its more than 700 long-standing community gardens by public auction, public outrage and media led to a deal.[56] The Trust for Public Land (TPL) saved 62 of the gardens for $3 million, or roughly one-quarter of their value, and the New York Restoration Project purchased the remainder. Since then, TPL has created three separate nonprofits to own and manage its gardens — 16 gardens in the Bronx, 34 in Brooklyn/Queens, and 14 in Manhattan.[57]

More recently, as we entered the twenty-first century, Los Angeles created a community garden land trust for different reasons. Community gardens were scarce or nonexistent in LA, but city leaders saw that gardens could help revitalize low-income neighborhoods and decided to create new open space for them. The city gave seed money to found the Los Angeles Neighborhood Land Trust (LANLT) in 2004, and now LANLT is creating urban park oases in seven traditionally underserved neighborhoods.[58] LANLT vaunts community gardens as sources of community pride and part of the answer to poverty, crime, poor nutrition, and obesity.

South End/Lower Roxbury, like some of its counterparts, was also formed in response to development pressures. In the 1960s era of urban renewal, blighted and dilapidated properties were torn down with the promise that the South End/Lower Roxbury neighborhood would be rebuilt — but that didn't happen. In the 1970s residents began creating community gardens in empty neighborhood spaces, often just starting them and only later asking for permission or obtaining a lease from the city, says Betsy Johnson.

But when the city began implementing its plan for redevelopment in 1986, bulldozing a community garden and building an affordable housing project on the site, the community gardeners woke up. People who cared about the gardens pulled together. And these gardeners then joined forces with a group of public housing advocates, who were just as appalled at the city's plans. Together they agreed that the city should preserve both its community gardens *and* affordable housing. At an important public meeting, they showed up some three hundred strong. City officials were forced to rethink their plan, Johnson says.

Over a period of five years, the garden group and the city pieced together a deal. In 1991 the City of Boston transferred ownership of eight gardens for $1 each to the newly formed South End/Lower Roxbury Open Space Land Trust. Finally, 20 years after vacant spaces had been claimed for gardens, South End/Lower Roxbury gardeners knew they would be protected.

"But setting up a land trust is not a panacea," says Johnson. "There is no such thing as *permanent* protection, because the government always has eminent domain." In other words, if a property is essential for a project perceived to be for the public good — usually a new road or water project but also private development — the government can invoke its right to eminent domain and snatch the property away from the owner, without the owner's consent, though it must provide the owner with monetary compensation.

Johnson worries about whether SELROSLT's efforts to protect South End/Lower Roxbury's community gardens will hold up over the decades and centuries ahead. But SELROSLT has two things in its favor, she says. First, it owns the deed to the land on

12 of its gardens, which not all community gardens do. Another garden sits on top of an underground parking garage owned by a condo association, and SELROSLT has two consecutive 99-year leases on that location that Johnson believes provides more legal protection than would ownership through the condo association. Three other SELROSLT gardens are still owned by the city, but the city has voted to transfer the parcels to SELROSLT. Second, the city has zoned the gardens as "open-space community gardens," which suggests that local government will not allow it to be taken for development.

In the end, Johnson says, preservation of community gardens depends on four things: a mechanism for long-term protection, such as ownership or easement; the long-term stability of the owning organization; maintaining the gardens in such a way that they are seen as neighborhood assets; and strong community

Southside Community Land Trust

In Providence, Rhode Island, another neighborhood land trust – perhaps the nation's first – wanted to grow community along with community gardens. When the Southside Community Land Trust was founded in 1981, some neighborhoods in Providence were considered so blighted that the city was selling lots for only $1. With a local philanthropist's donation of $5,000, SCLT purchased nearly 5 acres of vacant lots that, today, would be either developed or unaffordable. Though land values have steadily risen, SCLT has used its creativity and community connections to secure 13 community gardens and a ¾-acre commercial and educational farm in the city center. To augment these gardens and the farm, SCLT has developed a range of education and training programs.

Of the 260 SCLT community gardener families, 75 percent are people of color, reflecting a diversity of nationalities from all over the world, especially Southeast Asia, Central America, and Africa. Many immigrants have brought their own seeds from their homelands, as well as their culture's knowledge of how to grow food. Today, at garden potlucks, many share their produce and heritage. The SCLT City Farm, modeling biointensive agricultural practices, annually grows over 4,000 pounds of food on only 5,000 square feet, generating $20,000+ in farmers' market and restaurant sales, as well as enough food to contribute to SCLT's daily communal staff lunch and a neighboring soup kitchen.

Even more, SCLT now manages a state-preserved 50-acre farm that serves as an important incubator for new farmers. The Urban Edge Farm, purchased by the state in 2002, hosts seven farm businesses.[59] The farmers use organic farming practices and collaborate to manage the overall farm's upkeep, repair, and operations. Though Urban Edge was originally conceived as a four-year program for launching new farmers, SCLT learned through listening sessions that farmers who worked on bringing the dilapidated farm back into productivity didn't want to have to leave to start over somewhere else. "So the farmers can stay year by year as long as they comply with a lease agreement that they cocreated with us," says executive director Katherine Brown. "The farmers and SCLT are especially proud of the amazing collaborative model we have developed to incorporate the farmers as partners who, working together, manage the farm." This unusual partnership has allowed SCLT to run a productive urban farm without funding a farm manager staff position.

SCLT has plans – big plans – for the future. It has established the Urban Agriculture Resource Center website to share information about high-yield urban gardening, lead testing for urban sites, and other topics specific to the challenges of growing food in urban settings. With assistance from the Greater Providence Urban Agriculture Task Force, a 50-member citywide coalition founded by SCLT in 2004, the group has hired a coordinator to assist it in expanding by adding 16 new gardens to the network, with space for four hundred more family plots. And to ramp up its work with schools, especially those with low-income children, SCLT is more than doubling the number of visiting schoolchildren from three hundred to eight hundred.

– coauthored by Regine Kennedy

relations. If SELROSLT were to disintegrate, Johnson says, all of its community gardens would automatically revert to city ownership. And that could be a problem, because even if current city leadership wants to protect the gardens, there's no guarantee that its future leadership will. So the land trust itself must be stable if it is to ensure long-term protection of its gardens.

Managing the Challenges of Diversity

Community gardens, by their very nature of bringing together people in a closely shared space, are a perfect breeding ground for conflict. As more people want to garden, and the wait list grows, people begin to question a host of things: Is the assignment method fair or democratic? How is the money managed? Are the rules working? Are shared tools being cared for properly? Is water being conserved? Should people be able to compost or not? What if a gardener isn't taking care of his plot, or worse, what if someone is found taking from another garden?

But gardens are also a place for bringing people together to build understanding. "The beauty of the garden is that it's a place where differences may be neutralized," says Carol Bonnar, volunteer manager of South End/Lower Roxbury's second largest community garden on Worcester Street. With training and experience in project management in the world of art, Bonnar was eager to help the Worcester Street garden become all it could be.

She quickly discovered that managing a community garden is not like other projects. Diversity in a garden is a strength, giving people a place to reach across social and economic boundaries, and enabling people to learn from one another's heritage and traditions. Many of the Worcester Street gardeners are Haitians living in subsidized housing, and they bring to the garden seeds from their homeland. Because of the gentrification of the neighborhood, others, like Carol, come from predominantly white suburbs. There can be stark differences among the gardeners, and it's one reason Carol is drawn to the work.

Yet diversity in a garden can also present unusual challenges. Democracy is dependent on participation,

Bonnar says, and on people understanding the issues. So managing a garden based on democratic principles can be frustrating when some people don't speak English or when an entire culture typically steers clear of structure or governance. Bonnar readily admits that her life experience hadn't fully prepared her for the challenge of helping a diverse community learn to govern itself. She's learned that there's a fine line in taking leadership. "You need to know when you can act on your own," she says, "and when you need to call a meeting of the entire garden, to at least give all the gardeners an opportunity to participate." This, she says, is "due diligence," or doing the research and taking all the necessary steps to make an informed and appropriate decision.

Most community gardens have evolved a common set of principles that help them navigate the difficulties of diverse personalities, cultures, races, ages, gardening experiences, and even schedules. A garden operates more smoothly when gardeners sign a clear agreement that outlines their responsibilities and the garden's expectations of them. Community gardens usually require gardeners to agree to use only organic or no-spray methods. Many specify the kinds of materials that can and cannot be used.

Running a smooth community garden also involves deciding how plots will be assigned, often a highly contentious issue in older, more established gardens, where longtime gardeners may have accumulated more than one plot. In the South End/Lower Roxbury land trust, 11 of the 16 gardens have gardeners with multiple plots, and one gardener even accumulated eight garden plots, using different family member names.

Equally important is deciding when and how to evict gardeners who are not abiding by the rules. Most SELROSLT gardens are open during the day to the public but locked at night to prevent theft and vandalism. Internal theft — when a gardener is found taking from another garden — is usually grounds for immediate eviction. And most gardens have a system for warning a gardener to clean up his or her plot, and then, if the plot is not cleaned up within a month, notifying the gardener that the plot is being reassigned.

Carol Bonnar believes that every community garden should devise its *own* set of standards, through discussion with its gardeners. That way each garden will develop its own concept of equity and parity that reflects the values of its community.

Dealing with Soil Contamination

For many years the Orange Line train rumbled through the heart of Boston and Roxbury. In the mid-1980s, as part of a massive urban renewal project, the city moved the train underground and gave away the railroad ties to community gardens. When news of the arsenic and other toxins from the ties spread, Bonnar says, "we were told not to plant root vegetables near them."

Bonnar heard about Boston's Jamaica Plain project, where the Boston Natural Areas Network, the Boston University School of Public Health, and the state worked together to haul out all the toxic soil from a tiny garden, clean it, and then truck it back in. It was hugely expensive, she recalls, but it led her to contact the groups involved. The School of Public Health came to take soil samples at the Worcester Street Garden and concluded that the contaminated soil there was so extensive that it couldn't be hauled away for cleaning.

Like so many substances once used for good reason and later found to be a carcinogen, creosote was used to help preserve railroad ties and only later was discovered to leach into surrounding soil. As if that wasn't bad enough, a Boston University School of Public Health associate professor of environmental health, Wendy Heiger-Bernays, suspected there might be another toxin associated with creosote and decided to investigate. She was right. She and her colleagues found that polycyclic aromatic hydrocarbons (PAHs), which can harm immune and reproductive systems and cause cancer, were also leaching from the railroad ties into the soil.[60]

Heiger-Bernays recommended that SELROSLT gardens (and two other sites in Boston) remove soil underneath and around the timbers. Because the cost of replacing that contaminated soil with new soil would have been prohibitive, Heiger-Bernays created an affordable recipe for the garden: two parts donated city compost, tested and completely clean; one part old

soil. The clean compost, Heiger-Bernays said, would help break down any PAHs remaining in the soil.[61]

Worcester Street community gardeners and over a hundred volunteers set to work, tearing out all the old ties and toxic soil in 2008 and bringing in cleaner soil in 2009. The effort was massive, but the next step was equally massive. Through the winter the gardeners considered options for new bed borders. Trex (a recycled material used as a substitute for wood) was expensive. Pressure-treated wood, even though considered nontoxic, was abhorrent to the now risk-averse gardeners. They landed on bluestone, when a company that Bonnar approached agreed to help. The large paving stones would have cost about $40,000, but Bonnar raised $8,000 and the company generously donated the rest.

With the entire garden ripped apart, the gardeners decided to take the opportunity to redesign wider walkways and create high raised beds for the handicapped. Now the garden is a network of paths of bluestone dust, and the bluestone pavers create beautiful — and completely inert — borders. Bonnar says the soil will be tested again soon, to see how the remediation "recipe" is working.

Another SELROSLT community garden shares a legacy of contamination by PAHs. The Bessie Barnes garden is set among a neighborhood of homes historically owned by middle-class African-Americans. The PAHs are a group of one hundred chemicals, according to Heiger-Bernays, formed during incomplete combustion of organic substances such as oil, garbage, and coal.[62] When asked how the PAHs got into the Bessie Barnes garden soil without the presence of railroad ties, Betsy Johnson sweeps her hand into the air and says, "They're from burning coal and urban living."

In this garden, instead of removing the soil, the EPA decided to "cap" it, meaning they covered it with concrete. But the good news is that they didn't "pave paradise to put up a parking lot," and instead built a safe food garden. Tall raised beds filled with compost now grow an array of tomatoes, beans, squash, cucumbers, corn, and even grapes, pears, and nectarines. The Bessie Barnes garden proves that food can be grown in the worst of conditions — on top of concrete.

A Catalyst Role

The importance of community gardens can be a touchy and controversial subject. "Lots of people give community gardening flak," says Betsy Johnson. "They say community gardening isn't worth anything in our food security. . . . But it *does* make a difference," she says forcefully. "Particularly for low-income and immigrant gardeners, the food production is phenomenal."

Johnson has seen it, and so have others. I recall the story of residents in a Portland, Oregon, housing development who began digging up little plots of land outside their apartment entrances, not because they thought it was a cool new fad, but to put more food on their family tables (page 17). Or when the Puerto Rican population of Holyoke, Massachusetts, left jobless when textile and tobacco factories shut down, began growing food gardens because they didn't want to starve (page 123).

But are community gardens "the answer" to world hunger? Are community gardens for everyone?

Johnson readily answers, "No! It's hard work, and time consuming. So it's not for everyone." But Johnson makes a case for why critics are missing the point. Nobody is envisioning a future where every morsel of food is produced locally, she says, or where every family tends its own garden. And nobody is arguing that community gardens can feed the world, or feed all the hungry, or meet all of a community's food needs.

What community gardens can do, however, is in some ways far more interesting and powerful than the mere production of food. Betsy Johnson argues that community gardens are a first step — or a stepping-stone — to *other* food projects. She's seen it over and over again, as former interim executive director of the American Community Gardening Association. Johnson begins to rattle off the stories.

In St. Paul, Minnesota, she says, a simple children's garden led to a community garden, which then attracted Hmong immigrants, many of whom were deaf because of the bombing they experienced in Southeast Asia. The Hmong eventually became more intensive in their food production, which led them to expand the garden and find other sites. Then the Hmong wanted to be able to sell their produce, which led to a farmers' market. But because the Hmong were deaf and didn't speak English well enough to sell their produce, the situation led to a youth education program that taught youths to sell the Hmong produce.

Johnson barely takes a breath and moves on to another example. She reminds me of the small Bessie Barnes garden in Lower Roxbury, built on the concrete cap. This small garden has produced an abundance, she says, and the gardeners now want to sell their extras. So they are establishing a farmers' market in an underserved neighborhood. And now, to make the market successful, the gardeners are recruiting other farmers to sell at their market. The individual gardeners are not allowed to profit from their sales — a common rule for community and public gardens — but they can reduce the overall garden costs by donating their proceeds to the garden and thereby reduce their garden plot fees.

And Johnson keeps going, ticking off other projects. The Greening of Detroit, led by Ashley Atkinson, started as a community garden and has evolved into a comprehensive food project. Just Food in New York City began with community gardens in the Bronx, which led to a farmers' market that would sell produce from the gardens. Nuestras Raíces in Holyoke, Massachusetts (page 123), started as just one small community garden and very quickly transitioned to much more — youth empowerment, women's empowerment, and a farmer and small-enterprise incubator.

A community garden is a catalyst, Johnson says. It increases the community's awareness and interest in a host of shared concerns seen through the lens of food: economic development, social justice, nutrition, and public health. Growing awareness and interest, in turn, often lead to other important steps. Small grassroots projects begin to use food to help specific populations — shelters, food pantries, youth, seniors, and immigrants. And these, in turn, lead to interest in new farmer programs, new distribution networks, and new training programs. And then people get interested in the marketing, through local food guides, tours, and events. Whether it happens slowly, over many years, or very quickly, in a matter of only one or two years, Johnson says, it's a natural evolution — all spurred by community gardens.

A Long Tradition of Community Gardens

Today's community gardens are descendants of a long tradition that dates back to the "Progressive Era," from the 1890s through World War I, when well-intentioned women of the upper and middle classes considered it their duty to aid those less fortunate than themselves. Then, community gardens were seen as stepping-stones for the poor. They created jobs, educated, alleviated hunger, and served as a gateway for moral reform and assimilation. Gardens were planted largely by public service organizations in reclaimed vacant lots and at homes and schools, with the first recorded school garden established in 1891 – in Roxbury. These gardens gave way to the "war gardens" brought on by two World Wars. Home food gardens and canning were characterized as important tools for every citizen to help fight the enemy. The federal government even established the National War Garden Commission in 1917, supported with technical assistance by the Garden Club of America, and enlisted children into a "School Garden Army" in 1918. And during the Depression in the 1930s, gardens were an important stopgap, funded by the Federal Emergency Relief Administration, until greater relief measures could be put in place.[63]

But it was the famous victory gardens of World War II that revealed the amazing potential of community gardens – more than 20 million gardens in urban and rural communities produced over 40 percent of the vegetables grown in the United States.[64] Just a mile away from the South End/Lower Roxbury, the thriving Fenway Victory Garden is one of the few survivors of this bygone era. Home gardening and canning certainly continued after the war, but at a much-reduced intensity and often as a hobby more than a necessity.

The 1970s ushered in a brand-new energy for community gardening. Unlike in previous eras, when gardening was promoted by civic or government agencies, Betsy Johnson says, "*this* time it was from the bottom up!" The gardens of this era were grassroots and community-driven. Cities were being sickened by urban flight and disinvestment, and the flight of grocery stores to the suburbs meant that immigrants seeking a better life in our cities couldn't get healthy or culturally familiar foods. So, Johnson says, city dwellers began growing their own food. And these efforts grew into outspoken activist, nonprofit organizations that began to exert political and social influence. Community gardening became a movement. And this movement gained sufficient momentum in 1979 to create the American Community Gardening Association (ACGA).

Another sea change occurred in the mid-1990s, Johnson says. People began to see the broad benefits of gardens and wanted to use them to help specific populations. So community gardens started to become more top down and more programmatically driven. Instead of being started by community residents, now a nonprofit would decide to start a garden to help local youth, the hungry, or the homeless.

Johnson isn't sure this is a good trend. The successful gardens are those started by gardeners, for gardeners. But the trend has grown more extreme, she observes. Now, as the food movement is gaining steam, even larger organizations want to jump on the bandwagon. Johnson says some corporations and large nonprofits think they can just start a community garden by throwing a little bit of money at a community. She gives the example of a corporation that wanted to start six community gardens, for $1,000 each. Johnson doesn't have to vocalize the obvious questions: Does the community even *want* a garden? And where are the leaders who will lead and sustain this effort?

Another corporation was more savvy, in Johnson's opinion. A community garden, she says, drawing from her decades of experience, takes time, dedication, and leadership. This corporation decided it would commit $10,000 per city just to support the leadership development needed to get a community garden off the ground. To develop this leadership for community gardening, the ACGA offers a two-day workshop. And the workshop is not about gardening – it's about community organizing, meeting facilitation, fund-raising, coalition building. The workshop staff tell people, "If you've come here to learn about gardening, you've come to the wrong place." The reason is simple: the difference between a successful community garden and an overgrown lot that was once a garden is leadership — leadership from within, from the ground up, not the top down.

Of course, community gardens are not the ultimate or only answer. But they are a shared community experience that often catalyzes a cascade of expanding community awareness and network of programs. Where it leads depends on the specific community culture and needs. For Nuestras Raíces (page 123) it has led to youth and women's empowerment and a farmer incubator. For Janus Youth (page 105) it has led to a community store that will sell community garden produce and value-added products and provide jobs for local residents.

In all of these communities, it might be argued, a community garden may also be a stepping-stone to a new community culture — a culture that values food for more than the nutrition it contains. This emerging community culture sees food as an end *and* a means to a healthier community — a more diversified and locally sustainable economy, new job and small enterprise opportunities, better eating habits and nutrition, lower health costs, fewer days lost on the job — and to empowerment, self-esteem, and social justice. These may sound like heady claims, but projects around the nation are demonstrating their reality. "It's not like a light bulb goes on for *everyone*," Johnson says, "but a community garden is usually one of the first stepping-stones."

Lessons Learned
Running a Community Garden

Community gardens must begin with the community.

The desire to help is noble, but not sufficient. Time and again, the experiences of food projects across the nation point to the same guiding principle: if a project such as a community garden is to truly help a community of people, it must be theirs from the beginning. They must own it, design it, build it, and maintain it. Jamescita Peshlakai, lead coordinator for the Navajo Nation Traditional Agricultural Outreach program of DINÉ Inc. (page 177), says she is most effective in helping to develop community gardens when she articulates that her role is to *support* the community, helping it develop its own vision and goals for the garden. Leadership must come from within the community, she says.

In the low-income community of New Columbia in Portland, Oregon, Eca Etabo Wasongolo, a Congolese immigrant and recently hired community organizer for Village Gardens, a program run by Janus Youth (page 105), agrees. He says, "Never go into a community and say, 'Let's start a garden.' . . . First you have to meet with people, tell them about the nature of your program, and learn from them what they'd like to [have]. Maybe they'll say, 'Better health.' So then maybe the conversation leads to the idea of a garden. You need to be a guide, always asking people, 'How do we do this? Do you think we will be able to do this?'" Referring to the expansion of Janus Youth's Food Works teen farm into poultry production, he asks, "Do you think if they said at the beginning, 'We want you to grow chickens,' do you think anyone here would have supported that?" He laughs, then says, "But, through natural evolution, allowing the community to make the decisions, here we are, beginning to grow chickens!"

Similarly, the Virginia organization Lynchburg Grows (page 116) built its first community gardens slowly. It didn't rush into the neighborhood with a plan to "help." After purchasing abandoned city lots where it hoped to create neighborhood gardens, Lynchburg Grows did nothing for four years other than mow the grass on the lots and ask curious neighbors what they might like to see happen on there. Executive director Michael Van Ness is convinced that this organic approach explains why the eventual groundbreaking successfully

bridged all racial, cultural, and generational divides in the community — and why the neighborhood immediately wanted to expand their garden.

A group of committed community leaders is essential.

If a single person approaches the city with an idea for a new community garden, Richard MacDonald, manager of Seattle's P-Patch program (page 60), will tell him or her to return with five or six more people. One person is not sufficient to steward a new garden into fruition. "When they have a good *group* on board, then we sit down and create a strategy or map of how they will move the garden forward," says MacDonald. Key items to discuss are:

- Identifying the skills and tasks required and recruiting other people for these roles
- Making sure other community groups have opportunities to be involved, such as nearby food banks, native plant groups, or green building groups
- Making the planning group inclusive, to reflect the neighborhood's multicultural population

Rely on collaborative decision making.

In our conversation, Village Gardens' Wasongolo stresses one thing: all decisions about the garden must be made jointly, by all members. As an environmental mediator, I know that the strength of a group is related to its commitment to collaborative decision making, and so I can't help asking him, "What happens if people disagree?" The other garden leaders standing nearby all turn to look at me as if I had just sprouted horns. "But they must agree," says Wasongolo, surprised. "It's their garden and they have to reach a decision, and they must do it together."

Schedule regular meetings for decision making.

Village Gardens has found that the community gardens it supports thrive when they hold regular meetings for decision making. One community garden holds meetings on an unusually intensive schedule: once a week, every Saturday, without fail. During these two-hour meetings, the garden community plans events and workshops, prepares for visitors and tours, and discusses garden sign-ups, plant distributions, community workdays, and new program opportunities. This might seem like an intensive commitment, but program director Amber Baker believes the meetings feed a sense of community: participation is open to all members and is completely volunteer, and people keep coming.

A grassroots support group helps sustain a city-managed community project.

Through the decades, as Seattle's budget and politics shifted, an important force for sustaining city support for the P-Patch garden program has been a grassroots volunteer-driven group, Richard MacDonald says. Initially formed in 1979 to support the program, the citizen "P-Patchers" evolved through the years from an advisory council into a more formal group and finally into a formal 501(c)(3) nonprofit known as the P-Patch Trust. Without the long-term support of this citizen support group, which has worked diligently to secure land and raise funds for Seattle's community gardens, the P-Patch program might not have been able to weather the vagaries of city budgets through four decades.

Establish clear rules and guidelines for the garden.

Carol Bonnar, manager of the Worcester Street Garden in Roxbury, Massachusetts, for the South End/Lower Roxbury Open Space Land Trust (SELROSLT; page 62), urges community gardens to establish rules and guidelines in collaboration with the gardeners, to increase their sense of ownership of the garden and also to make sure that the rules will reflect their values and goals. Typical items to cover in the rules and guidelines are the process of allocating plots, fees, use of tools and watering equipment, maintenance requirements, and a system for allowing the garden to evict someone who isn't using or tending his or her plot. The Worcester Street Garden uses a system of flags: If, for example, weeds are overgrowing a plot and invading other gardens, a first warning flag is posted. Two weeks later, if the problem has not been addressed, a second flag goes in, of a different color. Two weeks later, or a full month after the first warning, the final flag tells the gardener that the plot is going to be given to someone else. (There is an appeals process that allows the gardener to be put on the waiting list for new plots.)

Management of the garden plots is not the only factor to consider when drafting community garden guidelines. At the Janus Youth farm and gardens, where members felt that ensuring safety, discouraging disruptive behavior, and facilitating emotional development in youths were priorities, all members are required to sign a "community contract" that is in essence a code of conduct (see page 107).

Learn from others.

For those wanting to start a community garden, Village Gardens' Amber Baker says, one of the best ways to avoid costly mistakes is to learn from others in the field. Visit other projects and meet with people who have experience with different types of projects. Then share what you've learned with your community so that you can tailor the project to your community needs.

To bolster community support, keep the garden looking good year-round.

Community gardens are always beautiful and lush in the summer. But the key to obtaining community support, says SELROSLT director Betsy Johnson, is whether the gardens also look good for the remainder of the year. Simple things like having an attractive entrance, such as a pretty iron gate or colorfully painted fenceposts, can help. And, of course, clear guidelines for keeping garden plots maintained are essential.

Don't be afraid to ask local businesses and organizations for help.

When SELROSLT discovered that the old railroad ties that had been used to create raised beds in its Worcester Street Garden were potentially leaching toxins into the soil, they were left scrambling to find something to replace them with. Carol Bonnar, the garden manager, went to a local business, McVey Stone Company, and asked if they would help – and they were happy to make a donation. The results are raised beds made of beautiful (and inert) hunks of bluestone, laid deep into the ground for stability.

SELROSLT knew that the task of digging up the old railroad ties, installing the donated stone, and replacing the soil would be immense, so it asked for help. It found volunteers in City Year's Young Heroes, a local leadership program for middle-school children, which brought more than a hundred volunteers to assist.

3: Urban Farming
Growing Food in the City

Agriculture has not always been banished from our cities. In fact, an argument might be made, however unlikely it might now sound, that history will prove the twentieth century more of an anomaly than the rule with its proclivity for "denaturation."

During the twentieth century, as cities edged toward greater sanitation, they also edged toward more sterile, rigid landscapes, stripping cities of their connection to nature. Streams, now in the way of development, were hidden beneath paved streets and parking lots. Cattle, goats, pigs, chickens, and bees, no longer suitable for sophisticated city sensibilities, were considered unsanitary and a threat to property values. Moving people efficiently by cars meant that roads were designed without consideration for people traveling by bicycle or foot, often making it perilous to do so. And in the interest of car safety, road design commonly prevented median and roadside plantings of edible fruit trees and shrubs, as they could attract wildlife and create hazards. Car-based lifestyles and community design conspired to create the unintended and bizarre consequence of social food amnesia, with generations of urban children unaware of where their food comes from.

Now, as the twenty-first century ushers in a new sensibility for urban life, communities of all sizes are revitalizing their core by softening their urban edge — planting trees down Main Street, "daylighting" streams that were hidden from sight for decades, redesigning roads to accommodate bicycles and pedestrians, planning and planting greenways and green infrastructure. And reconnecting with food is no less important, as communities blur the edge between urban and rural by fostering the formation of weekly or year-round farmers' markets, local food festivals, community gardens, rooftop gardens, school gardens, farm-to-institution programs, and urban farms.

Some cynics might dismiss these trends as just that, short-lived and superficial. Yet the configuration of today's pressures on communities is forcing leaders and citizens alike to rethink what they mean by the word "community." Communities are now seeking to foster qualities that will enable their economy, ecosystem, and social fabric to be *resilient* — that is, to be able to continue thriving in the face of both expected and unexpected events or catastrophes. Post-Katrina, the New Orleans Food and Farm Network created the Community Food Charter, which declares that "a vibrant food system is central to the health of our families, our economy, our schools, our health care system, our culture, our environment, and our sustainable future." Like New Orleans, communities of all sizes are turning to food initiatives for neighborhoods, schools, senior centers, at-risk youth — finding that these programs increase neighborhood aesthetics, reduce crime, build community, combat nature-deficit disorder, and improve public health.

Now, if the twenty-first-century urban farmers you'll meet in this chapter are right, urban farms are here to stay, becoming as common a sight in small towns and large cities as ball fields and parks. Urban farms may never grow enough food to feed their community in entirety, but the food they do produce contributes a measure of resilience to the local food system and economy, while also providing a compelling educational presence to feed community spirit and connection with the natural world.

Practicing Intensive Agriculture in an Urban Setting
Growing Power

coauthored by Regine Kennedy

At last count the nonprofit Growing Power included 12 working farms, both within and outside the **Milwaukee** city limits; a retail store; a CSA market-basket program that delivers to restaurants, schools, and institutions; and 12 regional training centers across the country. Its mission: to inspire communities to build sustainable food systems that are equitable and ecologically sound, creating a just world, one food-secure community at a time.

Will Allen, founder and CEO of Growing Power, grew up on a farm in Maryland. After a 10-year run as a college and professional basketball player and several years in corporate marketing, he returned to his roots, commercially farming some family land in Oak Creek, a suburb of Milwaukee. In 1993 he purchased the last remaining farm in the city of Milwaukee, totaling just less than 3 acres, with several run-down greenhouses. He started an urban farm on the site in 1995 when he agreed to help the local YMCA with a youth garden training project. Today, approaching 20 years from these humble beginnings, Growing Power now employs a staff of 50 people who work both on- and off-site in a variety of capacities, including growing food, advising on food policies, and developing renewable energy systems for more than 70 projects worldwide.[65]

Growing Power offers the full farm experience, not only growing vegetables and greens, but also raising fish and a menagerie of goats, chickens, ducks, bees, and turkeys. Growing Power sells its produce through a market-basket program for low-income families as well as from its on-site retail store, which features produce from a co-op of three hundred farmers. The group's warehouse is staffed by local youths who, in addition to learning about building hoop houses and growing vegetables, dedicate two hours each day to academics to help improve their reading and writing skills.[66]

Allen has seen the difference that connecting with the soil can make in children's ability to learn. He describes how kids can come to the farm edgy, hyperactive, and unable to focus, and how, when they dig their hands into the soil or the piles of worms, they quickly settle down. Allen suggests that the farm has an ability to reach and teach children in different, and sometimes more effective, ways than classrooms.

In keeping with his sports background, Allen views the Growing Power organization as a franchise of sorts where he is the team captain, but everyone is an essential partner and part of the team. In keeping with his farming roots, Allen believes urban farming is the answer to many of today's issues — jobs, education, economic development, equity, and social justice.

Allen has come to this vision slowly, after many years working with Milwaukee's North Side neighborhoods. Though he began by farming in Milwaukee's suburbs, "he was drawn to doing something in the *city*," says Jerry Kaufman, president of Growing Power's board of directors. Allen wanted to work with youths, to teach them how to grow food, because he felt they had little understanding of where food came from. "He had no great vision and no great plan at the beginning," says Kaufman, "but he bought the dilapidated greenhouses and slowly started cleaning up [the site], working with youngsters and getting some recognition for his work."

Today Allen's persistence and patient willingness to learn from his mistakes have earned him — and the Growing Power team — national and global recognition as a leader in urban farming. Yet Allen insists that what he is doing is as old as the hills: "Nothing at Growing Power is new. People have been growing food inside cities forever." Through his work and successes, he has come to believe that urban farming is about far more than food. "We have to grow farmers so we can grow healthy people and sustainable communities," he says.[67]

In 2002 Growing Power branched out to Chicago, where it helped establish community gardens in low-income neighborhoods, model urban farms, training programs for farm interns, and educational initiatives for neighborhood kids. Allen's daughter, Erika, who runs the Chicago operations, explains why Growing Power expanded and the reasons she shares her father's passion: "We just see food as a really powerful organizing tool. It deals with land, housing, transportation, economics, everything. For us, it's really a tool for transformation."[68]

In 2008 Allen's originality and dedication were recognized when the MacArthur Foundation awarded him a prestigious MacArthur Fellowship, their "genius grant," in recognition of his practical approach to fighting inner-city hunger through a holistic farming model that incorporates growing and distributing food in urban areas. "Allen has spent the last 16 years developing his business as an organic farmer, dedicated to growing and providing healthy food to low-income areas," writes Shannon Sloan-Spice. "He passionately teaches and encourages sustainability to communities all around the world, and in so doing is fighting economic racism."[69]

Growing Vertically

Allen believes that the key to boosting urban farms is paying attention, not to social justice concerns, but to the bottom line. "Everybody just says, 'Oh, I can't grow organic food cheaper,'" says Allen. "But we don't have to pay $10 a pound for food — I charge $2 a pound and make money. You could be a millionaire."[70]

The secret, Allen says, lies in growing vertically. A vertical farm can produce $200,000 on an acre, which is what Growing Power produces through intensive growing in its Milwaukee greenhouses,[71] as opposed to $500 an acre in a typical rural farm setting.[72]

At Growing Power in Milwaukee, vertical farming is best exemplified in the low-cost, high-yield aquaponic (hydroponics and aquaculture) system. Here farming is taken to new heights. Literally. Allen's aquaponic system is vertically integrated, consisting of three stages: two levels of plant troughs, typically tomatoes and salad greens, stacked above fish tanks stocked with yellow perch or tilapia. Water from the fish tanks drains down to a gravel bed, where beneficial bacteria break down the ammonia in the fish waste, converting it to nitrogen. Watercress in the bed helps filter the water. From there, the water is pumped up to the top level, where it fertilizes tomatoes and other plants, row by row, from one end of the platform to the other. The now-filtered, clean water flows back into the fish tanks on the bottom level, and the whole process starts over.

All the products from this system — tomatoes, greens, and fish — are cash crops for Growing Power, sold through its diverse outlets.[73] The system is meant to replicate the workings of a healthy creek system, and at an affordable cost in a community setting. At only about $1,500 for the simple materials, this fish-producing aquaponic system cost about one-tenth the amount of a conventional aquaculture system that uses chemical processes to clean the water.[74]

Allen works with the Great Lakes WATER Institute (GLWI) at the University of Wisconsin in Madison to raise yellow perch, a native Lake Michigan species that's seen a catastrophic decline since the 1980s.[75] According to GLWI scientist Fred Binkowski, fish farming is part of a trend to grow safer, better food closer to home. It is also a lucrative way to fill the gap between consumer demand for fish and what can sustainably be caught in the wild.[76]

Though lake perch require cold-water systems, the tilapia in Growing Power's tanks require warm-water systems. In the greenhouses with tilapia tanks, Allen says, simply heating the water for the fish provides enough heat for the plants in the greenhouse.

Bright Idea: A Cooperative of Small-Scale Farmers

In 1993, well before the idea for Growing Power, Will Allen joined four other farmers to create the Rainbow Farmer's Cooperative. Together the rainbow of farmers – Hmong, African-American, and white – pioneered what was, at the time, a new kind of farmer cooperative. Each farmer grew a percentage of each vegetable product, the cooperative pooled and packaged the products, and, collectively, they were able to market larger quantities to supermarket chains. The cooperative covered the transportation and marketing costs, and farmers were ensured sales. And by cutting out the middleman distributor, the farmers also received a higher percentage of sales dollars in their pockets.[77]

Today the Rainbow Farmer's Cooperative has grown to a dynamic enterprise comprising about three hundred small-scale farmers and sellers across nine states in the Midwest and Southeast, offering an array of vegetable, dairy, and meat products as well as honey, jams, and other items.[78] This kind of cooperative fills an important gap – providing the transportation, cold storage, and marketing for small-scale farmers who often don't have the time, staff, or desire to engage in this part of the business.

And some of the tanks are set in trenches, creating a form of geothermal heat.[79]

Composting: Black Gold

To make compost Growing Power processes approximately 500,000 pounds of organic matter between its urban and rural sites each year. Local businesses provide a steady stream of materials for creating this "black gold." Food waste is collected from a number of local partners, including Children's Hospital, Rockwell Automation, Maglio Produce Wholesaler, and University of Wisconsin–Milwaukee residence halls. Each week Growing Power picks up 300 pounds of coffee grounds from Alterra Coffee and Stone Creek Coffee, while Lakefront Brewery provides 20,000 pounds of brewery wastes. Added to this mix are eggshells, cardboard boxes, grass clippings, moldy hay, leaves, and wood chips. All told, Growing Power is able to collect, every week, more than 100,000 pounds of waste from its community partners, diverting more than 2,000 tons annually from Wisconsin landfills and turning it into rich compost for growing food.[80]

Heifer International initiated Growing Power's vermicomposting systems in 1996, when it brought in the first worms.[81] As worms "eat" their way through organic matter, they produce castings (feces) that are, as the Growing Power website says, "a nutrient-rich, organic fertilizer" that can be used to amend soil directly or to create compost tea, which is applied as a liquid fertilizer. Not all worms are created equal for

vermicomposting; Growing Power uses red worms (*Eisenia foetida*), sometimes known as red wigglers. The worms are raised in bins or added to the long compost windrows. The castings are used in Growing Power's own gardens and also packaged and sold in the group's retail store.[82]

The main function of compost is to improve soil fertility, but at Growing Power it also serves as a vital heating system for greenhouses. Long windrows of compost are piled against the outsides of greenhouses, and small compost piles and worm bins are kept inside. As the compost decomposes it generates heat, warming the greenhouses. Compost is also the foundation for the greenhouses' raised beds. A layer of 12 to 18 inches of partially finished compost, topped by another layer of completely finished compost, generates radiant heat from below, keeping the raised beds sufficiently warm to grow plants — even through the cold Wisconsin winter. This innovative heating system requires no oil or natural gas and proves that growing food through the winter is possible and cost-effective and can even be accomplished with green energy.

Going Green with a Biodigester

As a leader in urban farming, Growing Power is continually looking for ways to push the envelope to make urban farming more energy-efficient (and therefore more cost-efficient). One of its more recent endeavors is experimenting with a biodigester to turn waste into electricity. Biodigesters are airtight containers that use bacteria to turn organic waste, such as animal manure or food waste, into a mixture of methane and carbon dioxide, also known as biogas. Pilot projects are cropping up across the country to test biodigesters in transforming animal waste into energy, particularly in dairy states such as Vermont and California.

Allen is one of the forerunners in using this method to test the transformation of urban food waste into energy. The food waste that Growing Power receives from local partners is transformed through a process called high solid anaerobic digestion. Simplified, this process uses biological microorganisms to turn organic waste into acetic acid, which, with the addition of other biological microorganisms, is then turned into methane gas, which can be used to run a generator or generate electricity.[83] Allen plans to expand the collection of organic material, for use in composting as well as for fuel for the biodigester, in a pilot program that targets local residences.

For many farmers one of the greatest costs in expanding the food production season is the cost of heating a greenhouse. Allen has already shown how greenhouse food can be grown through cold Wisconsin winters with no heat source other than compost. Now he is exploring other ways that communities can contribute to energy-efficient urban farming. He hopes to prove to his city that residential organic waste collection makes sense and is essential to growing food in urban areas.[84]

A Good System Is Replicable

Allen has the unusual ability to inspire and convince others that what Growing Power is doing in Milwaukee is not unique, is not difficult, and can be replicated anywhere there are people with the desire to do so. And he is intentionally using this talent to spread the word, traveling to communities throughout the nation, indeed the world, often at a punishing pace, to share his conviction that urban farming is part of our future and to grow a new crop of urban farmers. From his first modest effort to help local youths with a garden in the 1990s, Allen has expanded Growing Power's educational initiatives to workshops

on urban gardening, renewable energy, vermicomposting, community food systems, aquaculture, beekeeping, community food project design, year-round greenhouse production, hoop-house construction, mycology, and cheesemaking.

Allen says he likes to "teach and grow food," which is how the workshops began. But that doesn't convey his core passion, which, combined with his self-deprecating humor and imposing presence, quickly warms his audience to open their hearts and minds to learning from his decades of experience. Now people from all over the world, from diverse backgrounds and professions, attend Growing Power workshops and its five national conferences about agriculture and food systems.[85] In his workshops Allen uses a sophisticated presentation to show the history of Growing Power — how it evolved from a raggedy place with glass missing from the greenhouse windows to the dynamic place it is now, teeming with plants and animals. Allen explains how the original worm bins, now about 13 years old, were made out of whatever wood was available and how the aquaponics systems have evolved from independent systems to the integrated, multilevel systems they are today — the vertically integrated system that he teaches workshop participants how to build.

Time and time again Allen comes back to the importance of healthy soil. "You don't need a green thumb," he says, "you need good soil." Composting and vermicomposting get the process started, and from there, according to Allen, things happen. At each step along the way, Allen stresses the importance of connections and community — connections for food waste for the composting systems, connections to schools for the youth corps, connections to the community for selling the vegetables and other products grown and raised on the farms. Systems are also one of his themes, whether he's speaking about the vertically integrated aquaponics systems or the network of suppliers, producers, institutions, and organizations. A connected and thriving community — or an effective community system, if you will — is the ultimate outcome for a successful urban farm.

Lessons Learned
Farming in the City

Reach out to neighbors to prevent perceptions of nuisance and conflicts.

Operating on Milwaukee's North Side, in a modest neighborhood with affordable state and federal housing projects, the nonprofit Growing Power (page 74) learned that not everyone likes the idea of living near a farm. Some neighbors were uncomfortable with the idea of living near chickens, ducks, goats, beehives, and a significant composting operation that might give off occasional odors.

"In the early stages, we had to work to gain the trust of surrounding residents," says Jerry Kaufman, president of Growing Power's board of directors. "They were unfamiliar with what a farm is all about. So we invited them to come to the site." Once the neighbors could actually see what was happening on the farm, their concerns diminished. Also, when Growing Power staff learned that odor was a concern of the neighbors, they did some research on the topic and learned that turning compost frequently generates more odors. Therefore, instead of turning the compost frequently, Growing Power staff now lets the worms do their work.

These and other lessons have led Growing Power to believe that urban farms are best created only where they are desired. When Growing Power is asked to assist in establishing an urban farm, one of the conditions that must be met is that the community must fully support the concept.

Build one thing upon another — figuratively and literally.

One of the highest priorities for Will Allen, founder and CEO of Growing Power, is to grow good soil, which allows everything else to follow. "Once you're able to grow healthy soil, you can grow healthy food," he says.[86] Composting and vermicomposting are key soil-building methods. Teaching others in the community how to compost builds connections. And community connections build the capacity to collect organic waste to be used for composting and the biodigester. Add vertical farming, a dynamic, integrated system unto itself, and more products are available for distribution and sales, which in turn builds more community connections. Growing Power's increasingly complex system of interconnected pieces is built on the simple foundation of healthy soil and community connections.

To gain political and community support, promote the advantages of urban farms.

Advantages of locating a farm in an urban area are many. Rather than spending money on packaging, lengthy refrigeration, and transport of produce to market, you'll find the market is right there, all around you. Both Philadelphia's Greensgrow (page 96) and Virginia's Lynchburg Grows (page 116) say their customers come from every section of the city, not just the immediate neighborhood.

Also, urban farms have ready access to labor. This may seem counterintuitive, as city people are not often considered prime candidates for farm labor. Yet Mary Seton Corboy, CEO of Greensgrow, notes, "Greensgrow has found that there are a lot of very smart, motivated, and hardworking people who want to work in urban agriculture."[87]

The biggest gap for urban agriculture, according to Corboy, is lack of available training. Of course, organizations trying to assist new farmers in rural areas often acknowledge the very same gap.

Earn neighborhood support with transparency and an open-door policy.

Everyone in the neighborhood should always know exactly what's going on at an urban farm, says Greensgrow's Corboy. Transparency is vital to earning the community's trust and support, particularly if the site was formerly contaminated (or otherwise problematic) and has a legacy of harm in the community. Corboy recalls a time when Greensgrow's neighbors were worried about a fluid that was running out of the farm into the streets. Greensgrow immediately took a sample to be tested and showed the neighbors that it was just water with soap in it, not a pollutant.

Another time, she says, one of the Greensgrow hives swarmed. A neighbor came to her office to tell her in whose backyard the bees had landed. Because of all the small ways she has shown that Greensgrow is trustworthy, the neighbors trust her. "I've been here 11 years," Corboy says. "They know I wouldn't raise bees in their neighborhood if I didn't think it was a good idea or safe."

Leverage both the advantages and disadvantages of an urban location.

Troy Community Farm in Madison, Wisconsin (page 41), has worked to leverage the advantages and disadvantages of its urban location to its benefit. Unlike a typical rural farm, which may be able to provide housing but experience difficulty in finding workers, Troy Community Farm does not offer on-site housing but has access to a huge labor pool. Students, interns, and community volunteers can simply bike or bus to work at the farm. Also, because it is easy for the city community to engage with the farm, Troy Community Farm can turn its attention to education, through internship programs, high school classes, and university courses.

A potential obstacle is that land and infrastructure may be more expensive for an urban farm than they would be for its rural counterparts. Zoning, use and building regulations, and soil contamination are other urban considerations. To address these challenges, it may be important to create broad partnerships – following the model of Troy Gardens – to piece together the funding and governance mechanisms for land, infrastructure, and future development. While working with partners can be a slower process than going it alone, for a farm in the heart of a community, the experience of Troy Gardens suggests that the benefits far outweigh the costs.

Prepare a business plan.

If you want backing from city government for your farming endeavor, prepare a solid business plan. Michael Porterfield of Gladheart Farms in Asheville, North Carolina (page 87), believes that cities are becoming more interested in hosting farms and gardens, and a business plan will build confidence in your chances of success. If you don't have an organization like the Appalachian Sustainable Agriculture Project (page 243) to help you develop a business plan and marketing tools, as Porterfield did, seek help from your local chamber of commerce or economic development commission.

Test soils for contamination, and take appropriate precautions.

Many urban soils suffer from lead contamination. The Food Project in Massachusetts (page 136) has worked to build awareness of the dangers of lead exposure in Boston's Roxbury and Dorchester neighborhoods, where lead is often found at concentrations of 1,000 parts per million or higher – more than double the 400 parts per million the U.S. Environmental Protection Agency has designated as harmful to human health. Exposure to lead disproportionately affects urban, poor, and minority communities, and Boston's Roxbury and Dorchester communities are prime examples. Of all the areas that make up the greater Boston region, Dorchester has the largest African-American population, and 23 percent of the community lives below the poverty line. Not coincidentally, Dorchester also has the highest percentage of children with elevated blood levels of lead.[88]

Certainly it is possible for wealthier neighborhoods to have lead contamination in their soils, but statistics suggest that community gardens being developed in neighborhoods with similar demographics should be especially aware of the greater likelihood that their soil may be contaminated with lead.

In a series of innovative experiments, the Food Project has been working with researchers at Wellesley College to explore the potential of phytoremediation for soils contaminated with lead. Beginning in 2003, they tested three "hyperaccumulators" – mustard greens, sunflowers, and collards – to see if they might work as biological soil scrubbers. Three years of research yielded disappointing results: it would take seven to ten years for these plants to reduce the lead by 300 parts per million, not fast enough to be a practical remediation method. The Food Project is continuing to work with Wellesley College to explore a number of other avenues, including using other hyperaccumulators such as geraniums; incorporating acidic fruits (oranges or lemons) into the soil to lower its pH, which would increase lead accumulation in hyperaccumulators; and using phosphorus to help organic elements in compost bind the lead in the soil, thereby making the lead less bioavailable to plants.[89]

The results of the research completed thus far indicate a couple of practical steps to take limit exposure to lead when working with contaminated soils.[90] The first is to garden in raised beds and use compost. While raised beds do not themselves remediate the lead, they limit plants' exposure to the lead-contaminated soil. The bed frames should be built of untreated lumber, and the bottom of the beds should be lined with landscape fabric to prevent roots from coming into contact with contaminated soil. The beds should be filled with at least 2 feet of clean soil or compost. Compost neutralizes the pH of the soil and contains organic elements that bind with heavy metals such as lead, which makes it less bioavailable to plants.

The second step is for gardeners to protect themselves against inhaling fine particles or allowing contact of the fine particles with their skin or clothing. Simple guidelines offered by the Food Project include:

- Do not work in the garden when it is dry and dusty; work in the garden only when the soil is moist or damp.
- Wear gloves while gardening.
- Wash your hands after gardening and before eating.
- Wash and scrub all vegetables before eating or cooking.
- Remove gardening shoes before entering the home to avoid tracking excessive dirt indoors.

Food Justice in the Heart of West Oakland
The People's Grocery

In **West Oakland, California**, an inner-city neighborhood that's dealt with economic inequities and uncertainties for decades, a thriving social justice movement has taken root. Not least among the players here is People's Grocery, a community-based group working to build and support a local food system — and the concomitant local economy, health system, and social fabric — through grassroots activism, community engagement, and advocacy.

While some urban food-justice organizations focus on growing food in city lots, People's Grocery takes a different tack, focusing on building movements: building relationships between community members and leaders, which in turn leads to projects and campaigns that improve the health and wealth of the community. But that's not to say that they don't farm. Indeed, in the parking lot behind the once legendary California Hotel (now a housing development for low-income residents), People's Grocery operates a microfarm and greenhouse, complete with raised beds, vermiculture composting, and chickens. Residents of the hotel participate in the greenhouse enterprise program, in which they raise plants to sell at farmers' markets. But the point here is less about production than about opening up space for people to build community together, through gardening. The farm, says executive director Nikki Henderson, is a demonstration project that shows how a local food-production system can build community.

With community engagement and grassroots activism as its foundation, the People's Grocery has a core commitment to leadership development. Its Growing Justice Institute, for example, works with eight West Oakland residents to support them in building community health. Each of these residents develops cohorts of another eight to ten residents, and over the course of two years they design community health projects and campaigns. People's Grocery staff support the program by offering workshops on topics ranging from the political economy of the food system to budgeting and fund-raising.

The Community HANDS project, on the other hand, begins with nutrition education programming. Individuals who have participated in People's Grocery's nutrition programming are eligible to become health and nutrition demonstrators, who receive stipends for offering nutrition demonstrations to the community at local events, area hospitals, and other venues.

People's Grocery also offers a CSA, known as the Grub Box program, in partnership with Dig Deep Farms, a Bay Area enterprise dedicated to employing local residents (with a focus on youth and people of color) to convert blighted land to agricultural purposes, growing produce for local consumption. As with most People's Grocery programs, the Grub Box program is focused on community engagement; the idea is that customers feel like they are participating in a reclamation and empowerment movement when they order a box.

Perhaps most interesting is People's Grocery's willingness to test the effectiveness of their programs. The group is collaborating with Highland Hospital in East Oakland — the nearest public hospital to the West Oakland neighborhood — in the "Bite to Balance" case study, wherein 15 families from the pediatric clinic whose children have diet-related problems will receive weekly Grub Boxes over the course of six months, along with health and nutrition demonstrations. The hospital will track the children to determine whether the availability of fresh produce affects their overall health.

Top: The People's Grocery's urban agriculture program is based in the transformed parking lot of the historic California Hotel, now a subsidized housing facility for very-low-income residents. *Bottom left:* Nyota Koya, a community health and nutrition demonstrator, leads a cooking demonstration for staff members. *Bottom right:* Residents like Mickey Martin help in the gardens and at the markets, bringing fresh, healthy food to low-income populations throughout Oakland.

Transforming Abandoned City Land into a Community Farm

Beardsley Community Farm

coauthored by Megan Bucknum

The Beardsley Community Farm in **Knoxville, Tennessee**, is an urban demonstration garden with a focus on education, sustainability, and community empowerment. Its mission: to promote food security in Knoxville's low-income communities through self-sufficient practices focusing on vegetable gardening, food preparation, and preservation techniques.

In 1997 a community food assessment of Knoxville found what some might consider the Sahara of food deserts: a neighborhood consisting of two subsidized housing developments, an abandoned middle school, and a park with a high crime rate, with no grocery stores, restaurants, or other sources of food. In an effort to help thwart the violence and drug activity occurring in the park, and to bring fresh food to the neighborhood, the city gave 7 acres of the park to the Knox County Community Action Committee (CAC) to build a community farm.

With funding from the city and the United States Department of Agriculture (USDA), the CAC built a greenhouse on the site, and in 1998 staff and volunteers grew 21,000 plants to give away to one thousand local families through the "Green Thumb" program, which promotes home gardening by providing plants, soil, tools, and gardening assistance to low-income families. The project then evolved into a farm of swimming-pool raised beds planted with vegetables to feed the neighborhood. Faced with highly compacted and low-nutrient city soils, the CAC used the above-ground swimming pools to demonstrate an easy way to garden at public housing, where residents are not able to till up a garden. To be eligible for this food, a community member would need to attend an educational class offered by CAC in life skills or nutrition. (This program continues on a small scale, though now the

CAC also simply donates food to four food pantries and kitchens.)

More than 10 years later, Beardsley Community Farm has evolved into a successful urban demonstration garden that grows produce (vegetables, fruits, and berries) and raises honeybees and free-range chickens. The farm offers education on urban organic gardening, raising free-range poultry, honeybee management, and vermiculture, and it has even developed a small-scale aquaculture project. Farm staff also work with elementary schools and after-school programs to teach students about healthy food and nutrition. The food grown at the farm goes to local food pantries and soup kitchens. And adjacent to the farm, 30 garden plots with on-site water are available to community members free of charge.

This remarkable evolution has been achieved by people power. The farm is run by a dedicated farm manager and as many as six AmeriCorps staff, supplemented by more than a thousand volunteers. By employing Knoxville low-income teens, aged 15 to 17, in a summer leadership program funded by a government grant, the farm has been able to build several demonstration projects teaching sustainability — small-scale vermiculture and aquaculture demonstrations, a straw-bale shed that serves as an entrance to the farm, and a water catchment system.

Building an Off-Farm Presence

One reason Beardsley Community Farm has been so successful in attracting volunteers is that it has established a solid off-farm presence. In order to be truly a *community* farm, it is imperative to have a presence in the community, says John Harris, director of the CAC's Green Thumb program. This means representing the project at community events. It also involves looking into other areas of the community to find those that could use the project's support, rather than focusing on how other community members or organizations can help develop the project. These partnerships can be mutually beneficial, while creating a stronger network of community members and organizations. Beardsley Community Farm, for example, partners with the University of Tennessee, as well as local businesses, such as two restaurants and a grocery store, that assist with the CAC's two or three annual fund-raisers.

Harris also advises that a basic level of "branding" is important, as it helps build name recognition in the community. Beardsley Community Farm has a distinct logo and, for the Internet savvy, an easily navigable, well-designed website. While these may seem like unnecessary frills for a grassroots urban farm project, the Beardsley farm experience suggests they are critical for a twenty-first-century farm to succeed.

Connecting with the Neighborhood

One of the underlying goals of the Beardsley Community Farm is to help expose the neighboring urban community to agriculture and local food. So how do you attract a group of people who may not have any experience with how farms operate and what they look like? Beardsley's answer is to hold festivals and events that offer familiar activities and fun to attract people to the farm, where they will also learn about the potentially unfamiliar world of agriculture.

The farm's popular Farm Fests feature bands, potato prints, corn-husk dolls, face painting, cake walks, and other carnival-like attractions. Also at the event are activities that teach people about food and gardening, and people can see the crops, chickens, and bees. With these festivals, farm staff have been able to address the challenge of finding ways to connect with the immediate neighborhood community.

Lessons Learned
Working with Volunteers and Interns

Make sure interns know that farm work is hard work.

Without interns and volunteers, Asheville, North Carolina's Gladheart Farms (page 87) would be hard-pressed to get the work done. "The key to our success is not financial," says Gladheart's owner, Michael Porterfield. "It's our human resources."

Gladheart Farms offers internships through Porterfield's community of faith network and also through World-Wide Opportunities on Organic Farms (WWOOF). To ensure a successful experience and avoid freeloaders who may think that being on a farm is "sitting in a lawn chair eating carrots and drinking beer," Porterfield says it's important to establish clear expectations in the very beginning. The main thing is to tell interns is that they're going to have to work hard. "If they know that," Porterfield says, "when you say 'you could work harder' it's not a surprise to them." During a typical summer day at Gladheart Farms, farm staff and interns rise at 5:30 a.m., eat breakfast at 6:00 a.m., and work until 9:00 or 10:00 at night — with breaks along the way for lunch and an occasional swim in the nearby river.

To attract volunteers, establish name recognition in the community.

Beardsley Community Farm in Knoxville, Tennessee (page 84), boasts an astonishing volunteer force of more than a thousand community members. Staff attribute their success in recruiting volunteers to their focus on a basic level of "branding," or building name recognition in the community. They staff tables at many community events, maintain a lively Web presence, and support energetic outreach efforts.

Insist on liability forms.

Volunteers are often the heart and soul of nonprofits, but even so, misunderstandings, mishaps, and accidents can happen. Regardless of how well intentioned the volunteers, it is important to clarify the risks involved, so volunteers are at least informed about the tasks and potential liabilities. Michael Van Ness of Lynchburg Grows in Virginia (page 116) says it is vital for all volunteers to sign a liability form, no matter how small the task. This could save you from a financial crisis.

Give your volunteers as much as they give you.

One of the biggest mistakes that organizations make is to undervalue their volunteers, according to Lucy Harris, executive director of SEEDS, Inc. in Durham, North Carolina (page 140). Most organizations, she says, don't give their volunteers half as much as they get from them. Giving to volunteers can mean anything from hands-on assistance or education to inviting them to take samples and sharing the bounty. It should also mean a volunteer appreciation day.

Learn to manage volunteers so that they bring real benefit.

A typical complaint is that volunteers can take more work than they're worth. The Food Project in Boston, Massachusetts (page 136), proves that it's all a matter of learning how to manage them. Based on its extensive experience working with thousands of volunteers, the Food Project has created the *Volunteer Program Manual*, an invaluable, detailed guide to recruiting and working with volunteers (it's available for free on the project's website; see the resources section). The manual's guidelines are too numerous to list here, but a sampling of their key lessons in working successfully with volunteers are:

ESTABLISH REGULAR SPECIFIC WEEKLY TIMES FOR VOLUNTEERS. Volunteers take time and focus from staff, so having them come whenever they want is unwieldy. Instead, work them into your plan in advance. At the Food Project, volunteers can come on Tuesdays, Thursdays, or Saturdays, between 9:30 and 12:30.

SHOW VOLUNTEERS HOW TO DO THE BASICS. Don't assume that volunteers will automatically understand the purpose of what they're doing or how to do it. The Food Project gives volunteers hands-on lessons in tasks.

MAKE IT RELEVANT. The Food Project emphasizes the importance of connecting volunteers to the larger mission, so they understand where the food will be going and how many people it helps feed. Rather than thanking them for weeding a row of squash, for example, thank them for increasing the yield of squash so that there will be more of it to feed people at the downtown soup kitchen.

Establishing a Farm on City-Owned Land
Gladheart Farms

Gladheart Farms is a certified-organic farm just minutes from downtown **Asheville, North Carolina**. Through a creative leasing agreement, the farm cultivates riverside city-owned land and operates a CSA, as well as a wholesale distribution business, networking with other organic farmers from across the country to bring fresh organic produce to local markets.

In Asheville, North Carolina, a young couple is proving that — with a little bit of help — new farmers can find land and develop a farming business in the heart of a community. In the spring of 2006, at the height of the real estate boom, finding land to farm would have seemed an impossible proposition to most. But Michael Porterfield, with the support of his wife, Michelle, was determined. He started looking for land and struck black gold — 7.5 acres of land just within Asheville city limits. Finding acreage so close to the city's downtown was a real opportunity. But the property was a hot ticket, on the edge of fairly dense city neighborhoods, and several developers were interested in the land as a potential building site. Because he didn't want to lose it, Porterfield snatched it up at the asking price.

Porterfield had one big advantage: the support of his community of faith, in the form of funding for investment and 10 to 15 people who live at the farm and assist with labor and management. His faith community had also supported Porterfield's former organic tea business, which gave him solid business experience.

His first year on the farm required preparation — building a greenhouse, digging a well, installing a walk-in cooler, clearing away overgrown vegetation, and picking enough rocks out of the field to fill a quarry. In the second year Porterfield planted his first crops and, selling directly to stores, he sold out. Porterfield credits this success in large part to Peter Marks of Appalachian Sustainable Agriculture Project (ASAP; see page 243), who helped him write a business plan and develop marketing tools.

Opening the Door for Urban Farms

The market was so strong in his first year, Porterfield says, that he could have sold "ten times more." So he started looking for more land. Once again he struck black gold — prime bottomland along the Swannanoa River that was unused, derelict city property adjacent to a city park. Gaining permission to lease the land was a six-month baptism by fire into the intricacies of ordinances, regulations, and multijurisdictional decision making. A soil and water erosion ordinance designed to govern development on slopes caused the city to stipulate that his fields couldn't lie fallow for more than 30 days without being seeded with a cover crop; a wetlands regulations led to a request for a 150-foot buffer that would have taken half the land out of production; and a requirement to conduct a study to determine if using the land would harm an endangered bird's habitat might have stalled the process for another year if the city hadn't already conducted a study to explore the possibility of using the land for a ball field.

Porterfield says that his involvement with ASAP helped him negotiate these hurdles, because ASAP's work had created sufficient interest in supporting the growth of a local food system that the city was willing to listen to his needs and negotiate conditions that were mutually acceptable. In the end the Asheville city council unanimously passed a resolution to allow unused city land to be farmed for organic vegetables, opening the doors for other urban farmers. Porterfield signed a six-year renewable lease, and the city waived his first two years of fees in exchange for his work in clearing the land. In 2009, his second year of growing

Bright Idea: Tailgate Markets at Assisted Living Communities

Gladheart Farms sets up a small farm stand at a nearby assisted living community, once a week for two hours. Porterfield says there's usually a line waiting for his wife to open the farm stand. "We sell more in two hours than we do selling all day at other markets," Porterfield says. The retirement community did not want a full farmers' market; they just wanted to deal with one farmer. The Appalachian Sustainable Agriculture Project brokered the relationship, and Gladheart Farms was thrilled for the opportunity to serve seniors. A small idea like this can provide large benefits — increasing access to fresh produce for seniors and also providing an eager market for the seller.

crops, Porterfield expected to gross $100,000 from selling 50 different vegetables through a CSA, a farm stand at a senior community, and a wholesale operation. Because of the up-front investments, he didn't expect to immediately turn a profit, but Porterfield is a businessman and says the farm must be "pushed out of the nest" and become self-supporting in its third year.

Wholesaling to Support Local Growers

Through Gladheart Farms, Porterfield was restoring city land to productive food production, and he wanted to help other farmers do the same. With his business experience in selling organic tea, he was immediately drawn to the idea of selling his organic produce wholesale. The Appalachian Sustainable Agriculture Project connected him with distributors, and in 2009, he says, 95 percent of his business was selling wholesale to restaurants, groceries, and catered events.

But Porterfield also soon learned there was demand for products he couldn't grow — organic pineapples, bananas, citrus, avocados, mangoes, ginger, persimmons, dates, nuts. So he has begun working with farms across the country and selling their products. Gladheart Farms transformed a former bank and warehouse into an organic commercial distribution center, with three rooms kept chilled for storing produce and a commercial kitchen for producing organic products. And he has brought in people with specific marketing and inventory management skills to help his distribution business grow.

Most organic farms in the Asheville region sell retail through CSAs and tailgate markets; few sell wholesale. Porterfield believes that larger farms that might wish to transition to organic would be encouraged by the presence of an organic wholesaler. He knows that organic cultivation is challenging in the mountains of western North Carolina. He hopes Gladheart Farms Distribution will reduce the risks of growing organically by offering opportunities for selling quantities of organic produce wholesale.

Lessons Learned
Starting a New Farm

Identify your customers early.

No business can survive without customers, whether it's a nonprofit or a for-profit enterprise. And even if the real goal of an urban farm is education and empowerment, it will not be able to go about educating and empowering anyone until it knows who can use the food it will grow. One of the first rules of business is always to know your customers. So whether you plan to donate or sell your food, spend time in advance lining up the potential recipients and buyers.

One radically simple idea is to show up with food and see who comes. Mary Seton Corboy of Greensgrow Farm in Philadelphia (page 96) suggests that you just buy local produce from community gardens, fill up a pickup truck, and take it to the site where you plan to establish a farm. Whoever shows up, she writes, "those are your customers!"[91]

Don't overcapitalize.

"New farmers are susceptible to the exact same traps as old-time farmers." Corboy has earned the right to say this, as she has grown a successful urban farm that now tops $1 million in revenues. One major mistake she's glad that Greensgrow Farm *didn't* make was overcapitalization. By staying lean and not investing in expensive equipment they didn't really need, Greensgrow was able to stay flexible and make the changes needed to stay successful. She and her partner had invested only $20,000 into equipment for growing hydroponic lettuce, so when they realized they needed to move into raised beds and a nursery, they could just pull the equipment apart and store it, without losing their shirts.

Corboy is passionate about this issue. She says there's a "psychological drama playing itself out," and new farmers end up convincing themselves that they need a tractor to be taken seriously by the big boys. But these decisions can ruin new farms, tying them unnecessarily to a particular mode of production. In her case Corboy is proud that she successfully resisted making a major capital investment in a forklift for nine years — the tipping point was when she *knew* she would use it every day of the week and it would finally pay for itself. A large investment is a "noose around the neck," she says.

There is little, if any, expensive equipment at Greensgrow Farm, proving that you don't need a huge capital investment for an urban farm. The farm office is an old trailer that Corboy rescued from going to the landfill. Refrigeration is no exception. Every CSA needs a good walk-in cooler. But instead of an expensive commercial model, Greensgrow did it on the cheap: they built several small sheds that they insulated with thick foam and installed small window air-conditioning units. Corboy says her cooling sheds are much less expensive *and* more efficient than commercial brands. And the components were rescued from someone who was going to throw them out.

Corboy also sees a time when urban farms could share or jointly own equipment. In rural areas of yesteryear, she says, one farmer had the tractor, and he went to other farms to bail their hay. Why couldn't this happen in our cities, too?

Find the right economy of scale for your operation.

Is it better to grow 3,000 pounds of vegetables that sell at $3 per pound or 5,000 pounds of vegetables that sell at $2 per pound? Or is it better to grow 50,000 pounds of vegetables that sell at $1 per pound? You might think the answer is obvious: more vegetables bring in more money. But revenue alone is not always the best measuring stick. Are the economies of scale worth the additional investment required to be able to produce on a larger scale? And, for people who care about which markets they sell to, will economies of scale mean you're no longer selling to the people you wish to reach? Put simply, how big is too big?

"This is one of our bigger challenges," says Edwin Marty, executive director of Jones Valley Urban Farm in Knoxville, Tennessee (page 158), during a discussion about the future of the farm. The reality, he says, is that larger grocery stores will tell you what they're willing to pay for your product, while smaller stores allow you more control of the price. "Grocery stores are generally willing to pay about half of what you can get at smaller stores," Marty says. On the other hand, if you sell only to smaller stores, then you can quickly saturate their capacity. So Marty advises that it's important to think strategically: decide where and how to market – and how big you really want to get.

Learn the standards of your intended business.

To be successful at livestock production today requires business savvy. It is not enough to raise high-quality pastured beef, pork, lamb, goat, or chicken. You also need to know how to market it. What this means, says Amy Ager of Hickory Nut Gap Farm in western North Carolina (page 234), is that the farmer needs to learn all aspects of the meat business, from how to butcher a beef to which cuts restaurants like to work with and the quantities and cuts that groceries will require. When you have this knowledge, your customers – particularly the wholesalers – treat you like a professional. Ager admits that she and her husband have made plenty of mistakes, some fairly costly because meat is expensive. But she also suggests that farmers – like all aspiring businesspeople – need to be prepared to make mistakes and consider them simply setbacks. Overall, the best strategy for learning the business without too many costly mistakes is to start small and grow slowly.

Begin with relatively easy crops.

The experience of Appalachian Sustainable Development (ASD) in southwestern Virginia (page 253) underlines the importance of new farmers – whether new to a particular project, such as growing organic vegetables, or new to farming altogether – beginning with less risky crops in terms of both production and perishability. Hard squashes are relatively easy to grow in southwest Virginia; they aren't bothered by many pests, don't demand daily attention, and have a long shelf life. On the other hand, hard squashes are also less lucrative. The next step up in difficulty from hard squashes might be bell peppers, which are not susceptible to many diseases but require more care, more attention, and weekly picking. Even riskier crops for this region

tend to be cucumbers, tomatoes, zucchini, and lettuce. Cucumbers and zucchini require daily picking to meet the grading standards for commercial grocery stores. Tomatoes require a lot of attention with staking and pruning. And, because of breaks and slugs, it is particularly challenging to get leaf lettuce into the box as a "pretty head."

ASD suggests that farmers identify crops that will work well with their experience, soils, growing conditions, and available time commitment. Weekend farmers, for whom the farm is supplemental to another job, might be advised to avoid crops requiring daily attention or picking. Conversely, farmers with more time and experience might be advised to step up to riskier crops that offer potentially greater profit.

Find a sister project.

Staff at Beardsley Community Farm in Knoxville, Tennessee (page 84), suggest that new community agricultural projects should find projects already established in the same temperate zone with a similar culture, to share knowledge and resources. Although not every project will face the same challenges, former farm manager Ben Epperson says, there will be some common hurdles, and he suggests that it is beneficial to collaborate with those projects instead of trying to reinvent the wheel by yourself. "Don't be afraid to copy what works," he says.

In the case of community gardens, challenges that all gardens share include recruiting volunteers, working with city ordinances, and water procurement. Taking advice from existing gardens that have already solved some of these problems will save a new project a great deal of time and energy, says Epperson.

Similarly, working with a sister project in the same temperate zone can help new agricultural projects decide when and what to grow. A project in Vermont might gain some insight from a California project, certainly, but would likely learn even more from a northern-climate project about appropriate rotations, cultivars, and techniques for extending the growing season.

Plan for the long term.

New projects can risk running into problems if the start-up momentum translates into moving forward too quickly. Try to think 10 years down the road to build a strong foundation for a project, advise Beardsley Community Farm staff, because trying to grow too quickly, both figuratively and literally, can lead to unsuccessful results.

In an urban setting, in particular, in its first year a community garden may need to concentrate on soil remediation or building good soil and may not even be able to grow produce during this time. Though this step might generate less community excitement, it is crucial if the garden hopes later to grow an abundance of food.

Thinking 10 years down the road can help contribute to a project's overall sustainability. For instance, fruit trees, nut trees, grape vines, and berry bushes take years to come into production, so it's a good strategy to invest in them at the beginning of the project. Beardsley Community Farm did plant an apple orchard and blueberries between 1998 and 2000, but most of these trees and bushes fell victim to vandalism and were destroyed. A solution could be to make the plantings a community project, so members of the community feel a sense of ownership toward the trees and might therefore be more likely to monitor them and prevent vandalism.

Seeds want to grow.

Michael Porterfield's trajectory to success with Gladheart Farms in Asheville, North Carolina (page 87), didn't come from book learning or classroom knowledge. He says he's never read one book about farming or gardening, and probably never will. Books, he says, would scare off anyone in their right mind, with all that information about diseases and pests. Farming can be intimidating, Porterfield says, so a lot of people don't even try.

But people have been growing food for thousands of years, Porterfield argues, and there's no reason people can't get out and do it today. He encourages people to grow just two tomato plants, to see how easy it is. Something as simple as growing food doesn't have to be hard or complicated, he says, and too much emphasis on book learning is a way to crush a person's creative force. Sure, mistakes are going to happen, he admits, but farming is a lifetime of learning. "The premise of farming," Porterfield says, "is that seeds have life in them and want to grow! So you've got that going for you — it wants to happen."

Porterfield passes on advice that someone once gave him: start with something that's easy to grow, so that you'll be encouraged by success and want to grow something else.

Porterfield says he has learned the most from people who've been successful and have been generous with their time, so his doors are always wide open to anybody who wants to come learn from him. And he's not afraid to experiment, even when he's told "you can't grow onions here." Sometimes they're right, he says, but sometimes they're not. He grows beautiful onions.

Cultivating Food and Farmers
The Intervale

by Tim Beatley

northeast

The Intervale, an agricultural haven within the city limits of **Burlington, Vermont**, is home to an unusual and highly productive array of activities: one of the largest CSAs in Vermont, a unique incubator program for new farmers, a native plants nursery, community garden plots, a large retail garden store, educational displays and research plots, a municipal composting facility (which has done much to rebuild the fertility of Intervale soil), and a city power-generating station. And it is also the site of new thinking about urban farming, imagining how waste from a city might be repurposed to meet the food, energy, and recreational demands of that same city's population.

A remarkably short distance from downtown Burlington, Vermont, that bastion of progressivity, lies one of the most unusual experiments in local food production anywhere: the Intervale. "Intervale" literally means a low-lying place between two hills, and that characterizes well the feeling of this landscape. Long neglected, a former dumping ground and industrial farm, it has risen like a verdant and bountiful phoenix, employing an incredible mix of creative farming ideas to serve as a model for how to sustainably occupy land at the outskirts of a city.

The nonprofit Intervale Center owns and manages the grounds, encompassing about 350 acres located within the city limits of Burlington, along the floodplain of the Winooski River. This area has a long history of agriculture, dating back at least five thousand years to the Missisquoi Abenaki. It is now home to a diverse array of activities, including 12 commercial farms that lease their land, a large retail garden center (with demonstration gardens), a major composting facility, a wood-fired electricity-generating plant, community gardens, a heritage farm and farmstead (the original farmstead of Ethan Allen), a conservation plant nursery, and even a network of nature trails that take visitors to the banks of the Winooski River.

One of the larger farms is the Intervale Community Farm (ICF), Vermont's first CSA and one of its largest. Unlike many urban-based CSAs, ICF requires its shareholders to come to the farm to pick up their shares, something possible given the close proximity of the farm to Burlington. And it's not just a pickup but an assembly — shareholders essentially put their boxes together themselves and are then allowed to venture out to the fields to pick certain crops (such as flowers) each week. There are abundant toys, even a kid's garden, to entertain accompanying children. Andy Jones, the farm manager, feels that this pickup process helps build an understanding of farming and a connection to the farm that may be lacking in the usual urban CSA. And in an unusual twist, the farm is actually owned by its members and operates as a consumer cooperative (though not all CSA participants are owners). Joining the Intervale Community Farm Cooperative provides additional benefits (such as automatic renewal of CSA membership, rebates, and profit-sharing) and allows participation in the governance of the farm. The cooperative element serves to strengthen the economic viability of the farm and to further engage members in its mission.

The Intervale Center runs its own (smaller) multi-farm CSA and also undertakes a variety of other food-based and agricultural activities, including research and consulting.

Fostering Farmers

One primary goal of the Intervale Center is to foster the start-up of new farms and agricultural enterprises. Through its Farms Program, the center leases land, equipment, and storage facilities to commercial farmers operating in one of three categories: *Incubator* farmers receive subsidized land leases, technical assistance, and mentorship from experienced farmers, for up to three years. *Enterprise* farmers have been in operation for at least three years, while *mentor* farmers have been farming for at least five years and agree to provide advice and guidance to the newer farmers. There are no residences on the farms, so the farmers live off-site. The idea is that as these farmers become more experienced and profitable, they will "graduate," finding land elsewhere and making room for new incubator farms.

The Intervale approach significantly reduces the up-front costs of farming, making it easier for new farmers to get going. Most importantly, because farmers can lease land, they don't have to come up with the capital to acquire good farmland, which can be prohibitively expensive, especially near an urban center. Farmers are also able to participate in an equipment-sharing cooperative, which can substantially reduce the costs of start-up and operation. Having access to tractors, greenhouse space, and other specialized equipment means farmers need not invest in big-ticket items that they need only at certain times of the year.

Glenn McRae, formerly the executive director of the Intervale Center, believes that the main advantages to farmers participating in the incubator program are the mentoring and daily interaction with other farmers. This social factor shouldn't be underestimated, and it's something that rarely happens in more rural areas where farms and farmers are more spread out. The brand-name value of producing in the Intervale and the ready access to an urban market are also clear advantages.

Building Farmers' Community

Farming is often thought to be of necessity a solitary endeavor, with large tracts of remote land and the farmer spending most of the day with animals, crops, machinery. The Intervale model makes clear that farming can be a community-building project — and this may be one of its major lessons.

The project brings 10 to 15 farmers into relatively close proximity. No single farm is very large, and all are within a short walk of one another. It is telling indeed that when visiting the Intervale it is possible to walk, at a leisurely pace, from farm to farm and cover the entire site in the course of an afternoon. Reaching out to other farmers is easy in this context, and individual farmers tend to see each other over the normal course of a day. Farming in the Intervale is not a remote and isolated endeavor at all; quite to the contrary. As former executive director Glenn McRae observes, there is much more interaction among farmers than would usually be the case. "They're next door ... you see them every day, whereas if you're on your 100-acre farm out

Firsthand Feedback

Adam Hausman runs Adam's Berry Farm, a pick-your-own and wholesale berry farm, one of the Intervale's 12 farms. When asked about concerns those interested in getting into farming might have about the Intervale's model, he says that the Intervale approach does help farmers overcome the major problem of the cost of land, especially around cities, as an impediment to starting up a farm. Hausman himself was grandfathered in to the Intervale and not subject to the five-year time frame. He wonders whether all farmers would accept the limitations of leasing for restricted time frames, particularly in cases (like his) of substantial up-front investments in berries or fruit trees that are not portable (his berry bushes "will outlive both of us," he says). Nevertheless, even for a perennial fruit farmer, being able to share the costs of equipment – in his case, renting walk-in freezers – is a tremendous help. He also confirms the value of the social dimension, noting that the sense of community and camaraderie immeasurably enhance the quality of the farming experience.

there, you've got to plan to get together, which doesn't happen very often."

This closeness makes the equipment-sharing system more practical, of course, and has led to an informal work exchange between farmers. McRae offers this example: "I'm going to disk my field; do you want me to do yours as well, and later you can do something for me?" While such offers can certainly be seen in more conventional agricultural settings, the proximity of the farms at the Intervale make mutual aid much easier and much more likely.

Andy Jones, who runs the Intervale Community Farm, stresses the importance of this set of peers, a community of farmers to interact with and whose collective experience and wisdom can be drawn upon when needed. It is pretty clear here that the individual farmers know each other, have a real sense of a farm community, and profit from it.

Proximity to the city of Burlington is another important element of the Intervale's farming community. City residents are able to interact with and get involved with the farms to a degree not possible in other traditional farming operations. McRae refers to this as "farming in public," and he notes that any member of the public can come and visit the farms, something the Intervale's farmers simply accept. At the Intervale, farming is not something that happens far away and out of sight; it is very much a part of the public experience.

And that experience is intentionally extended to the public. In the summer the Intervale Center hosts a Thursday-evening outdoor party, with live music, food (including fresh, oven-baked pizza from Burlington's own American Flatbread), and various talks and activities. In June there's a conservation nursery tree walk and an Abenaki youth dance, in July a talk by the Vermont Beekeepers Association. Indeed the design of the spaces at the Intervale Center amounts to the creation of a public green, suitable for picnics and many other events such as these.

Incubating Agricultural Enterprises

One reason the Intervale is so unique is the almost dizzying array of activities and uses that occur on the site. The composting facility at Intervale Compost Products, for example, accepts all the organic waste from Burlington. In its early years the composting facility had been moved from site to site within the Intervale, part of the process of reclaiming the degraded land and restoring its fertility. Now, too large to be moved, it sits in a permanent location and provides a highly salable product, coveted by farmers and gardeners, that is packaged in specific "blends" for different markets.

The facility has not been without controversy, however; in recent years it ran afoul of several major state environmental permitting programs. In response to these issues, the facility is now owned and operated by the Chittenden Solid Waste District, a state agency for the county.

The Intervale Conservation Nursery, which is owned by the Intervale Center, consists of greenhouse and propagation space in which some 50,000 native riparian species of trees and shrubs are grown. These plants are used largely in wetland restoration projects around the state.

Gardener's Supply Company, founded by Will Raap (who also happened to found the Intervale itself), is an anchor at the Intervale, with both a large retail store and the main offices for the company. Just about any gardening product can be found at this store, from bird feeders and gardening gloves to large greenhouses, along with various educational materials and demonstrations. The company maintains a series of research plots that are testing different planting regimes and products.

And there is the Tommy Thompson Community Garden, where residents of the city can lease small 25-foot-square plots to grow their own vegetables. This combination of activities means that the Intervale has become a vibrant and sometimes busy community farming center, with mutually supporting enterprises.

Evaluating Success

Early on in the development of the Intervale, founder Will Raap set the goal (no one is quite sure how or where this number came from) of the farms there providing 10 percent of the food needed by the city of

Burlington. It's not clear what percentage the farms do provide, but there is no doubt that the amount of production is impressive — estimated at more than 1 million pounds per year.[92] And as Raap notes, the Intervale continues to serve as a catalyst for community farming initiatives and projects around the state and nation. And finally, the approach of incubating new farms and farmers reflects the growing sense among many (including those in the land conservation community) that simply saving the land itself is not enough — that it will be increasingly necessary to train and nurture the next generation of farmers.

There are certainly some pieces missing at the Intervale, and some limitations to the Intervale model, and McRae is quick to acknowledge them. The emphasis has been on growing food, and while there are some added-value products, there is not really a processing component. This is one direction McRae believes will be important in the future. Indeed, there are plans to construct the Intervale Food Enterprise Center, which would consist of a 20,000-square-foot food-processing facility (for small processors) and 20,000 square feet of greenhouses. Most interestingly, the facility would be heated with the waste heat of the nearby McNeil power station.

Success of the Intervale is due to many factors, but strong support by a string of Burlington's mayors has been key, beginning with Bernie Sanders (then mayor, current congressman). And location in a city with a long history of support for sustainability doesn't hurt either, of course. Raap likes to point to the importance of the composting operation as a key factor in the Intervale's success, noting that restoring the soil and fertility of the degraded site was an essential precondition for everything else. As Raap notes, "The leaves and yard waste, and subsequently the food waste, from Burlington became the source of the fertility to rebuild the soil from essentially the ravages of twentieth-century industrial agriculture." Starting with the soil, it seems, is good advice for farming and as well for growing new and innovative food and farming institutions such as the Intervale.

Changing the Way Urban Communities View Their Food

Greensgrow Farm

with research by Robin Proebsting

In eastern **Philadelphia** Greensgrow is a vibrant and profitable urban food farm that occupies a formerly quarantined, contaminated city block. It serves its community through a CSA, an on-site farmers' market and nursery, and demonstration projects, workshops, and community events. Its mission is to "be a profitable, urban, green business dedicated to growing the best products, people and neighborhoods."

Mary Seton Corboy and Tom Sereduk wanted a better tomato. As chefs they wanted to delight friends who ran Philly restaurants with fresh, deliciously flavored tomatoes picked the very same day. With this wild vision they decided in 1997 to start a farm.

More than a decade later, Greensgrow Farm is one of the leading and longest-lasting pioneers in urban farming. The farm is a colorful oasis amid the gray and brown of sidewalks and rows of crowded brick townhomes. Unlike many other urban farms that couldn't survive without volunteer labor, Greensgrow is a profitable, growing business. Its CSA, nursery, garden shop, and farm market, along with small grants, feed a budget nearing $1 million, with 19 paid staff. Greensgrow is also a powerful model for our nation's cities, proving that contaminated urban sites can be fully rehabilitated for immediate community benefit. Whether an urban farm is a temporary or long-term use of a contaminated site, food farming offers new potential for the urban landscape to become a source of green jobs and local, fresh food.

How Greensgrow arrived here is rich with lessons for others who may wish to try their hand at urban farming. Corboy and Sereduk began their venture outside Philadelphia, in Jacksonville, New Jersey, and brought produce into the city. But this didn't last long. They lived in Philadelphia, and commuting *out* of town to the farm, to bring fresh produce back into town, just didn't make sense. So they searched for land where they both

lived, right in Philadelphia. "I wasn't out to change the world, I just wanted cheap land," Corboy says.

But finding cheap city farmland wasn't easy. Most available and affordable land in Philadelphia was blighted by former industry or dilapidated buildings. Instead of giving up, the partners adapted to the challenging land conditions and began thinking outside the box, revising their ideas about how, where, and what they might grow.

Corboy and Sereduk found an abandoned city block in Kensington, one of the oldest Philadelphia settlements that now is home to one of its lowest-income neighborhoods. Corboy talks about Kensington's heritage as a fishing village on the edge of the Delaware River and how it was home to industrial milling for items such as Stetson hats and light industry such as Dutch Boy paint. Nobody could have guessed that the abandoned, concrete-capped, fenced-in Superfund site contaminated by a steel-galvanizing factory, just minutes from the I-95 traffic corridor, could become an urban farm. A series of open tanks containing lead, cadmium, zinc, and arsenic had been abandoned when the industry left town, and Corboy says the tanks eventually filled with water and offered a tempting place for local children to swim. People started getting sick. "A lot of people died young around here," Corboy says.

When the neighborhood fought to have the lot cleaned up, the Environmental Protection Agency declared it a Superfund site, removed the tanks, and contained the remaining contamination with a three-foot layer of concrete. Then it erected a fence, locked the gate, and left. When Corboy and Sereduk found the site, it had been abandoned for eight years, and though it was stigmatized by contamination, it was affordable. The partners developed a plan. They could grow vegetables hydroponically, without any soil. Later, they could build raised beds and fill them with organic soil. They would transform a derelict ¾-acre city block in an old, low-income neighborhood into a food-producing sanctuary.

Piecing together a deal took creative gumption. The city's local community development corporation (CDC) was pursuing a "Clean Green" project and wanted to clean up the privately owned Kensington lot, which was just collecting trash. While the city was happy to have the site farmed, it wanted to buy the lot itself so that the neighborhood would feel some ownership. "The neighborhood was screwed by all previous owners," Corboy explains.

So, as a complicated purchase deal was slowly put together by the city and the CDC, Corboy and Sereduk obtained a short-term lease from the private owner in 1998 and set to work on establishing the farm. When the city's purchase of the property finally went through, it passed on management to the CDC, which continued leasing the land to Corboy and Sereduk at the same rate of $150 a month. Based on her experience, Corboy suggests that leasing city property for urban farming may be the easiest and most affordable route for new farmers trying to get started.[93] Corboy and Sereduk also envisioned a need for future expansion and bought a small parcel across the street, bringing their farm to 1 full acre.

After finding the land, money was another issue. Corboy had $25,000 to invest, but they needed more. They approached a venture capital group, which agreed to loan them $47,000. Last, they went downtown to obtain a business permit to grow food, where people outright laughed at them. "This was a rough neighborhood," Corboy says. "The idea of two white kids wanting to grow hydroponic lettuce here was ludicrous. If anybody wanted to do something in that neighborhood, the city would support it. But they

didn't think we would last. They had no clue what they were permitting — and even typed it wrong — 'hydrophonics.'"

The beginning was brutal. They worked 15 hours a day, seven days a week. "What we thought was all incorrect," Corboy says. Two years later the U.S. Environmental Protection Agency took notice and awarded them a $50,000 Sustainable Development Challenge Grant to assist them in developing a model that could be replicated. But after three years of mind-numbing stress, Sereduk left — something Corboy calls "an arranged marriage gone bad."[94]

Corboy wants people to understand that urban farms are *real* farms, growing real food for real people. Under her leadership the Greensgrow CSA reached 375 shares in 2010, providing fresh local food to some five hundred households. Overall, Greensgrow serves more than five thousand customers with its garden nursery, small garden shop, and twice-weekly farmers' market. Greensgrow also holds educational workshops, makes biodiesel for its farm vehicles, and manages honey-producing beehives.

What's more, if Corboy is right, the end is not in sight. In 2009, she reports, Greensgrow experienced a remarkable 25 percent annual growth. The city block in Kensington will always be Greensgrow's home, she says. But as it reaches maximum production capacity, Corboy is dreaming of expanding her operation to numerous farms. She confesses that this may be difficult, "like holding onto a handful of nails." But Corboy is convinced that multiple farms will be a better *financial* model.

Strategies for Sustainability

One leg of sustainability is financial stability. And diversification, argues Corboy, is important for financial stability. To achieve long-term sustainability, she says, you may need to abandon your personal goals, even do some things you don't want to. Multiple sites, for example, may be that compromise. And the crops you grow may be another.

Growing greens, Corboy explains, is easy from June through October. But that's not a long enough season to be profitable. Extending the season is important not only for additional profits but also to "extend

Bright Idea: CSA Farmers' Market

One corner of Greensgrow houses the cute CSA pickup shed, where "co-owners," as Corboy calls those who have bought a CSA share, come to collect their weekly share of food. Nearby, so shareholders will be able to shop when they pick up their CSA basket, Greensgrow hosts a farmers' market.

Corboy believes farmers' markets do better when they have more competition. So in a counterintuitive business move, Corboy decided to open the Greensgrow CSA farmers' market to other farmers, and in 2009 it was hosting 80 producers. She suggests that customers believe competition will give them a better deal and that they'll find better quality for less money. So the more vendors and products, the more people come.

the interest in who's watching and patronizing you." This was a key reason that Corboy started a nursery business, which now provides as much as 30 percent of Greensgrow's income in the first two to three months of every year.

Corboy laughs, remembering the early days of her nursery business. Like everything else used on the farm, the greenhouse was junk to someone else but gold to her. She rescued it from someone who was going out of business. When Corboy and Sereduk brought it back to Kensington, they changed their minds. Instead of using it as a season extender for their hydroponic greens, they would use it to get the attention of serious Philadelphia gardeners. They would grow flowers.

This strategic decision reflected their frustration. Corboy says the local horticulture clubs, who are major opinion leaders in the city, didn't understand that Greensgrow was growing production-level hydroponic lettuce *every week*. She says they thought Greensgrow folks were lousy growers, because all they could grow was lettuce. So when Greensgrow began to grow flowers, it was a major breakthrough, finally gaining the attention — and respect and patronage — of the horticulture community.

Shortly after this, Greensgrow began a CSA with 25 shares. As usual, Corboy took a different tack. Instead of overwhelming customers with 25 pounds of Swiss chard and zucchini at one time, she wanted to offer a range of different foods, so she enlisted the help of a host of regional growers and businesses. A Greensgrow weekly basket can include an array of local products — locally produced cheddar, chèvre, and mozzarella cheeses; locally raised chicken, lamb,

or chorizo sausage; locally produced potato pierogi or seitan; a choice of local yogurt, butter, eggs, or milk; and seasonal items such as bread, cider, honey, beer, or soaps. She describes it as "a picnic basket for a young urban couple who wants to eat more at home but who may not even know how to cook, even though they subscribe to *Gourmet*."

As a former chef Corboy is still excited by what they've done. "We teach them how to cook!" she says proudly. The day before the pickup, shareowners receive an e-mail about what's in their weekly basket, the history of the farms providing the foods, and recipes for how to use the produce. And when people pick up their baskets, staff are on hand to show them how to prepare the food. Greensgrow also hosts recipe sharing on its website. By 2010 the Greensgrow CSA boasted 375 shareholders, a testimony to Corboy's winning recipe.

Changing Ideas about Urban Farms

Farms *do* belong in cities, Corboy believes. She is trying in many different ways to change how people view urban farms, by being a good neighbor, delivering knockout weekly baskets, and conducting countless educational tours. And she pays attention to small details, too. She fought for state approval of a "farm truck" license plate for the Greensgrow delivery truck. While it may seem unimportant, the fact that she had to fight for this license plate is just one reflection of how urban farms are still not viewed as "real." For Corboy having that license plate meant that everywhere Greensgrow makes truck deliveries it will be taken more seriously as a farm.

Another issue, she says, is that rural people think a city is unappealing, dirty, and full of graffiti. So Corboy had local schoolchildren paint a colorful mural of Greensgrow on the side of its delivery truck, so country folks could see the city farm in a different way. She points to the mural. "There's our industrial past right over there. Now we're making beautiful things grow and feeding the people of Philadelphia. I hope it changes how people from rural areas think about cities — that we're not as nasty as they think." Most important, says Corboy, "I want them to recognize that urban farms are real farms."

Greening the Farm with Biodiesel

Greensgrow Farm offers all kinds of delightful lessons in sustainability — from green roofs to a solar-powered composting toilet — and all are explained by interpretive signs for children's education. Perhaps the most impressive, however, is its production of biodiesel.

Emissions from delivery trucks, tractors, and other farm vehicles can be an important consideration in a "green" farm system. Depending on the source of biodiesel, use of this alternative fuel may cut vehicle emissions by about half. At Greensgrow 21-year-old Cody Richter manages the biodiesel production. Richter says that emissions are only part of the story. Considering the full life cycle of biodiesel, he says the corn, soy, or canola that produce it actually absorb more air pollutants during their growth than are emitted by the trucks during its combustion. This is somewhat akin to creating wind energy that feeds the grid and watching your electricity meter run backward. So, Richter says, biodiesel is contributing to what he calls "negative pollution."

A tiny garden shed houses the most rudimentary but efficient biodiesel production equipment. Richter built the system and runs it single-handedly. Using vegetable oil from three city restaurants, he creates biodiesel in 40-gallon batches. With this small system he produces up to 250 gallons each week for about $2 per gallon, less than the pump price of gasoline. Of this, Greensgrow uses 50 to 70 gallons each week in its farm truck and tractor.

Richter likes to do this work, he says, because he's producing a product that isn't nearly as dangerous as petroleum products: biodiesel is biodegradable. And the biodiesel that Greensgrow doesn't use Richter refines to a cleaner grade, producing about 250 gallons every three or four weeks. Because the cleaner grade can be used by more demanding engines, Greensgrow sells it to the city's water department for its "green" boat that trolls the Delaware River to suck out plastic bags and other garbage. Corboy says the city is eager to use biodiesel for its vessel to avoid a calamity; if the boat ever springs a leak, now it will release just vegetable oil into the water, not a poisonous petrochemical oil. Corboy is proud to show off Greensgrow's biodiesel operation because it demonstrates that biodiesel doesn't need to be a stand-alone operation and can be incorporated easily and effectively into a larger operation such as a farm.

Richter is somewhat like Corboy — not content to do what others do. It's not enough for him to recycle used restaurant oil. He doesn't want his production of biodiesel to pollute the local storm-water system with by-products, one of which is glycerin, or soap. So Richter collects the glycerin filtered from Greensgrow's biodiesel, along with glycerin from others who are making biodiesel in the city, and runs it through his system for final purification. The purified glycerin, he explains with a big grin, can be made into "local" soap.

The key to successful biodiesel production, says Richter, is finding good-quality restaurant oil, with a good hydration rate — meaning a low level of fatty acids. The more a restaurant reuses its oil, the more it accumulates fatty acids. Anything below a 3 on a scale of 0 to 6 means it is acceptable. Richter says that one Philadelphia restaurant has a hydration rate of only 0.5 — "off the charts!" He says it's not surprising that this same restaurant is known for the "best French fries in the city." Corboy chimes in, "They're known for their fries, so they're going to use quality oil. It's part of their business."

Richter says that restaurants that give him their oil believe it's good for their business. They advertise to customers that Greensgrow recycles their cooking oil into biodiesel, which makes them "greener." Most restaurants don't recycle their oil, Corboy explains, instead storing the used oil for pickup and disposal by the same company that delivered it. But some restaurants, she says, are willing to go the extra mile. To participate in the biodiesel program, the restaurant must purchase a special tank to store used oil for later collection by Richter — which is a real commitment from the restaurant.

Diversify for Agri-Edu-Tainment

At Greensgrow raised beds are rotated quickly through different crops. And every year the crop selection changes, to give city people an opportunity to see how different food is grown, says Corboy. In addition to the basics, such as 30 varieties of heirloom tomatoes, for which there is high demand, Greensgrow also makes a point of raising food that people in the neighborhood might never have seen growing, such as corn. Corboy points to a long raised bed of beautiful, tall corn. "This is for them. We can get all the corn we need — we have 80 producers in our CSA market. So [this is] just for showing people how it grows." She chuckles, recalling the prior summer: "Kids come here with their parents, who ask, 'How tall are you — are you as tall as corn?'"

Some things, Corboy says, you have to do for show. "We do things here that I'd rather we weren't involved in. Urban people have such a short attention span, you have to entertain." So Corboy is constantly dreaming up playful ideas for "agri-edu-tainment." Children tour the farm by following fun, hand-painted interpretive signs, learning about soil, bugs, plants, the green roof, the composting toilet. All the signage is ostensibly authored by Blanche, the black farm cat. If children follow Blanche's farm tour and can then answer questions about all of Blanche's educational signs, they receive a Blanche coloring book.

In the winter the farm sells Christmas trees and firewood in ready-to-transport Greensgrow wood bags. In 2009 Corboy planned their first summer farm-to-table dinner for 50 people, with one long table set up smack in the middle of the farm. For a modest price of $45, the dinner sold out quickly. These are just a taste of Corboy's many inventions to educate people while entertaining. Corboy is using the farm as her creative tableau, and you can be sure by the time you read this that she will have dreamed up many new ways to attract people of all ages to her farm.

Honey in the Hood

Tucked against a wall behind the Greensgrow biodiesel shed, four beehives are humming. Corboy, their keeper, tells me that bees fly around 2 miles a day, coming and going all day, entering the hive at the bottom. The age of the hive can be determined by the number of rectangular box frames stacked one on top of the other. Some older hives are stacked 10 frames high. The queen and her eggs inhabit the bottom frame, nearly 10 inches in height. Above her, the smaller, nearly 6-inch-tall removable frames house the honey, and each may contain between 20 and 40 pounds of honey. Greensgrow sells the honey at its gift shop, along with "Honey in the Hood" T-shirts.

As is the case for just about everything else at the farm, Corboy says you don't need to be an expert to start beekeeping. She is a firm believer in learning by doing. "You just do it and learn the hard way. Sometimes you screw it up, and sometimes you get it right." She is allergic to bee stings, but she says, "If you don't bother them, they won't bother you."

Greensgrow doesn't maintain any special plantings for the bees, but Corboy says there's surprisingly plenty in a city for bees to eat, including clover and all kinds of trees. "The honey is delicious," she says. "People clamor for it."

Fresh Food for All

Corboy believes it is important to bring fresh, nutritious food to all people, regardless of income — to cross the barriers that contribute to deteriorating community health. But as with everything else she does, she approaches the challenge of improving food access for lower-income populations with fresh and innovative ideas. Greensgrow has learned how to accept coupons from the federal Farmers Market Nutrition Program (FMNP), which provides both low-income mothers and seniors with supplemental funds for purchasing fresh food. Greensgrow also holds a special market at a local senior center, bringing fresh produce to seniors who can't come out to the farmers' market.

And in 2010 Greensgrow launched the Local Initiative for Food Education, or Philly LIFE, a CSA that provides food shares to low-income families eligible for the federal Supplemental Nutrition Assistance Program. Greensgrow makes participation in the CSA easy by charging on a biweekly basis and automatically withdrawing the member's fee from his or her SNAP account; if the member receives maximum SNAP benefits, the weekly basket represents less than 12 percent of his or her total SNAP allotment.[95]

"People [in lower-income neighborhoods] believe that fresh food is unavailable and that fresh food is too expensive for them, and [LIFE] wants to make the point that it is available — that fresh food is available within our area, it is not too expensive, it can be very affordable like anything," Corboy said in an interview with a local newspaper.[96] And reflecting Corboy's belief that education needs to be integrated into every aspect of the farm, when people pick up their shares each week, they participate in a cooking class to learn how to use the foods included in their weekly basket.

Farms as Education Outposts

Some people see a potential future where cities could grow significant food for their residents. Activist Malik Yakini of Detroit, for example, suggests that, with its vacant 6,000 acres transformed into farms, Detroit could produce as much as 25 percent of its own food.[97] For others the point of urban agriculture is not to catalyze a fundamental change in the complex fabric of our food system. Rather, urban agriculture can offer an important window of education and access into our food system for the 84 percent of Americans who are metropolanites.[98] As our older, rural-based generation dies, these 233 million people have dwindling connections to our food production and probably less understanding of how or where our food comes from than any previous generation in our history.

As a society we are discovering that the consequences of this gap between people and their food may be profoundly costly. The gap has been filled by fast, fatty, sugary foods that are fueling a national epidemic of obesity, which, looking at just direct medical expenses, was found to have a total annual cost of $161.3 billion *per year*.[99]

Now the local food movement is trying to fill that gap with something else: fresh, healthy, nutrient-dense food. And many of the strategies to achieve that outcome involve education. The real point of an urban farm, says Mary Seton Corboy, is to create an "outpost" where people can learn about their food. We need more informed, discerning, concerned, and savvy consumers, she says. And one way to get there is through urban agriculture. Urban farms should be centers for education and access, she argues, enabling people to make better choices about their food and also expanding access to fresh healthy food to people with less means.

Corboy remembers the natural-foods movement of the 1970s, when local food and community gardens were all the rage. That movement, like many passing fads, soon petered out, and Corboy is worried that this movement, too, will fade. But she hopes that our nation will reach a tipping point, a point of no return, where the national consciousness about our food becomes so deeply ingrained that it cannot go away. She likens this to what happened with feminism.

Back in the heyday of feminism, Corboy remembers, women thought it was here to stay and couldn't imagine a time when the very term itself would seem archaic. But the term now seems outdated, she explains, only because the concepts were successfully ingrained into our laws and regulations. There are now entire generations who have no memory of the years before, for example, Title IX, which in 1972 guaranteed women protection from discrimination on the basis of gender in educational institutions.

Corboy's vision is startling and challenging. She is suggesting something quite different — and even more radical — than many other leaders in the local food movement. She is suggesting that a future measure of success could be when "healthy food" finally sounds antiquated or tautological — when maybe the local food movement has worked its way out of a job, and healthy food is no longer a subject for national debate — because sufficient measures have been taken to ensure that America's food *is* healthy and that *all* Americans have easy, affordable access to it. This is certainly not a vision of no coffee or chocolate or ice cream, heaven forbid, but it is a vision for shifting the preponderance of our diet from cheap, sugary, fatty foods to healthy, fresh, nutrient-dense foods. Women's rights in our nation are so protected that most everyone today takes these rights for granted. Could a day dawn when fresh, healthy food is also so prevalent, so protected, that it, too, is taken for granted?

Lessons Learned
Finding Land for Farming

Look for urban land that's not being used, even if it's contaminated.

When people think of a farm, they usually think rural, bucolic scenery. One of the biggest challenges today in fostering new farmers to replace the rapidly aging population of farmers is the prohibitive cost of land. New farmers usually can't afford to buy prime farmland outright. But urban areas offer another concept of farmland that may be more accessible and affordable than commonly thought. Suitable land in the city may include formerly contaminated, remediated brownfield sites, such as the industrial Superfund site that Greensgrow Farm leases (page 96) or the former brownfield site that Lynchburg Grows purchased (page 116). In fact, says Greensgrow founder Mary Seton Corboy, one of the best interim uses for a brownfield site that cannot be immediately redeveloped is an urban farm.[100]

If you're interested in farming a piece of land that you suspect or know is contaminated, contact your local government office of planning. This office can help you determine what kind of zoning or legal permissions may be needed, whether a site assessment has been completed, and current clean-up status or guidelines. You may also want to contact the state agency responsible for working with brownfields. The state agency can conduct a full site assessment, and this cost may be covered by the federal brownfield program, devoted to assisting in the assessment and remediation of contaminated sites.

If the soil on the land is not suitable for farming, you can excavate and replace it, as Lynchburg Grows did, or you can simply farm over it, in raised beds, as Greensgrow Farm did.

Consider vacant or abandoned city lots.

Suitable urban land may include vacant or abandoned lots, such as the land turned into community gardens by Nuestras Raíces (page 123) in Holyoke, Massachusetts, or the land farmed by Jones Valley Urban Farm (page 158) in Birmingham, Alabama, or the 6,000 acres of abandoned lots in Detroit. And sometimes suitable urban land may include overgrown, unused land that nobody has ever considered for farming, such as the riverside property owned by the city of Asheville, North Carolina, and now farmed by Gladheart Farms (page 87).

Community gardens are prime candidates for abandoned urban land. While some community gardens are intent on protecting their land in perpetuity, such as the South End/Lower Roxbury Open Space Land Trust (page 62), such gardens may not need to own the land to be successful. Begin by asking the city government what spaces it owns that might be used for community gardens. Next, enlist the city's help in contacting the owners of abandoned lots to learn whether they would give permission for use of the space. The absentee lot owner may actually appreciate the idea of community garden because it keeps the space neat and clean, says Julia Rivera, organizing director for Nuestras Raíces.

When I ask whether it's worth the effort to develop a garden in space that might one day be lost to development, Rivera doesn't seem overly worried. She would rather transform a space and take her chances. "If we don't utilize those spaces, those spaces are doing nothing — and if we don't use them as gardens, they'll become something else. One place was being used as a dump for trash, and a place where drugs were being used. How are you going to feel about yourself if you live in a place like this?" she asks. When I press harder about this issue, she admits that, because Nuestras

Raíces does not own the community garden spaces, the spaces could be taken away. But, she argues, "if you're going to think negatively about it, then you'll never get started." The best insurance against losing a community garden is to make it a collaborative effort, involving as many local groups and individuals as possible, and to also help the community learn to exercise its political power. "They're finding political power by coming together," says Rivera. "And by bringing politicians and the community together, we can make things happen." She is also clear that the gardens are about far more than food. "Transforming these spaces is changing how people feel about where they live."

Consider public parkland for community gardens.

Marra Farm, a historic farm in South Park, Seattle, offers an interesting model for a community garden in a public space. Owned by the City of Seattle's Department of Parks and Recreation, the farm hosts a variety of interconnected projects, including a community garden in the P-Patch program (page 60), a large plot for the city nonprofit Lettuce Link (which grows food for low-income residents), a subsistence garden for low-income Mien gardeners from Laos, and a pilot market-gardening project.

The market-gardening program in particular is a new paradigm for using community public land. While public lands in the West have long been leased to private ranchers, communities have often balked at the idea of allowing private gain from farming public lands. Moving beyond this barrier, Seattle's market-gardening project is designed to help low-income people grow food on a larger scale than that of a P-Patch garden plot and to sell that food to supplement their income. People eligible to rent a market-garden farm plot must be from the local neighborhood, low-income, and experienced in organic gardening, at a P-Patch garden or elsewhere, and they must have a rudimentary business plan or idea of where to market their produce. If the program goes well, says Richard MacDonald, supervisor of the P-Patch program, he believes Seattle will expand the program to other sites. "The key is that these individuals have a real incentive to do something with the land because they're getting something out of it," he says.

Begin by leasing the land.

Mary Seton Corboy suggests that the best route for a beginning urban farmer is to lease the land, rather than purchase it. To begin with, the often exorbitant price of urban land puts it beyond the reach of most urban farmers. But leasing makes sense for other reasons. The liability associated with ownership of a contaminated site, particularly a brownfield site, can be prohibitive. Ideally, Corboy suggests leasing for a specified term of five years or more, from either a nonprofit or a city redevelopment group that owns the property and would like to see it in use.

In some cases simply taking over management of derelict land is enough of a benefit to the owner that the lease is token. Tricycle Gardens, a community gardening association in Richmond, Virginia, for example, searched for many years for a site to create an urban farm, says its director, Lisa Taranto. Finally, because of the tanking economy and development grinding to a halt, Tricycle Gardens negotiated a deal with a private developer to lease a half-acre undeveloped lot beginning in 2010 for $5 per year.

Jones Valley Urban Farm leases a 3-acre city block in downtown Birmingham for only $1 per year, on a five-year lease, in exchange for agreeing to provide food directly to people who live in the downtown and to provide garden-based nutrition education to children living downtown.

Always prepare a written agreement for land use.

No matter how much you know or trust a property owner who has agreed to let you farm on his or her land, whether for free or for lease, write your agreement down on paper. When the Jones Valley Urban Farm first got started, the owner of a piece of land in the city agreed to let them farm it for free. Then he decided to charge rent, asking for $250 per month. Six months later, the owner filed for "back rent" for the years they had farmed the lot for free. Though JVUF successfully fought this suit, and it has sinced moved on to other land, it learned a lesson the hard way: always create a written agreement.

4: Empowerment
Food Movements in At-Risk Communities

The stories in this chapter are full of surprises. I'm reminded of Julia Child, whose intense love for food led her to places she never dreamed — Le Cordon Bleu, a masterful cookbook, and *The French Chef*, a television show that inspired millions.

Just as Child's passion led her to unexpected places, each of these community food projects began with a fairly simple passion that led to something much bigger, with broader implications and outcomes than could have been imagined at the outset.

In Portland, Oregon, a woman's passion for finding ways to help runaway youth led to a very small garden at a shelter known as "Harry's Mother." And in turn this small patch of soil caused an unfolding of remarkable events at Janus Youth, with repercussions for countless people in at-risk communities across the city. Even the organization's executive director, Dennis Morrow, shakes his head in wonder at how a regional social services nonprofit organization has come to be the steward of neighborhood gardens and a certified organic farm that feed the community body and spirit in a host of unexpected ways.

In Holyoke, Massachusetts, a more basic passion — the desire to have food for their tables — motivated poor Puerto Rican immigrants to negotiate the transformation of an abandoned lot into a neighborhood garden. None of these people could have predicted that their simple act of hope, in the face of despair, would someday lead to a nationally recognized project, Nuestras Raíces, that incubates new businesses and farmers and empowers women and youths with special programs — all while celebrating Puerto Rican food heritage. The garden was the first step toward the resurrection of an entire community of people.

In the Boston area, the Food Project's desire to connect youth with land and community seemed a fairly simple proposition: get some land, and bring youth together to learn to grow food. Yet it led to one of our nation's premier examples of effective food farming through creative youth training and empowerment. Something so simple, yet so good!

In St. Louis two women wanted to bring affordable fresh food to inner-city residents, while reducing "undesirable activity" on abandoned lots. Today Gateway Greening supports more than two hundred community, youth, and school gardens, as well as an urban farm that offers jobs and training to those who may have nowhere else to turn — individuals who are dealing with homelessness, substance abuse, mental illness, and recidivism.

In Virginia the travesty of one man's garden being bulldozed catalyzed a cascade of unexpected events. The result today is a thriving urban farm, Lynchburg Grows, which has transformed nine nursery greenhouses that were destined for destruction into a productive CSA and education center. Yet farming for Lynchburg Grows is simply a means to a greater purpose: providing a place for people with disabilities to work, gain skills, and feel a part of a larger community.

More than inspirational, these projects are a message to all of us. And the message is simple. As John Muir pointed out in *My First Summer in the Sierra* (1911), "When we try to pick out anything by itself, we find it hitched to everything else in the Universe." Food, it appears, in the hands of people who are trying to make the world a better place, is never *just* food. It is hitched to everything else we care about: nourishment, health, skills, jobs, leadership, nature, community, economy, and, yes, even second chances, redemption, and empowerment.

Reclaiming Communities through Gardens
Janus Youth Programs, Inc.

with research by Robin Proebsting

Janus Youth Programs is one of the largest nonprofits in the Pacific Northwest, serving over six thousand high-risk youth every year in **Oregon and Washington** through 20 programs aimed at runaways and the homeless. One of Janus Youth's most preeminent projects is its urban agriculture program, which it offers because it believes that gardens have proved an effective tool in addressing key community issues such as hunger and food insecurity, community revitalization, environmental stewardship, and economic development. As part of the program, Janus Youth supports community gardens in low-income and immigrant neighborhoods, along with garden programs for youths of all ages.

Janus Youth Programs began in 1972 as a demonstration program of Multnomah County, Oregon's smallest and most populous county and home to Portland. As a program to help high-risk youth — runaways, homeless, children dealing with substance abuse — Janus Youth had nothing whatsoever to do with food. Now, however, Janus Youth demonstrates that a social services program can actually further its goals by entering the improbable arena of food gardening. More, it offers proof that a social services program with multicultural skills can steward community gardens with an unusual grace that transforms the gardens into catalysts for changing lives and even entire communities.

As Janus Youth's youth programs experienced success in the late 1970s, several different state agencies began to provide support for its work. Nearly 20 years later, the Janus Youth runaway program center — called "Harry's Mother" — started a tiny garden patch for its residents for strictly therapeutic value, to give the runaway youths something productive to do with their time. But the garden didn't stop there. It was a small seed that eventually germinated and grew into something quite large, bearing unexpected fruit.

At the helm of Janus Youth is Dennis Morrow, a dynamic man who appears to manage this multiservice agency and its $9 million-plus budget with grace and humor and who, according to one source in the field, has earned the reputation of being one of the best nonprofit executive directors in the nation. His willingness to be "on the edge" professionally is reflected in his personal life, with eight children, six adopted out of the Oregon child welfare system. A former Vista volunteer, Morrow has been with Janus Youth for over 30 years, and in his spare time he teaches nonprofit management at Portland State University and also teaches in the alcohol and drug counselor program at Portland Community College.

These details are a window into more than Morrow himself, as the culture and drive of nonprofits are usually shaped by their director, influencing who is attracted to the organization and the kinds of programs undertaken. "Who he is is who this organization is," says Rosalie Karp, who, as advancement director for Janus Youth, writes grants and finds way to advance the organization's mission.

When I visit Janus Youth, I expect to be taken straight out to the gardens. But this successful executive not only makes time for me, he wants me to sit still long enough to hear the *real* story of why something as improbable as food gardening has become so important to Janus Youth.

When Morrow arrived in 1980, the program had five residential homes. Now it has about 40 programs in 20 locations across both Oregon and Washington. Morrow believes Janus Youth is probably the largest service agency for runaway homeless youth and teen parents in the Northwest, providing service mostly in residences and shelters. Its shelter in downtown Portland, for example, offers 60 beds to homeless youth every night. In addition to these core programs, it provides programs for kids in the welfare system and residential programs for juvenile sex offenders before and after they go to the state training school, and it also operates an alternative school.

But the supreme irony, says Morrow, grinning wryly, is that Janus Youth now has more square footage in "dirt" than in residential programs and that it also has a certified organic farm. "This makes no sense at all, whatsoever," he says.

The gardens have never been a "natural" fit for Janus Youth. But Morrow wants to me to understand one thing: their success is directly related to the passion and tenacity of specific people. The person who planted the first seed was a staff person, Tera Couchman Wick. As a Master Gardener, Wick thought the runaway and homeless kids who came to Harry's Mother, with no solid footing in life, might become more grounded by getting their hands into the soil. So she turned an unused nearby lot, loaned by a local bank, into a garden.

When the bank needed its lot back to build a parking lot, Morrow thought the experiment was over. Janus would be out of the gardening business. But Wick had other ideas and approached Morrow about using the space behind Harry's Mother for a garden. Morrow replied that it was a paved parking lot. "But," he recounts, "she looked at me, in only the way she could, and said, '*But you could dig it up.*'"

So out in the asphalt parking lot, the second iteration of Harry's Mother's garden was built in the form of raised beds. Years later, in 2003, Janus Youth was hit hard by the recession and huge cuts in the Oregon state budget, and they were forced to sell the Harry's Mother property. This time Morrow was sure it was the end of the garden. But once again Wick persevered. She began to think bigger, not smaller. She imagined a community gardening project in North Portland, where Janus Youth knew the community, and wondered if she could apply for a USDA grant to fund it. When she approached Morrow about the idea, he didn't think Janus Youth would have a chance at the grant. Janus Youth had never worked with the USDA, and its work wasn't connected to food security. But it was no skin off his back to be agreeable, so his response was a quick "Of course."

But Morrow was astounded. They did get the grant. "Damn!" he says, slapping his hand on the table, remembering it as though it was yesterday. "Now what?" He looks at me, his eyes wide. "Now we had to do it!"

Reclaiming Community

Before the first community garden was dug, gangs, vandalism, drug trafficking, crime, and hunger characterized the North Portland St. Johns Woods public housing community. To stave off hunger, residents of the housing project — many from East Africa, Laos, and Vietnam — began to cultivate tiny garden patches outside their apartments; many had been farmers back in their homeland. But the gardens strained their relations with the public housing managers. The complex was served by a single water meter, and the managers were reluctant to foot the bill for watering the gardens. They threated some of the gardeners with eviction. Residents didn't have any other means to supplement their tables — no nearby community gardens where they could grow food legally. Upon learning about the garden that had once existed at Harry's Mother, an organizer for the St. Johns Wood community sought out Janus Youth for its help, and soon Tera Wick was working with the community to create a community garden.

At first the management company at St. Johns Woods didn't want such a strange commodity on its property. Now this same management company is one of Janus Youth's best references and wants Janus Youth to put community gardens into other public housing communities. What turned the company around?

The garden created by the St. Johns Woods community may have been small, tucked inconspicuously into a triangular corner on the south side of a building, but its power was immense. People came out of their apartments to see the garden, to work in the garden, and then, best of all for the community, to connect with each other. Almost miraculously, the more people were drawn out of their homes by the garden, the less crime and vandalism there was in the community. People were reclaiming their community.

The beauty of it, Morrow says, is that the immigrants now had an opportunity to replicate what they knew from their homeland: growing food. And too, the parents and grandparents — the community elders — now had an important role in the community: to share their agricultural knowledge.

This success gave Janus Youth ample reason to expand its work in community gardens — an unconventional approach to working with a high-risk neighborhood was yielding unforeseen bounty. Now, just nine years later, Janus Youth has a full array of programs that connect people and build community through food gardens.

Teen Empowerment

Food Works is one of the more remarkable — and unexpected — outcomes of Janus Youth's foray into the world of community gardening. Dennis Morrow says they began Food Works, a program for teen youth, at a time when Oregon had the highest food insecurity in the nation. As with all Janus Youth food projects, the idea for Food Works arose from the community itself. When community leaders began talking about an employment and empowerment program for teens, Janus Youth was ready to facilitate and support the realization of their vision. And the idea of a teen-managed garden, where teens could learn to plan, grow, and market produce, was born.

In the beginning Food Works was just a tiny garden, the size of a meeting table. Its teen managers began by growing salad greens because they're easy to grow and easy to market. When the teens were successful at the farmers' market — selling out within an hour! — they quickly decided their tiny garden wasn't big enough. They wanted a *farm*. Again Tera Wick went into action. She made a connection with the regional government, which owns unused green space, and was able to negotiate the use of a 1-acre plot of land on Sauvie Island, 7 miles upstream in the middle of the Columbia River.

Historically Sauvie Island had been farmed, and Rosalie Karp tells me that island families were glad to see the teens reviving that agricultural heritage. Also on the island was Sauvie Island Organics, an organic farm that offered to share equipment and knowledge with the teens. Within several years the Food Works teen farm on Sauvie Island had a well, irrigation, a greenhouse, and tractors, and when the teens began selling over $5,000 worth of produce per year, it became a certified organic farm. In 2009, Karp is proud to say, Food Works teens sold over $11,000 worth of produce. The money from sales goes back into the program, with guidance from the teens on how to allocate the funds, and a portion set aside to help them buy back-to-school supplies.

Of course, in addition to building leadership, job skills, and self-esteem, the gardens do help reduce hunger in the community. Whatever the teens don't take back home to their families or sell at the markets is donated to the community. In 2009 Food Works donated over 5,000 pounds of fresh, organic produce to over 750 individuals. As a reflection of Janus Youth's core belief in supporting, not directing, community leadership, it is the youth leaders who decide what portion of their harvest to donate — and they consistently agree to donate more than 50 percent of all produce from their farm.

Food Transforming Community

Morrow tells me, "Every one of these kids is facing a choice whether to be in a gang or not. And the garden mentors say they can tell when they walk through the community who are garden youth and who's not. The garden youth act respectful, they go out of their way to help." After involvement in Food Works, some youths are receiving college scholarships, excelling at sports, gaining recognition in different ways. Morrow says the impact is nothing less than transformative.

As he recounts the story, Morrow is clearly enjoying the fact that the joke is on him. Above all, he wants me to understand that Janus Youth's garden and farm program has grown *organically*. It started small with Harry's Mother, as a small garden out front, became raised beds in the parking lot, then grew into a community garden in North Portland at St. Johns Woods, where in turn it spawned community gardens in other neighborhoods, as well as new focused programs such as Food Works. These programs arose from the community itself. Janus Youth simply assisted. It was Tera Wick's conviction that real change happens only when it comes from the community itself that led to this unusual approach. To this day Janus Youth carries out Wick's commitment to community-led change. The community creates its own vision and goals, and Janus Youth supports the community passion by garnering resources to make that vision possible.

From the beginning the gardens of Janus Youth have been about creating a place where people from poorer communities can grow their own healthy food. That's the simple story. But Morrow wants me to understand that the real story is much bigger.

"The amazing thing," says Morrow, "is that it's *never* been about the food. It's been about the food transforming community." He shakes his head. "I don't know anything that compares with this. I write grants, I've done all this stuff, all of my career. And this — and there is not a lick of b.s. here — this is a *miracle*. This makes things happen like you cannot *believe*. Kids transformed, families transformed, communities transformed."

Expanding Circles of Leadership

The USDA grant that funded the St. Johns Woods garden and its larger successor program called Village Gardens is a multiyear but one-time event. Projects are never refunded. So when the grant was nearing its conclusion after three years, Dennis Morrow once again worried that the successful gardens would likely end. But as before, Tera Wick continued to innovate.

The Housing Authority of Portland had received a Hope VI grant to level its poorest housing development in North Portland, not far from St. Johns Woods. The grant would redesign and build a new mixed-use (commingling residences and businesses) and mixed-income community that would be called New Columbia. The Housing Authority very much wanted Janus Youth to create a community garden in this socially engineered community. And the community leaders from Village Gardens wanted to share their work with another community. So Janus Youth built on its success by creating a leadership program: the now-experienced gardeners from nearby St. Johns Woods would mentor the creation of the new garden in New Columbia.

The result of this leadership program was entirely new. People from St. Johns Woods were growing, developing, and transforming their community, and now they were elevated to a position of leadership in helping another community create its own garden. So like the proverbial loaves and fishes, the benefits of the gardens were multiplying — growing confidence and self-esteem within a community and growing connections between communities.

When I visit Seeds of Harmony, the New Columbia community garden, I am greeted by residents who proudly show me their food gardens. Though it is winter in Portland, there are still collards, leeks, and garlic sprinkled through the gardens. Eca Etabo Wasongolo, a slender man with a smile as wide as he is tall, is a political refugee from the Congo, where he earned a degree in social work. He worked for years in the camps for Rwanda refugees, until he, like others who worked in the camps, became considered a political threat. Now hired as a full-time community organizer at Village Gardens, he will be able to use his professional training for the first time in the United States. Agnes Sola, a stout woman from Nigeria, is beginning her third year in the garden. She says proudly, "I am a *good* gardener because I grew up on a farm, and I know how to make it easy for myself." She proudly points to the only raised beds in the community garden and to the narrow raised wooden planks she has laid across the paths, which enable her to tend her beds without compacting them.

Another garden leader, Michelle Hanna, a single white mother with one son, talks about the importance of commitment. She says, "These people are really serious, meeting every Saturday, and it's important to have that commitment." In return for the commitment, they all agree that the garden has given them specific job skills — "people skills" enabling them to work with a broad diversity of people, volunteer management skills, and project management skills — all of which improves their résumés for the job market.

Wasongolo tells me about the garden community and how the garden *is* community, referring to how the garden generates social capital — networks, trust, goodwill, and the potential for joint action. He points to Michelle and Agnes and says, "We would never know each other without the garden." They all chime in, talking about how people from all parts of the community meet through the garden. In just the first few years of the garden, they say, hundreds of people in the community have met and developed new relationships.

They are all in high spirits, and soon I learn the big news: they are designing their first community-managed corner store! Even something as simple as a corner store, in the hands of Janus Youth, is "edgy." The store will be managed and run by residents, and they envision it as a meeting place for residents to gather, learn, share information, and, of course, sell their produce. More, because the store will boast a professional kitchen and walk-in coolers, residents hope one day that they'll be able to process their own garden products for sale, in pickles, jams, and salsas. I briefly meet Pascal Ananouko, an immigrant from Togo, who is rushing to the planning meeting at the store. His son, Egvebado, started as a youth gardener, was later hired onto the Food Works team, and now leads a kids' garden program called the Little Apricot Garden Club. Ananouko's 18-year-old daughter has also been active in Food Works for years. The gardens are clearly important to his entire family, a way of life.

Later I'll visit the store and see its walls plastered with sophisticated charts and diagrams that form a detailed strategic plan. Ananouko and the store team have conceived five community teams to steward the store opening and its future management, with each team in charge of different aspects of the planning. The store will be operated as a nonprofit and is expected to at least break even after the first few years. The Housing Authority will support the store by sharing the cost of a project coordinator, leasing the building to the community at no cost in perpetuity, and underwriting all the utility costs. Michelle informs me, "The Housing Authority approached *us* about the store. I think that shows a lot of confidence in us and is something to be proud of, don't you?"

Because of their involvement with the gardens, residents are now reporting changes in their blood pressure, cholesterol, and diets. Michelle thinks her son eats more vegetables because of her garden, because, she says, they're more *interesting* when they come fresh out of the garden. Amber Baker, the Village Gardens program director, shares that when the first garden started at St. Johns Woods parents began coming to the office to ask, "What are you doing to my kids? Now they're asking for *salad*!" They were happily surprised to see their children finally eating vegetables! Nearly all the teens in Food Works report at year-end that they've increased their consumption of fresh vegetables. Baker also tells me of people who have lost weight and lowered their blood pressure because of working in the garden. Some people, she says, report that without the garden, they wouldn't even leave their house! To document these changes a grant from the Northwest Health Foundation will fund the planning for participatory community-based research into long-term health impacts of working in community gardens.

Janus Youth couldn't have predicted that growing food would become a powerful tool for helping at-risk youths or multigenerational diverse low-income communities, but its entrepreneurial and innovative spirit enabled it to embrace the unexpected harvest from its gardens. Maybe it's not the miracle that Morrow suggests. Maybe it's what we should expect when people connect with the land and food, put food on their table, and find they are competent at doing something to help themselves and others. Maybe there are psychologists who can cite detailed studies explaining why these results would be predicted. Even so, if it's not a real miracle, it's pretty darn close.

Lessons Learned
Working in At-Risk Communities

Give respect, and honor the people's voice.

Giving respect means giving people a real voice, then honoring that voice. Dennis Morrow, executive director of Janus Youth in Portland, Oregon, puts it plainly when he says, "You cannot b.s. this." If you go into a community with a "Let me help you because I know what's best for you" traditional power-structure mentality, people who have experienced this treatment their whole lives will see it, know it, and not participate.

Both Morrow and advancement director Rosalie Karp agree that the form of respect and engagement they're espousing is far from a social services mentality. They believe their approach has a spiritual context and that giving the community voice will have integrity only if rooted in the belief that people must be given a say in what impacts them. Many organizations do espouse these values, they say, but are not able to put them into practice. Karp says people in communities like St. Johns Woods can smell that lack of integrity a mile away and will be gone in a flash.

Karp tells me of one woman who took her aside at the garden and said, "I've always hated white people. But I've learned that I don't hate you, and I don't hate the people who work in the garden with Janus." She tells another story of an African-American man who stopped the program director one day and said, "I want you to know, I've been watching you. Lots of times, you white folks come out here, and you tell us what's wrong and bring some money, and then you walk away and leave us. You haven't done that. You've come out here, and you've listened to us. And you're working with us. And I like what I see."

Still, despite their program's success, Tera Wick, the midwife to the Village Gardens program at Janus Youth, worries that they'll become overconfident and comfortable. It is so easy to slip off the path of humility and to just *do* something without asking, she says. For the long term she believes it may be important to have a formal mechanism for accountability with the community, perhaps something like a community advisory council. This way, she says, when we fall back into making assumptions about what's needed and inadvertently making decisions *for* others, "there will be a pathway for people to let us know that our awareness has fallen short, and we can course-correct." Wick says her next project is to establish this formal feedback mechanism – designing and establishing it with the community, of course.

Ask the community what it wants, rather than telling it what it needs.

Janus Youth enjoys large success with its community gardens, and everybody from the executive director to its garden managers attributes this success to one thing: *attitude*. The attitude they all talk about, even if they give it different names, is one of humility. Put into action, humility usually means one thing: you and I don't have all the answers.

The lesson of *asking* is a logical extension of giving respect. Amber Baker, the Village Gardens program director, explains that their goal is to listen to the community, not to come in with their own agenda. Dennis Morrow says that Tera Wick, the founder of Janus Youth's garden projects, was an "absolute goddess" at this. Morrow explains that Tera Wick always told people, "I am here at your request. I am here to help *you* figure out what *you* want to do. I am here to bring in resources *as you need them*."

Wick demonstrated this humility when, at the beginning of the St. Johns Woods effort, she went in, she says, "with a lot of questions and no answers." Dennis Morrow is more blunt: "Nobody [from Janus Youth] ever went in and said, 'Us white folks want to tell you how to do this.'"

It is the Janus Youth position that, fundamentally, the attitude of "I'm here to help you" assumes that we know what you need better than you do. This is not only disrespectful but also deeply flawed. Instead, it is always better to ask:

- Ask the community what it wants and what it needs.
- Ask how we can help you make this happen (as opposed to "Thank you, we'll make this happen for you now").

Wick tells me a story of how the simple concept of asking plays out with the teen Food Works program. At the front end of the nine-month program, the crew leaders, who are required to have worked in the program the previous year and must apply and be interviewed for the position, spend a month with the adults in the program to design the entire year's program. Every year the adults ask the youths, what do we want to accomplish, and how can we make that happen? One year the teens said, "We want to learn cashiering!" Instead of giving the traditional adult response – "Okay, let's go find a good curriculum on cashiering" – the adults kept asking instead, "How do we do that?" The youths thought they should be tested on their skills. The adults pressed, "How do we do that?"

By continuing to be asked instead of being offered answers, the youths became very involved in developing a cashiering skill-building program. And the more they became involved in designing it, the more they became invested in the outcomes and the more time they put in and the more they stayed engaged through the entire school year.

Wick's story could just as easily have been about adults designing their garden, deciding on rules for the garden, and more. It is powerful in demonstrating how real community ownership is achieved through humility and asking, not telling.

Ensure that all decisions and approvals are made by the community.

This practice may seem like an obvious extension of the previous one, but it is not always, and perhaps not even commonly, practiced. At Village Gardens, with Janus Youth, collaborative decision making is applied in remarkable ways: Community members must approve all new community volunteers. Community members must approve all new Village Gardens staff. Community members must even approve the grant proposals. Rosalie Karp explains, "When we go home at night,

they're still here. This is their home, the garden is in their home, they live with the decisions, so they should be able to have final say." While this practice of collaborative decision making may seem daunting, it is perhaps the most important core practice and truly empowers a community to take ownership. The downside, of course, is that decisions take time – sometimes a *lot* of time. The upside of this practice, however, clearly outweighs the drawbacks. When the community makes decisions about the garden – and takes responsibility for those decisions – the garden and its future truly belong to the community.

In the Food Works program, giving the teens responsibility for making their own decisions has led to more positive outcomes – more engagement through the school year and increased skill-building and confidence. To make decisions they use the "fist to five" decision-making process, a form of consensus building in which a fist equals "no," five fingers means "yes!" and the degrees in between reflect how the participant feels about the issue at hand. Staff certainly has input into the decisions, but there are times when the teens have overruled the staff.

Create a safe environment.

Most people in housing developments, Rosalie Karp says, live under institutional poverty, institutional racism, and institutional exclusion from the public process. When Janus Youth started, the St. Johns Woods community had the highest *fifth-grade* dropout rate in all of Portland and the highest call rate to the police department. But within only eight months of putting in the community garden, the community began changing. The call rate to police dropped dramatically, and residents started taking ownership by picking up garbage, paying their rent on time, and taking community responsibility for what kids were doing.

At its core Janus Youth deals with people whose lives have been unsafe, whether because they are homeless, at-risk young runaways, poor, or immigrants. If runaway youths come into the downtown home and feel unsafe, says Karp, they leave, just like that. Likewise, if they come to work at a garden program and feel unsafe – psychologically, physically, mentally, emotionally — they'll take off. On the other side of the coin, the establishment of a garden may be precisely what's needed to help create a safer environment in the larger community.

Janus Youth works to create a safe environment in several ways. In terms of physical safety, the group has established a series of safety rules at its farm and gardens regarding tool use and physical behavior. For psychological and physical safety, community adults adopt and review annually a "community contract" that sets the standards for the Food Works youth and other gardeners. For emotional safety, in addition to a strong ethic of confidentiality, a series of guidelines establishes behavioral norms, including encouraging people to try new things; to realize that it's okay to disagree but it's not okay to blame, shame, or attack; and to take 100 percent responsibility for their own actions, while remembering the difference between intent and impact. Youth teams use skits to learn how to put these behaviors into action. And some teens even memorize the guidelines, using them to call each other out when needed.

Define your roles.

At all of the Janus Youth gardens, the titles given to staff are a matter of "walking the talk." Since the organization believes the community should own the project, Janus Youth personnel believe they are more appropriately called "Invited Staff," while community residents are given the title of "Community Staff." These titles are meant to show and constantly remind people of their respective roles and, I suspect, their relative power.

As a self-described "outsider," Amber Baker, the Janus Youth Village Gardens Program Director, says that someone who isn't a resident of the community needs to be very attentive to figuring out her role. "There may be times," she explains, "when I have the skills to do something for the community project, but it may not be *appropriate* for me to do it." Even if residents take longer or struggle with something, certain things are better done by them, for themselves. Baker says a lot of well-intentioned people jump into the idea of a community

garden with enthusiasm, not realizing the importance of building relationships and listening to what the residents want and need. When she works with AmeriCorps volunteers who come in gung-ho to work in the garden, to write grants, to *do* things, she often needs to remind them that their most important job is to build relationships. Time meeting with community residents is never time wasted.

Baker likens a community garden project to being invited to someone's house for dinner. "If someone invites you to dinner," she says, "you don't invite friends along without asking." Similarly, if you were invited into a community, you wouldn't bring in just anybody without asking. She drives home her point one more time: "It's not your house, so you don't just begin cooking in the kitchen."

Build on community strengths.

"You may be poor, you may be an immigrant or part of a minority, but you do have skills and talents," says Rosalie Karp. The approach of Janus Youth is "strength based" — a buzz word more common to the realm of mental health but equally appropriate and powerful for the new community gardening movement. A strength-based approach focuses not on what's wrong with someone or a community but on what's right. Tera Wick says, "If we started with 'what's wrong,' we could spend all day on it and make no progress." Instead, the success of the positive approach involves assisting the individual or community group to identify strengths and what they need to heal, to grow, to change, to move forward.

For example, strengths in one neighborhood might be that there are a lot of grandparents who know a lot about the kids, because they're not leaving for work during the day, and also that many are immigrants who used to farm in their country of origin. A strength-based approach would then ask how the grandparents and immigrants could play a role in the community garden. "The program belongs to the community," says Karp. Janus Youth simply "provides a *platform* for people to be a part of the decision process."

Coming Together to Garden
Seeds of Harmony

The Seeds of Harmony community garden serves the mixed income residents of the New Columbia housing development in **North Portland, Oregon**. Part of Janus Youth's Village Gardens program (see page 105), the garden not only provides food for the table but also fosters relationship building and social integration in this very diverse neighborhood.

Like all Village Gardens programs, the Seeds of Harmony Garden focuses on empowerment by and for the community. Here, the garden is run by the gardeners, with support as needed by Village Gardens staff. The gardeners themselves set the guidelines and standards for the garden, which basically boil down to two principles: do the best you can to keep up your designated plot, and be respectful of other people.

Perhaps as a result of this consensus-building structure, the Seeds of Harmony Garden has been able to build a sense of connection and relationship in a community that varies widely in age, income, ethnicity, and culture. The New Columbia housing development is a mixed-use, mixed-income facility, bringing together businesses and residences, subsidized housing and luxury condos, Seattle urbanites and newly arrived immigrants, in a mish-mash community that, surprisingly enough, seems to work.

And the garden may just be an important pillar for this development, supporting residents in common cause to grow food, connect with the land, and step outside the confines of their homes to truly associate with their neighbors.

Training and education are active elements of the garden. The garden has designated advocates and mentors, who come from among the community gardeners themselves, with training provided by Village Gardens. With an eye toward encouraging future gardeners, the community has designed one section of the garden as the children's garden, where kids can get their hands in the dirt and explore gardening on their own. With these and other endeavors, the Seeds of Harmony garden has truly engaged its gardeners, both drawing inspiration from and giving inspiration to an intercultural collaboration that builds community even as it feeds its members.

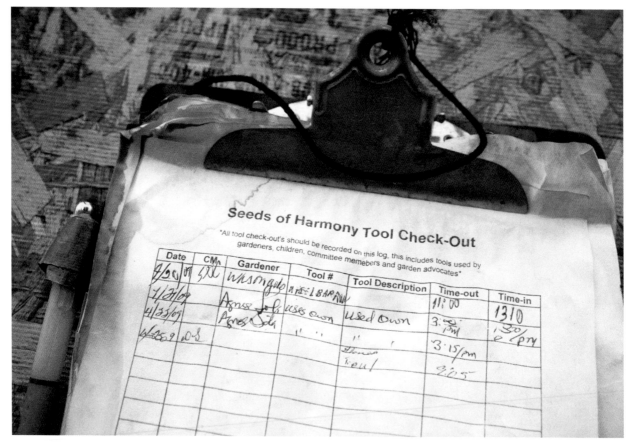

Seeds of Harmony Tool Check-Out

All tool check-out's should be recorded on this log, this includes tools used by gardeners, children, committee memebers and garden advocates

Date	CM	Gardener	Tool #	Tool Description	Time-out	Time-in
8/20	SM	Whsongulo	4 PEE 1 BAR AW		11:00	
4/21/09		Agnes Sola	uses own	Used Own	3:00 pm	1310
4/22/09		Agnes Sola	" "	"		1:30 / e pm
4/25/09	D-S		" "	"	3:15 pm	
				Rauel	2:05	

Members of the Seeds of Harmony community garden come from incredibly diverse cultural backgrounds but are linked by the common denominator of food. The 71,000 square feet of garden plots set in Portland's New Columbia and Tamarack housing communities provide food, education, and a place to meet.

Empowering People with Special Needs
Lynchburg Grows

with research by Jessica Ray

Lynchburg Grows is an urban agricultural center in **Lynchburg, Virginia**, that grows food, offers educational programs, and provides a haven for job and skills development for people of all ages with special needs, including a vocational training program for disabled and low-income individuals. In its nine historic greenhouses, the farm grows vegetables, herbs, and tilapia fish, while also managing beehives and chickens for eggs. A portion of this food is donated to local food pantries, and the remainder is sold through a CSA, at a farmers' market, and to restaurants. Through its educational outreach, Lynchburg Grows assists neighborhood gardens and hosts workshops and programs at local schools and community centers.

In the heart of Virginia's southern Piedmont, south of Lynchburg's old downtown shoe manufacturing center, tucked into a corner of a vintage 1920s neighborhood of tidy homes, an extraordinary transformation has occurred. A large abandoned industrial site of 6.8 acres — once burdened with contaminated soil and nine glass greenhouses nearly one hundred years old — is now a lush, thriving urban food farm.

Inside these historic greenhouses are vegetables, flowers, and herbs growing in neatly maintained raised beds, historic roses climbing higher than could be imagined, and water burbling in newly built aquaculture structures for tilapia and perch. Outside, the clucks and scratchings of chickens are mixed with the laughter of children who are learning that Dr. Seuss wasn't kidding — chickens can indeed lay green eggs. Bees are pollinating and making honey in the hives, and worms are wriggling and creating rich compost that will fertilize the raised beds.

With only three full-time and one part-time paid staff, Lynchburg Grows achieved this remarkable transformation in just under five years, thanks to tens of thousands of hours put in by volunteers from nearby schools, numerous youth and civic groups, and the local community. Now Lynchburg Grows is putting fresh, healthy organic food on tables throughout the community and reaching people at all income levels. It sells produce at a weekly farmers' market, delivers 120 CSA shares from May through October (including a number of "scholarship" shares), and provides weekly deliveries to Lynchburg Daily Bread, a shelter and soup kitchen that serves nearly 50,000 free meals each year.

But Lynchburg Grows is not content with just tending its own gardens. It believes in the power of growing food to transform and empower people, and it is finding ways to sow this seed in other fields. Its three leaders — Michael Van Ness, Dereck Cunningham, and Scott Lowman — believe that the simple act of growing food can build technical and overall life skills, offering people an important path to self-reliance and food security. Their core passion is to offer people with special needs and disabilities, who are often institutionalized or housebound, a place to venture out, be with others, work and learn job skills, and laugh and build self-esteem.

The Birth of an Idea

When Michael Van Ness first saw the abandoned site populated with nine neglected glass greenhouses, he was astounded. It was 2004, and the greenhouses hadn't been watered for five years. Yet roses were climbing 20, 30, 40 feet, stretching up and out, through greenhouse roofs and vents. At one time, back in the 1920s, the greenhouse complex had been the only large-scale source of cut flowers in Virginia. Later, in the 1950s, H. R. Schenkel Sr. bought the site and converted the greenhouses to growing roses exclusively for Lynchburg Wholesale Florist, which distributed his roses throughout the mid-Atlantic states, accounting for one-fourth of Virginia's flower industry. Some claimed these were the "best roses east of the Mississippi,"[101] and indeed they "decorated the White House and crowned Kentucky Derby winners."[102] When Schenkel's finally closed its doors in 1998, unable to compete with the cut-rate prices of flowers from South America, nearly 100,000 rose bushes were flourishing and producing about 20,000 cut roses daily.[103]

Five years later, after sitting idle — unwatered and untended — the greenhouses had become a dense thicket of roses gone rogue. Somehow, without water, without heat, without cooling, the roses had survived unmitigated summer and winter temperatures in the greenhouses.

No sensible person would ever have deemed these greenhouses salvageable. All who had viewed the property considered them a lost cause. But Van Ness was on a mission. He was accused of "not being able to see the thorns [for] the roses." Together with friends Dereck Cunningham and Scott Lowman (and a third friend, John Wormuth, who later left the project), Van Ness was seeking a place for a very special garden.

Their search had begun nearly six months earlier, in November 2003, when a local newspaper reported the story of Bedford resident Paul Lam. Suffering from a severe speech impediment as a preschooler, Lam had been misplaced in a mental institution for the severely disabled, where he languished for nearly 40 years. Then, as deinstitutionalization swept the nation in the 1990s, Lam was transitioned out into a group home, where he spent every day possible outside tending his very own small garden. One day, as Lam watched helplessly, a large dozer blade flattened his garden, the result of a bureaucratic muddle.

As news of this travesty spread, so did the desire to right the wrong. Together, not knowing they were beginning a far longer journey, four men in different professions who happened to know each other teamed up to find Lam a new garden. Within months the four friends were able to provide Lam with a larger garden, complete with tiller, compost, and seeds.[104]

One of these men, Dereck Cunningham, a stained glass artist, was especially sensitive to people with special needs. Born with spina bifida, and not expected to live beyond six months, Cunningham had not only survived against all odds, he had learned to thrive, thanks to years of working beside his grandfather in his garden. His grandfather was a double-leg amputee and would wheel Cunningham to the garden on his lap, where they would work side by side among the vegetables. He imbued Cunningham with a sense of possibility.

Now, influenced by Lam's situation and by Cunningham's own journey, their aspirations grew. They all saw that growing food was transformative. Growing food could not only feed people, it could also change lives. It was about to change theirs.

They decided they wanted to create a place where disadvantaged people of all stripes could enjoy the healthy benefits of gardening, obtain new skills and concomitant self-esteem, and have access to the space to do so. When the partners toured the Schenkel property, Van Ness saw an unusual opportunity. He was an environmental lawyer with deep roots in land conservation, so he was able to structure a land-purchase deal. He and his three partners would turn the abandoned greenhouses into an urban farm. The Schenkel family was thrilled. An urban farm would honor their greenhouses, their family history, and especially their roses.

Reclaiming a Contaminated Site

Nothing in life is easy. The partners were going to create an urban farm with a core program for people with special needs. It was only fitting that the site itself had special needs. The city of Lynchburg had already completed an initial assessment of the Schenkel site and a neighboring plot, an equally abandoned and overgrown manufacturing site, and the city's preliminary data indicated that the sites were contaminated by both pesticides and lead. The neighboring site was also contaminated by barrels of paint solvent that would definitely need to be cleaned up.

Despite the preliminary report of pesticides on the property, "I could tell the Schenkels were good land managers," says Van Ness. "It was clear that they applied just the specified amount of pesticides, when and where they were supposed to, and not more. This can't be said for all land managers." Van Ness was trying to explain that contaminated soil doesn't necessarily mean a previous farmer was a poor steward or misguided. Farmers and greenhouse growers did what they believed was right for the land and their crops, and only later did we begin to learn the consequences. DDT, for example, was once considered a World War II miracle drug for delousing soldiers and killing mosquitoes that carried deadly malaria, and entire communities would actually welcome the fine mist as it was sprayed by planes flying overhead. Only years later would we learn that DDT doesn't biodegrade like other chemicals and persists in the soil for untold decades, a carcinogen waiting to be taken up in food crops for human consumption.

Van Ness pored over pages of soil tests. He and his partners were especially concerned about the possible persistence of DDT. They worried that children and people with developmental disabilities would likely have greater sensitivities to chemicals. Even if there was only a one-in-a-million chance that something in the soil might be harmful to someone who passed through or worked in the gardens, that one-in-a-million chance was too much. They wanted clean and safe soil. Period.

With these pressing concerns the new owners of Lynchburg Grows took several actions. First, they sought assistance from the Virginia Department of Environmental Quality (DEQ). They learned that they could receive cleanup assistance by entering a voluntary brownfields remediation program. This program, federally funded by the Environmental Protection Agency, would help them clean the site to support the desired community uses. A key benefit of joining the program would be that, if all the assessment and remediation hoops were successfully passed, the new property owners would be formally relieved of all long-term legal liability for any consequences of the contamination. Lynchburg Grows applied and was accepted into the program.

The DEQ then conducted a full site assessment. For each greenhouse and the remainder of the property, it assessed computer-generated randomly selected sites over a comprehensive grid. Just to obtain the raw data, an assessment in 2009 cost on average roughly $10,000 per acre. The cost for assessing the Lynchburg Grows site came in at just over $80,000, and all of it was covered by the federal program. The news was good: Levels of arsenic — a residue of rose pesticides — were not as high as had been believed and were actually within "safe" background levels. Lead was found in a narrow strip along one of the greenhouse drip walls, but nothing would ever be grown there, so a simple barrier cloth would suffice for containment. The old pesticide shed was a problem, however, with the soil contaminated by pesticide residues, so the new owners enforced a "no dig" stipulation on the shed and covered the floor with a gravel cap to keep the pesticides contained.

At the same time they received advice from an emerging leader in urban farming, Will Allen, founder of Milwaukee's Growing Power (see page 74). Allen visited in 2005 and recommended that they remove some of the greenhouse beds to enable easier access for disabled people inside the greenhouses. He also suggested that they replace all the soil, if only to ensure peace of mind.

Bright Idea: Farmer for a Day

This is one of Lynchburg Grows' innovative youth programs. They invite schools to bring classes to the farm to learn about where food comes from and healthy eating and to participate in farm chores while learning about the different systems on the farm, such as composting, vermiculture, raised beds, and more.

After the DEQ conducted its initial assessment, the partners made a fateful decision. They would take on the Herculean task of replacing all the soil in the raised beds. They figured that it would take them nine years, replacing the soil in one greenhouse each year.

Thus began the bucket brigade. Nobody foresaw the magnitude of the task. Each of the nine 200-foot-long greenhouses would require 150 tons of new soil. The next summer, in 2005, sweltering under the glass in heat climbing to 105°F, the partners began cleaning out roses and replacing soil. Van Ness became a self-described "maniac." With his partners and more than a hundred student-athletes from Randolph Macon College, 250 students from Lynchburg College, and many students sent many times by the Virginia Episcopal School, they completed replacement of all the soil in two greenhouses.

"It was like eating an elephant," says Van Ness. "Some days you'd have to close your right eye, squint your left eye as hard as possible, and try to ignore the rest of what was around you. It was so overwhelming. I learned to say, I'm just going to do 10 feet today!"

During this effort the partners learned that the soil they had purchased was not worth its weight in pennies. It didn't have the necessary tilth, or the ability to hold water the way healthy soil should. So they decided to get into the soil-making business and launched a composting effort. Now, in a win-win partnership with the city, Lynchburg Grows receives about 2,000 tons of leaves and woods chips for composting each year, saving the city about $96,000 per year in landfill costs. It also takes in over 4,000 pounds of dining hall waste from a local college.

Winning Community Support

While it may be obvious that a nonprofit food project has different qualities from those of a for-profit one, one not-so-obvious factor is that the success of a nonprofit is directly proportionate to community support and feelings of ownership. Michael Van Ness, now executive director of Lynchburg Grows, is blunt: Don't try to "own" the project, he advises. The project has to be the community's — grown *with* the community, built on relationships and partnerships every step of the way. If you do this, the community will be there for you, helping at every step.

Today Lynchburg Grows is developing new outreach programs that reflect the community's needs. "Roots and Shoots" is an intergenerational gardening program that brings together the community's elderly and youth to share and learn the joys of gardening. Its colorful "Grasshopper" bus is a mobile learning center for youths, offering hands-on learning laboratories where schools can meet their SOL (standard of learning) requirements while integrating agriculture and farming into the school curriculum. Its "Elementary Education Garden" program has helped install gardens at elementary schools, early learning centers, and even a camp, to provide outdoor classrooms where children can learn about such topics as plants and healthy eating. Farm tours and field trips to nearby and distant schools focus on learning where food comes. For more in-depth field trips, it offers a day of "Farm Camp," during which schoolchildren can participate in farm chores while learning about composting, vermiculture, aquaculture, gardening, renewable resources, and animal stewardship. And it's converting five vacant, neglected city lots into community gardens.

Each one of these different programs increases the community's food security by increasing the supply of local food. Each one teaches practical knowledge about how to grow food to hundreds of community volunteers, students, and neighborhood residents. And each one teaches practical job skills to people who would otherwise be marginalized by society, while also fostering hope, empowerment, neighborhood, and a sense of community. Food is the medium for growing Lynchburg.

Growing Neighborhood Gardens

As a perfect example of growing the project with the community, Van Ness points to the five vacant lots Lynchburg Grows bought in its first year, in the hope of turning them into community gardens. For the next four years, all Van Ness did was mow the grass. Neighbors would approach to ask, "What are you planning to do here?" Van Ness would respond, "It'll be a garden. *Your* garden. What do *you* want to do with it?"

The first groundbreaking for one of these neighborhood gardens was in 2009, and Van Ness delights in recounting the story. It was multigenerational and multicultural, he says, bridging racial lines that had begun to emerge as a result of white gentrification of an older black neighborhood. But because they were breaking ground for their own community garden, all the neighbors came out together — black and white, richer and poorer — and many met each other for the very first time.

Within only four weeks of this groundbreaking for the neighborhood garden, Van Ness reports that neighborhood children had begun to start their own gardens at home. Members of the neighborhood approached Lynchburg Grows and asked for its help in cleaning up another neglected lot to make it into a park. Within only four weeks, because the neighborhood itself was excited, because the idea of their neighborhood garden had been grown *by them* over four years, the community now wanted to double the size of their garden. Van Ness says proudly, "This is what it means to grow a project with community."

Financial Stability through a CSA

The Lynchburg Grows community-supported agriculture (CSA) program provides more than weekly baskets of freshly grown food to shareholders from May through October; it also provides a steady, dependable stream of income for the nonprofit. The group produces all its vegetables, herbs, and flowers in its glass-enclosed greenhouses, even during summer months. The historic greenhouses are the group's greatest asset, and it has found a way to use them by draping shade cloth under the ceiling during summer months, installing drip tape for automatic irrigation — which paid for itself within one month — and fixing a "swamp cooler" misting system to keep the greenhouses cool. On a hot, sunny day, a greenhouse is a surprisingly welcoming environment, and workshops that meet in the broad middle aisle are treated to a pleasantly cool breeze.

After testing the CSA for two years at 60 full shares and working out all the kinks, Lynchburg Grows doubled its production to 120 full shares, which it calculates feeds about 218 people. Unlike most CSAs,

which charge for a whole season's worth of produce up front, Lynchburg Grows allows the purchase of shares in one-month increments, and while it guarantees a basket of organic food every week, it doesn't guarantee that all the food is grown by Lynchburg Grows. This has provided room for error, time to learn, and financial flexibility.

For those who are just starting out, it can be hard to also build a CSA on top of learning the ropes of running the farm. Lynchburg Grows' strategy was to form a partnership with a nearby established producer — Appalachian Sustainable Development (ASD; see page 253) — to ensure a reliable source of quality produce. This way, if a greenhouse crop doesn't do well or is wiped out by insects, shareholders can still be guaranteed delivery of a full basket of food. The farm commits to buying a certain amount of produce at wholesale prices, planning each month's purchases based on the number of shares purchased that month. Whatever isn't used in the weekly CSA baskets is sold at the farmers' market. Lynchburg Grows marks up ASD produce by about 30 percent, which is just enough to cover its packaging and marketing costs and far less than the average 80 percent retail markup, so their farmers' market customers still receive a bargain price.

Through the CSA Lynchburg residents have affordable access to healthy, organic food, and Lynchburg Grows gains a stable customer base that has confidence in its ability to deliver quality food and a critical, steady revenue stream of more than $4,000 per month.

Maturing into a Multifaceted Farm

Lynchburg Grows began as an idea, a garden project for at-risk youth and people with special needs. Now, however, it is far more. In the space of just five years — it's still a young nonprofit by anyone's standards — the group's annual budget increased from $8,000 to over $260,000. It has grown into a full-service urban farm for the entire city.

Lynchburg Grows' core mission is still to grow food for the community while providing opportunities to at-risk youth and people with disabilities and special needs. But it's also growing new ideas, each year bursting with new ventures. The nine greenhouses offer an amazing tour of the expanse of Lynchburg Grows' ideas. Greenhouse 3 features 12 varieties of the original Schenkel historic roses, once gone rogue and now restored to their former splendor in neatly tended beds, creating a small oasis of perfumed heaven. Disabled staff prepare small bouquets of these roses for donation to residents of nursing homes throughout the Lynchburg area. Greenhouse 4 is being used to study the potential of switchgrass and other crops as alternative fuel. Part of Greenhouse 5 will always feature the overgrown, untended roses climbing through the roof, a reminder of where Lynchburg Grows began, while the remaining part will be used to graft and clone the historic roses for eventual sale. Greenhouse 6 is dedicated space for nonviolent juvenile offenders to develop their entrepreneurial skills by growing, producing, and marketing salsa. Greenhouse 7 is being converted into an innovative production and education center for three-dimensional aquaponic food production, based on a system advocated by Will Allen of Growing Power (see page 74). In Greenhouse 8 the congregation of St. John's Episcopal Church grows food for the hungry, producing 14,000 pounds between 2004 and 2009. And last, Greenhouse 9 is now a rose-lined space available for meetings and special events; in 2009 it hosted its first wedding.

As Lynchburg Grows continues its efforts to restore all nine historic greenhouses to productive uses, it is also striving to stay true to its original goal: creating programs to help people with special needs and at-risk youth learn important life and job skills. A partnership with the local high school and community college brings special-needs "transitional" youth (so called because they are transitioning out of school), ages 18 to 21, to the farm once a week to work as "Farmer for a Day," during which they participate in a variety of farm chores. Work at the farm prepares them to enter the workforce and also builds their life skills and self-esteem. For those who are interested in working in agriculture after graduation, Lynchburg Grows arranges for them to work for two hours every day at the farm to develop their skills in farming.

Another partnership for job training brings at-risk youth to the farm to work side by side with people with disabilities. Youths who have been sentenced to community service through Lynchburg's Community Court system are assigned to work at the farm, in the hope that exposure to this sort of work and community support will help them stay out of trouble in the future. Lynchburg Grows reports that many of the at-risk youth enjoy this work, and several have expressed interest in becoming future employees.

Yet another partnership brings in at-risk youth who have just completed time at the juvenile detention center to learn how businesses are formed and operated. This "Post-D" program helps these young people develop business skills through agriculture. They plant tomatoes, peppers, red and yellow onions, and cilantro in Greenhouse 6. "The end-game," writes Aaron Lee, a blogger for Lynchburg Grows, "involves getting the group to grow enough veggies that we can have them make a salsa they could turn into a micro-enterprise (a product they can market). They're proving to be some of the hardest-working (even when they get the lousy dirt-moving jobs) volunteers we've had."[105]

Experience Before Power

Young and small nonprofits usually have roll-up-your sleeve "working boards," and their composition is often determined by who is willing to devote 20 or more hours a month to the cause. Larger and more mature nonprofits with one or more paid staff often appoint or elect policy and fund-raising boards, with people selected according to their connections, expertise, and ability to help the organization raise money. While some boards are willing to take someone who is relatively new to the cause, Lynchburg Grows takes the opposite approach. Anyone applying for a position on its board must have been involved in volunteer activities for at least six months. This ensures that that he or she fully understands the organization before being in a position to make decisions about the organization.

A fourth partnership is occurring at the Lynchburg farmers' market, one of the oldest indoor farmers' markets in the country. The Lynchburg parks and recreation department, funded by a grant the city received in 2006, has helped Lynchburg Grows develop a business plan for the market, become a market anchor, and provide job training for disabled individuals who run the market.

In these and many other ways, using an urban farm as the central pivot point, Lynchburg Grows is creating innovative pathways into the community for disabled people and at-risk youth to become, and be seen as, productive community participants. Like other community farms in the nation, Lynchburg Grows proves that a farm can grow far more than food.

Lessons Learned
Working with Special-Needs Populations

If you build it . . .

The experience of Virginia's Lynchburg Grows shows that if you want to open your doors to people with disabilities and special needs, you won't have to wait long. Scott Lowman, the director of stewardship and one of the group's founders, says, "We did not need to go looking; they came to us." Executive director Michael Van Ness, another of the group's founders, emphasizes that people who are marginalized, disabled, or at risk want to give and be a part of society; they want to do something important and feel that they are contributing.

Chris Matheson, who was born with Down syndrome, originally started as a volunteer and now enjoys coming to Lynchburg Grows as one of the stipend workers. His father, John, says, "Chris just loves it here. It gives him a place to anchor his life. It's low pressure, but he really feels like he's contributing here. And he'll eat anything that comes from the greenhouse — even if he doesn't like it!"

You don't need special skills to work with the disabled.

No special training is needed, says Scott Lowman. But working with the disabled does require compassion and kindness, as well as proper supervision. Perhaps the most important skill for success is the ability to describe tasks and state goals clearly and simply.

Find appropriate tasks.

Most gardening tasks can be accomplished by people with special needs. The key is to realize that everyone is unique: some can water and feed animals; some can help in the garden with weeding and harvesting. Lynchburg Grows president Dereck Cunningham says, "Be patient, and you will eventually find something for everyone." One woman, who is blind and has cerebral palsy, now spends her day competently pulling off dried rose petals, which will be used to make potpourri. But it took a year and a half of patient experimentation with different jobs to identify this as a task that she could successfully perform.

Expect to learn and teach empathy and respect.

Empathy, Michael Van Ness says, is not something kids are naturally born with. It is something that we need to teach, and he believes it is taught by working with people with special needs. He says, "That's why so many college volunteers keep coming back — because they see this is a creative, learning farm."

Van Ness tells the story of one student intern who asked another, "Have you learned anything from working here?" The intern looked at another nearby who had cerebral palsy and replied, "Normally, I would beat the s--- out of him at school. I would normally make fun of him." At which the intern with cerebral palsy also spoke up. "Normally," he said, "I would feel sorry for myself." Here, though, their sense of "normal" had expanded to include empathy and self-respect.

Reclaiming Community through Heritage and Food
Nuestras Raíces

with research by Benjamin Chrisinger

Nuestras Raíces is a grassroots nonprofit organization founded in **Holyoke, Massachusetts**, by Puerto Rican immigrants who sought to put their agricultural experience to good use by growing food for their tables in a community garden. As the organization worked to help residents celebrate their Puerto Rican culture while putting down roots in their new home, it naturally expanded beyond gardens and food into issues of social, economic, and environmental justice. Today, Nuestras Raíces' mission is to promote economic, human, and community development through projects relating to food, agriculture, and the environment.

"This mural means that we left everything behind to come to Holyoke," says Julia Rivera, organizing director for Nuestras Raíces and president of its board. She is pointing to a colorful mural painted by children featuring a large tree with a wide canopy, which she explains is symbolic of her homeland of Puerto Rico. The mural is a backdrop to a small neighborhood garden with a distinctly Caribbean flair. A quaint garden shed and covered stage, painted a bright robin's egg blue, are decorated with pots of brilliant petunias and geraniums, creating a gathering space with brick-red picnic tables. Nearby, individual garden plots are bordered by fences painted the same bright blue. And everywhere the garden is neat as a pin.

Julia Rivera is just one of thousands of Puerto Ricans who sought a better life in this small town in western Massachusetts, when the paper mill and tobacco farms here were booming in the 1950s and '60s. Their pursuit of the "American dream" was cut short, however, when the mills closed and tobacco farms went fallow in the 1980s, leaving the town reeling from a catastrophic loss of jobs. More than 30 years later, the town is still recovering; in 2009 the poverty level in Holyoke hovered near 31 percent, nearly triple the Massachusetts average.[106] The Puerto Rican population remains prominent; outside Puerto Rico itself this little town of only 40,000 has the highest concentration of Puerto Ricans in the United States — 37 percent, according to the 2000 census.[107]

Nuestras Raíces, whose name translates as "our roots," began in 1992 with one community garden, La Finquita. The garden was so successful that local community leaders came together to found Nuestras Raíces to preserve La Finquita and create more gardens.[108]

Since then Nuestras Raíces has steadily grown into a vibrant community organization seeking to improve *all* aspects of life in Holyoke, including reclaiming Puerto Rican rural heritage. The challenges in Holyoke remain daunting: high rates of teen pregnancy, crime, and welfare dependency and, according to the Holyoke Health Center in 2010, the highest rate of diabetes mortality in Massachusetts — and 89 percent of those who succumb to diabetes here are Latino/Puerto Rican.[109] Operating on a budget edging toward $1 million, with a staff of 25 full-time and part-time people, Nuestras Raíces works with young people, women, new farmers, new businesses — all anchored in the goal of empowering residents to increase their economic, environmental, and personal sustainability. "We're not just a community development corporation," says Daniel

Ross, former executive director. "We emphasize crops, environmental justice, and murals and arts projects to validate residents' rural and ethnic roots."[110]

Food Is a Gateway

As with so many other community food projects, Nuestras Raíces demonstrates that food is both a gateway and a catalyst for broader community health. "It goes way beyond food," Ross told the *Atlantic* magazine in 2008, "but it starts with food."[111] Julia Rivera says that Nuestras Raíces has made countless changes in the community, transforming vacant lots into beautiful gardens, changing how people view and understand food, helping people start their own farms, helping women start businesses, and teaching children about healthy food and gardening.

Women are particularly at risk in Holyoke, says Rivera. When she realized that many women were feeling trapped in their houses, often depressed and unable to do much for themselves, she was inspired to found a support group called Raíces Latinas to help them get their feet back on the ground. In 2009, five years after founding the group, Rivera tells me how Raíces Latinas has taught women about health, nutrition, obesity, and managing money and has even trained women in restaurant and school cafeteria work to help them become eligible for those jobs. "Some women are deciding to go back for GEDs," Rivera continues. "And some are excellent at making dresses, so we're helping them do business plans to start a business."

Another unusual outgrowth of the community food gardens is an educational campaign about environmental justice. Rivera recites all the environmental issues that residents face. Mercury contamination in the Connecticut River, on whose banks the city was built, means that residents need to be educated about not eating any fish they take from the river. The prevalence of brownfields means that abandoned industrial lots, where people may want to start gardens, may have contaminated soils that need to be tested and replaced. Rivera is working with young people to create a video about brownfields, which will, in turn, educate their families. Also, she tells me that the city's very low air quality, a problem discovered by the EPA, coupled with chronic pulmonary health problems, means that people need to be educated about how to clean their homes without worsening asthma and other pulmonary conditions. As an example, Rivera, whose son has asthma, says she was shocked to learn that she should never use bleach in her house. On just this topic alone, Rivera conducted 35 "Healthy Homes Workshops" over three months in spring 2009, to help educate people about how to ensure better indoor air quality.

Making a Difference

Rivera started working with Nuestras Raíces more than 12 years ago as a volunteer community gardener. "When I saw that gardening was good for the kids, I joined the board," she says. Next she started a garden for kids ages 7 to 12, called Cuenta Con Migo or "Count on Me." She began with only 12 kids, but the program grew to serve about 25 kids in just one year. "Now the graduating students are teaching other kids about gardening," she says proudly.

With 10 community gardens, two of which are dedicated for youths, Rivera is not satisfied. There is a waiting list for garden plots, and she thinks the community could easily support 20 to 40 gardens all over Holyoke. "I began to see how important the gardens are to people," says Rivera. "They come to the gardens at six in the morning. And how important it is to the *community* — there's no vandalism!" Joel Cortijo, another community leader, explained to the *Atlantic* magazine that everyone takes ownership in the gardens, and doing anything to damage them "would be like vandalizing your own car."[112] Overall, Rivera estimates that the gardens benefit about four hundred people and that a typical garden can produce almost $900 worth of vegetables in the summer. For a low-income family this might be the only reason fresh vegetables ever appear on the table.

La Finca: A Farm Incubator

There is a waiting list for a plot at La Finca ("The Farm"), a 30-acre farm along the Connecticut River. Nuestras Raíces owns 4 of its acres and leases the other 26 acres from the Trustees of Reservations, a nonprofit dedicated to preserving land of historic, ecological, or scenic value. While the Trustees share a strong commitment to urban farming, according to Tito Santana, interim codirector for Nuestras Raíces, they are unable to provide the group with a long-term lease, which can present some challenges when they're applying for grants for long-term projects. Nevertheless, Nuestras Raíces has proceeded with its master plan for the entire parcel, which includes an array of mixed uses that reflect and celebrate Puerto Rican heritage.

Just as food is a gateway for Nuestras Raíces' broader community work, the farm plots anchor much more than farming. Near the farm entrance are a farm store and gift shop (which offers irresistible homemade tropical ice creams); a barn and riding area for Paso Fino ("fine-stepping") horses, the pride and joy of the traditional Puerto Rican equestrian community; and a *lechonaria* ("pig roast") stand that offers weekly community pig roasts. A greenhouse here is a reminder that this is a real farm; not far from the entrance are the smallest plots, of one-eighth to one-quarter acre. Down by the river a colorful tree house and stage offer a shady area for summer concerts and theater, and nearby is the youth farm, where teens can farm for free on a larger scale than a community garden plot.

Farther downriver a large flat field graced with dark river-bottom soil is divided into farm plots varying in size from ¾ acre to 4 acres. This incubator farm has attracted Puerto Rican as well as other immigrants. The farmers grow a range of crops for their extended families and to sell for income at the farmers' market: corn, squash, peppers, pumpkins, okra, tomatoes, greens, onions, and — the prize of Puerto Rican cuisine — a sweet pepper known as *ajies dulces*, the base for the dish *sofrito*. On the slope above the farm plots are a petting zoo, an orchard, and pasture plots for raising goats, sheep, chickens, turkeys, rabbits, and geese. Overall, Nuestras Raíces hosts 20 to 25 farmers every year.

"I think thousands of people are affected by these farms," says farm manager Kevin Andaluz. Not only the farmers and their families but other community members benefit directly — people who come to visit the farms, or attend the music festivals, buy produce at the farm store, buy seedlings at the greenhouse. And now, he says, the farms are also providing food for the salad bar at the local school!

"This is for everyone — not just for people with money," he says. "We try to sell at reasonable prices, because most people don't have a lot of income. Most people who are served are low-income people." For those who think that local farms and food are a fad, Andaluz disagrees. "This is a project with very, very long-term impacts," he says. "When you survey people, people say they want more, not less."

The two hardest aspects of teaching people to become farmers in Massachusetts are planning for seasonal rotations and marketing. "In the Caribbean

Incubating Businesses

Nuestras Raíces has transformed an abandoned building and vacant lot in the center of town into the Centro Agricola, a space for meetings and small business development. Over the years these facilities have incubated a variety of new businesses, including a restaurant, bakery, greenhouse, *sofrito* sauce maker, shared community kitchen, and coral aquaculture producer.[113]

Marine Reef Habitat, where Gerard Ramos installs and maintains saltwater fish tanks and raises and sells tropical corals and fish, is one of these businesses. Ramos remembers free-diving for lobsters and fish with his father in the waters of Puerto Rico, where he developed a love of coral. "I went six years ago, and it was all dead. You couldn't catch a lobster," says Ramos.

Ramos's goal is to clean up his native waters so that, one day, the sea is once again a healthy, living habitat. And one way to do that is to repopulate the waters with coral that can survive and grow in the polluted waters. In 2009 Ramos grew 70 coral species, when he made an important discovery: at least one species is able to survive and even overtake invasive algae. Working with his brother, a diver in Puerto Rico, and with children they are teaching to become certified divers, Ramos hopes to repopulate the Caribbean reefs with this coral.

you can grow the whole year," says Andaluz. "In the United States you need planning. It's only six months for growing, so seasonal planning is very difficult." Marketing is equally difficult, he says, because many farmers don't like to go out and sell their produce. So Nuestras Raíces offers options to the farmers who don't want to go to market: they can sell their own produce at the Nuestras Raíces farm stand, or they can provide produce to Nuestras Raíces staff to sell at a farmers' market for a 30 percent fee. A lot of the farmers raise food just for their families, says Tito Santana, and many also do it in addition to full-time jobs.

Nuestras Raíces has had some challenges with its incubator model. Originally, the plan was for farmers to stay for no more than two years, following which, with help from Nuestras Raíces, they would move to their own land. However, the organization has learned how difficult it is for farmers to gain access to land.

To begin, most land is located at a distance from Holyoke, requiring transportation that the farmer often doesn't have, explains Tito Santana. And purchasing farmland requires access to capital or loans, both of which are not readily available to the Holyoke low-income community. Nuestras Raíces, itself, wants to be able to offer very-low-interest loans but hasn't been able to broker this deal with a financial institution. The group is working its way through these issues, and until it is able to find other solutions, farmers are able to stay at the incubator for longer than two years. This means turnover is smaller, averaging about 20 percent, or three or four new farmers each year.

While incubating new farmers remains a challenge, the training for new farmers is a clear success. Every year Nuestras Raíces offers a training course in which prospective farmers learn the basics in western Massachusetts and creating a business plan. At the end of the class, experts come in to review and score the business plans. Farmers who receive a high score are eligible for a plot at La Finca; farmers who receive a score of less than 50 percent need to either go through the course again and develop a better business plan or decide that farming is not for them. The course usually begins with about 20 people and successfully graduates 12 to 15, says Andaluz.

Evaluating Economic Impact

Community nonprofits focusing on food tend to be so young that their long-term impact on a community may be hard to gauge. A western Massachusetts research center has attempted to do just that for Nuestras Raíces, which, given its long tenure of nearly 20 years, offers sufficient history for study. In its February 2007 study, the Center for Creative Community Development (C[3]D) concluded that Nuestras Raíces has a total annual economic impact on the community of over $2 million. C[3]D found that this food-oriented nonprofit was having impacts not only in the most obvious way — through its own direct expenditures — but also in its indirect effects on real estate values, educational services, medical services, food services, and wholesale trade. Moreover, Nuestras Raíces was drawing about 2,500 nonlocal visitors to the town in 2006, who were spending on food, lodging, and retail stores. In turn, increased visitor spending was having its own set of indirect impacts.

The bottom line for Holyoke, according to C[3]D, is that annual expenditures of Nuestras Raíces of about $560,000 in 2006 were creating measurable positive economic impacts of another three times that amount. What's more, the future impacts of additional businesses, increased visitors, and capital improvement projects were projected to double that impact, to almost $4 million.[114]

Nuestras Raíces is just one grassroots nonprofit among many in the nation. For communities that are seeking ways to pull themselves out of the recession, the C[3]D study suggests that a community-based food organization can offer numerous localized economic benefits well beyond the food itself. Depending on the specific goals of the organization, a community might benefit in new jobs; youth educated in agriculture, building trades, and entrepreneurial arts; new businesses incubated; farmland preserved; additional dollars circulating in the community to support community health and professional services; and unique cultural offerings that can attract new visitors. Food is, in a sense, just the diving board into a much larger pool of economic opportunity and benefits.

Farming as a Gateway to Rehabilitation
Gateway Greening

Since 1984, Gateway Greening has been on a mission to green **St. Louis**, with an end goal of not only beautifying the city but stabilizing neighborhoods, improving community health, combating homelessness and poverty, and building food security. The group supports community and school gardens across the city, as well as its own City Seeds Urban Farm, where it provides job training and "therapy gardening" for at-risk populations.

Gateway Greening was founded in 1984 with a mission of making affordable fresh food available to inner-city neighborhoods, putting to good use the derelict vacant lots in the inner city, and bringing gardening programs to at-risk schools. In 2005 the group created the City Seeds Urban Farm in downtown St. Louis, on a patch of land between a high-rise hotel and a freeway on-ramp. The farm is, as described in its mission statement, "an urban agriculture initiative providing job training and therapeutic horticulture to individuals who are homeless and underserved. A collaboration of several local organizations, City Seeds Urban Farm produces and distributes affordable, healthy, locally grown produce."

The farm is tended by clients of the St. Patrick Center, a local faith-based service agency working with those who are or are at risk of becoming homeless, with Gateway Greening staff and AmeriCorps volunteers providing mentoring and support. The concept is simple — to provide basic training in farming, gardening, and landscaping skills — but the impact on the clients is profound.

Some City Seeds workers simply need a leg up. Herman Reid, for example, a war veteran, has no criminal record and no history of addiction, but he's been been down on his luck after an honorable discharge from the armed services and has been living at a local Salvation Army shelter. He's always loved the outdoors, he says, but when his grandparents had tried to interest him in gardening when he was younger, he'd resisted. Now, with his experience at City Seeds, he's learning to appreciate the process, he says, and he's been hired to work for a landscaping company with large, year-round contracts with clients like Washington University in St. Louis.

City Seeds trainee Jason Smith, too, has had a run of bad luck, but one might chalk it up in this case to different circumstances. Smith, who's been in and out of prison since the age of 16 for offenses ranging from assault to drug trafficking, says he would title his story "Convict Gone Green." In his early 30s, while in prison, he worked for the Missouri Department of Transporation pruning trees. He found that he enjoyed the work. His parole officer referred him to City Seeds Urban Farm, and with the skills and experience he's gained here, Smith plans to start a tree-trimming and gardening business, partnering with other City Seeds clients he's met. He likes the idea of maintaining community and home gardens for people who want them but don't have time to do the work themselves. He's tired of jail, Smith says, and wants to move on.

Stories like these abound. With its focus on empowerment and training, Gateway Greening has become a true gateway, transforming the lives of its clients one step at a time.

Preceding pages: The City Seeds Urban Farm is set in an unlikely venue between a freeway on-ramp and a high-rise hotel in downtown St. Louis. *Above:* Here workers like Anthony Woodfork gain work experience, helping them get past a range of challenges, including unemployment, homelessness, drug abuse, and a criminal record.

Top left: City Seeds farm manager Annie Mayrose (in the white shirt at left) runs training programs ranging from landscaping to vegetable farming. *Top right:* Ron Williams, on his last day at City Seeds, and nine months clean, was in charge of picking produce for the afternoon's nutrition demonstration. *Bottom left:* Jason Smith, in and out of jail since he was a teenager, is looking forward to starting a landscaping business. *Bottom right:* City Seeds sells some of its produce at a farmers' market in downtown St. Louis.

Top left: Trainee Eloise Williams, on her last day at City Seeds, shows off one of the onions she had planted on her first day with the program. *Top right:* Training at City Seeds prepares clients for jobs in many related fields, including agriculture and landscaping. *Bottom:* Veteran Herman Reid plans to parlay his newfound skills to improve his standing at a local landscaping company.

Mike Terry, a St. Patrick's Center client enrolled in City Seeds' therapeutic horticulture program, picks carrots at the farm. The program, aimed at those recovering from addiction, homelessness, and other issues, teaches clients how to grow food and improves nutrition literacy while fostering a strong connection to their recovery goals.

Growing Today's Food with Tomorrow's Leaders
The Food Project

with research by Robin Proebsting

The Food Project works in the **Boston, Massachusetts**, metropolitan region, providing internships and jobs to teens who work together on rural farms and inner-city community gardens. Through this diverse community of youths and adults who learn and work together, the Food Project is growing large quantities of food to distribute to food kitchens and pantries and to sell at farmers' markets. It aims to build a sustainable food system, produce healthy food for residents of the city and suburbs, provide youth leadership opportunities, and inspire and support others to create change in their own communities.

The Food Project is as dense, rich, and aged to perfection as a New England cheddar cheese. As it approaches its twentieth anniversary, the Food Project is known as one of the premier community food nonprofits in the nation, producing massive quantities of food for the greater Boston metropolitan area by also empowering and transforming youth into our leaders of tomorrow.

But it isn't until I visit the 21-acre farm in Lincoln, Massachusetts, that I fully appreciate and savor its unique qualities. At 8:30 a.m. the chatter and laughter begins. About 60 teenagers of all colors, sizes, and shapes are piling off buses and walking down the narrow dirt lane toward a shelter and its picnic tables. All are shouldering knapsacks over white T-shirts with pink Food Project logos. To get here, some spent a full two hours riding the rapid-transit "T," leaving Boston's low-income and working-class neighborhoods to arrive 18 miles outside downtown Boston for an all-day farm job. Others, coming from wealthier suburbs, were able to sleep in a little and travel a shorter distance by car.

One of the wealthiest towns in Massachusetts, where million-dollar homes are considered inexpensive, Lincoln is known for its lush tree-lined streets and curving stone walls, Walden Pond, the Thoreau Institute, and its colonial heritage. At first Lincoln seems an improbable site for this teen melting pot of race, culture, and agricultural hubbub. Yet as I learn more, Lincoln makes perfect sense.

Lincoln made a name for itself as "one of the first communities in the nation to establish a private land trust," the Lincoln Land Conservation Trust (LLCT), in 1957. The town's Conservation Commission, established soon thereafter, raised the bar for local government activism by pioneering wetlands regulation and has been gaining public support for land acquisition. The town has been off the charts in its commitment to conservation. Together the town and LLCT conserved more than 1,800 acres of agricultural and forest lands, and its recent Open Space Draft Plan boasts that it has permanently protected nearly 3,000 acres, or almost 40 percent of the entire town.[115]

So back in 1991 when local resident Ward Cheney approached the town with an unusual idea for a pilot nonprofit farm for youths, he found an unusually receptive audience. Cheney had all the needed skills — he was a farmer, an organizer, an educator, and an activist — and he had a vision. According to Food Project lore, Cheney felt that kids were not connecting with the land, with each other, or with meaningful jobs. Teens everywhere seemed isolated and frustrated, whether in the wealthy suburbs or the inner-city neighborhoods. And Cheney saw a way to change that.

Cheney convinced the town to lend him 2½ acres at historic Drumlin Farm, which headquarters the state Audubon Society. There he began the Food Project with 3 staff members and 18 teens. After two successful years of growing and donating 20,000 pounds of food a year to the hungry, Cheney decided to leave. But he had made his mark and proved a point: the combination of youth, land, and community worked.

In its first years Cheney's program was heavily influenced by Allan Callahan, an African-American pastor who later became a Harvard divinity professor, and Ché Madyun, president of Boston's Dudley Street Neighborhood Initiative. They contributed greatly to the development of three landmark features that are still a part of the program to this day: it offers teens a *paid* job, it intentionally *mixes* inner-city and suburban teens from different racial, cultural, and socioeconomic backgrounds, and all youths work together in *both* suburban and inner-city neighborhoods.[116]

Since then the Food Project has taken a rocket trajectory to become one of the premier food projects in the nation, with a budget edging upward of $3 million. It has expanded its reach throughout the larger Boston metropolitan area: it now works two farms of 10 and 21 acres in Lincoln, jointly known as the Baker Bridge Farm, which it leases for only $35 per acre from Lincoln's Conservation Commission; a 6-acre farm in Ipswich that serves the North Shore; a 2-acre plot in Beverly; and three city farm lots in Boston's Roxbury neighborhood.[117] And its accomplishments are impressive: in its 2008–2009 fiscal year, the Food Project employed 142 teenagers and hosted nearly 2,849 volunteers to produce over 200,000[118] pounds of 186 types of fresh vegetables, of which it donated 49,000 pounds to homeless shelters and soup kitchens and sold the rest through 492 CSA shares and four low-income city farmers' markets.[119]

But the Food Project is hardly content to sit on its laurels. Its staff have set five-year goals to double the project's impact on the regional food system. They want to increase the number of teens hired to 215, increase their annual harvest to over 350,000 pounds, create a local food policy council, and partner with others to make more healthy food available in the city.

All of these numbers, however, fail to convey the enormity and complexity of the Food Project endeavor. And they convey nothing whatsoever about the Food Project's *real* success, which has been to prove a radically new way of producing large quantities of food, while also giving kids a solid foundation for success.

What is truly impressive about the Food Project is that it does not sacrifice efficiency and production in the name of teaching youths. And its success is counterintuitive. People who are serious about food production rarely if ever think about teens as a vital, untapped resource. For serious production we think about cost and availability, which usually means migrant farm labor, green cards, visas, and minimum wage. And when we think about teaching and empowering youths, we often think of new educational models, alternative schools, charter schools, drama, music, dance, and sports — not farmwork.

The Food Project demonstrates a middle path for youth empowerment. Iris Ahronowitz, who spent a summer participating in and studying the Food Project, notes in her thesis for Harvard University, "The Food Project's philosophy of youth empowerment is viewed as conservative by some youth organizations, the most extreme of which hand full power and control over to young people." Instead, the Food Project

Bright Idea: Bounty Bucks

In partnership with the City of Boston, the Food Project has created "Bounty Bucks" for Boston's farmers' markets, which doubles peoples' purchases up to $10 per person. Like many similar programs in the nation, Bounty Bucks enables people to spend up to $10 of their Supplemental Nutrition Assistance Program (SNAP) dollars and receive $20 worth of fresh produce. A grant pays the farmers the difference, thereby supporting better health for low-income residents while also supporting a local farm economy.

provides teens with their first paid job, in a highly structured learning and production environment with clear guidelines and clear consequences. And as the teens "prove themselves through hard work, respect, maturity and commitment, they are given increasing opportunities to shape the organization."[120]

Tier One: The Summer Program

Teens are able to begin working at the Food Project as early as age 14. Many become hooked and stay with the program right through graduation and beyond. Structured into several tiers, the program pushes teens in physical, mental, and even spiritual ways. As teens move through the tiers over several years, gaining more skills and taking on more responsibility, they grow into young leaders in their schools and communities.

The first tier, the cornerstone of the Food Project, is working a paid summer job for eight weeks, from mid-June to mid-August. Those under 16 years of age cannot legally work for pay, so they receive stipends for an education and work program. The summer crew is intentionally split evenly by gender and split by 60 percent from the city and 40 percent from the suburbs. The teen diversity is broad: some speak another language at home, some were born in other countries, some are comfortable with race and diversity while others find the exposure to diversity quite "radical."[121]

For almost all the youths, aged 14 to 17, this is their very first job. And the impact cannot be overstated. For some the paycheck is an important motivation, as it will help pay the household electric bill or put food on the table. For others the paycheck is unimportant. More urban youths apply and are eager for the job than suburban youths. One city boy interviewing for a job at the Food Project said, "I need to stay out of trouble, and I know if I stay at home I'll get into trouble." Jen James, communications director, says with a snap of her fingers, "I could easily fill this program with just urban youth, just like that. But no transformation would happen." James says that mixing urban and city youth in both urban and suburban environments stretches them.

An independent assessment in 2008 found that the Food Project is distinct from other youth-development programs in three ways. First, it's a "phenomenal first job . . . a high quality, learning-rich experience" where the "work is 'real' and the responsibilities are significant." Second, the Food Project goes "beyond where most youth development programs tread" by "embracing friction" and structuring youth interactions in a sustained, meaningful, and authentic way that encourages growth. Third, the work itself is the core of the program. It is "working side-by-side — miserable at times, elated at others" — where the teens bond and form relationships across boundaries.[122]

The Food Project accomplishes all this with a highly structured curriculum that has evolved over nearly 20 years. They do not simply put kids to work out in the fields. The teens are formed into crews of 10, half from the city and half from the suburbs. The crews work together through the summer, rain or shine, weeding, planting, and harvesting, led by a crew leader who is usually an alumnus (age 20 to 28) and an assistant crew leader who participated in the program the previous year.[123] Every day includes a mix of education, work, and purposive fun. The teens begin the day with a group orientation, game, or presentation about the veggie of the week. The morning I'm there, the crew leaders seem like cheerleaders — high energy, positive, pumped — and it's not long before the sleepy teens are energized. Then the crews break off and head in different directions. Today some will weed the melon patch, plant a succession crop of broccoli, or trellis tomatoes. Other crews head to small shaded structures where they begin a hands-on mini workshop, learning about weeds or soils or insects. Midmorning, the crews will rotate, with those in workshops heading out to the fields and vice versa.

Suddenly, I notice that almost all of the crews headed into the field are actually *running*. It's a rare sight to behold — teenagers running flat out, so early in the morning, as if a prize awaited them. One was even barreling along at the head of the pack with a wheelbarrow. Nobody was lagging or drooping. I'm told that one crew leader, a former marine, began the practice with his crew, and the other crews wouldn't be outdone, so they followed suit. Clearly, something different was happening here.

At lunch the teens gather under the open shelter for more scheduled activities. Their afternoon is a mirror of the morning. Field work is scheduled for Mondays and Fridays, while produce is harvested on Tuesdays and Thursdays, when community and city volunteers join in. Wednesdays are set aside for community service, when the crews and their leaders meet in the city to work as volunteers at food pantries and shelters. In 2009, Food Project youths contributed 2,800 volunteer hours at hunger relief agencies.

The crews rotate between the suburban/rural plots and the urban plots in Boston's Roxbury neighborhood. The rotation exposes suburban and urban youths to the home neighborhoods of their counterparts and expands their diversity education.

Through the summer the program builds on a series of themes in a specific order: community, responsibility, service, initiative, commitment, hope, courage, and (again) community. In addition to agriculture, workshops cover diversity and personal finance. Diversity topics progress from stereotyping to group affiliation, community building, cultural sharing, and leveling the playing field. Personal finance may seem an unusual topic, but it is critical for helping youths learn the basic principles of managing their money.

"Straight Talk" is another hallmark of the curriculum, allowing youths and supervisors to talk in a highly structured one-on-one meeting about their experience in a way that fosters clear and open communication. Scheduled once a week, "Straight Talk" enables a supervisor to give individual structured feedback with "positives" and "deltas," meaning behaviors the supervisor believes could be improved.

To provide relief from the challenging, hard, sweaty work, time is carved out for games, skits, jokes, overnight camping, and at the end of the summer a celebratory feast, to which all of the teens' families are invited. Activities are designed to build public speaking skills through presentations, writing skills through a journal notebook, and interpersonal communication.

"The first day they arrive," says Julien Goulet, director of youth programs, "all the kids from the suburbs are on one side and the kids from the city are on the other." He points to different sides of the shelter and says, "The divide is huge. But six weeks later you can't tell where they're from, they're so integrated in each other's lives. That's the transformation I like."

Tier Two: The School-Year Program

A testimony to the summer program is that in 2008 over two-thirds of its youths applied to continue into the school year program,[124] known as Dynamic, Intelligent, Responsible Teens (DIRT). The 40 or so teens hired on to the DIRT crew meet after school and on Saturdays to lead community volunteer work at the farms, conduct special projects — like an analysis of the farmers' markets — and explore in greater depth workshop topics on homelessness, inequality, and hunger. In 2008 DIRT provided 1,400 hours of volunteer service in soup kitchens and food pantries.[125] These teens also recruit and interview youths for the next summer's program.

Tier Three: Assistant Crew Leader

The ideal track for a teen is to begin working in the summer at age 14, enter DIRT during the academic year, then return the next summer as an assistant crew leader. Learning from and working with more experienced crew leaders, assistant crew leaders share in all responsibilities — motivating and challenging their crew to greater efficiency, teaching their crew a new task, breaking it down into specific pieces, managing that new task for 15 to 30 minutes, then leading by example and hard work, all the while talking and educating and motivating their crew. Both crew leaders and their assistant crew leaders stay with their crews at all times, supporting their work, and are trained to resolve behavioral or other issues that may arise in ways that help build self-esteem.

The impact of the program increases with youths' tenure. One summer alone, says Jen James, has a very small impact. "We find we have the biggest impact if they're with us for a minimum of two years." And, of course, some youths continue with the program beyond tier three and may even return as crew leaders after they have graduated from high school or while in college or graduate school.

DIG in the Garden

Durham Inner-City Gardeners (DIG) was started in 2000 by a grassroots nonprofit, SEEDS, Inc., in response to a growing need for assisting teens from poor neighborhoods in Durham, North Carolina. DIG helps kids learn life skills while developing a "close-knit team and family unit," explains Kavanah Ramsier, a DIG co-coordinator.

A key for success is that DIG doesn't expect the teens to want to learn to garden. Instead, work in the garden is a paid job. When the youths arrive for work on their first day, says Ramsier, they are usually "uncomfortable with many things in the garden. Many don't want to get their hands or shoes dirty. Some don't want to see worms or spiders. Some are even surprised that food comes out of dirt!" But by *hiring* teens to attract them to the garden and teach them job skills, DIG is demonstrating in a very real way that it values their contribution.

One criterion for working in DIG is that you're *not* involved in a gang. "But," says Ramsier, "we actually talk about reasons why people get involved with gangs and how we can meet those needs here in the garden." DIG's method for creating the strong fellowship that teens may seek is to create leadership from *within*. It takes the Food Project model and kicks it up a notch, with students accepting ever greater levels of responsibility in a design of overlapping leadership teams.

Here's how it works: DIG hires five teens, the "year-round crew," who meet on Friday evenings and Saturdays during the school year, during which time they are engaged in intensive leadership training. Part of this crew's responsibility is to read seed catalogs, help prepare the orders, and even develop the application and interviewing process for the two summer crews. Lucy Harris, executive director of SEEDS, says that the program will never hire someone not recommended by the youth leaders. So interviewing is a heavy responsibility for the year-round crew, as they know that their decisions can make or break their summer. Do they hire someone who is fun or someone who will work hard? And which responsibilities should they assign to which kids? Making these hard decisions is how the youth leaders grow and gain confidence: when given responsibilities they become responsible.

Hired by and working alongside the year-round crew, the first summer crew of ten youths works for 12 weeks, followed by a second summer crew of another ten youths working for another 12 weeks. At the end of the summer, the year-round crew replaces its graduating members by selecting emerging leaders from the summer crews.

Part of the beauty of this system is that it allows youths to learn from each other. The new summer crew may be finicky about their clothes, but when they see their peer youth leaders excited about getting into the garden, it's easier for them to get down and dirty. Ramsier says it's important to view the garden more as an education center than a production center, so that you allow mistakes to be made and allow the youth to figure things out for themselves. As a rule, DIG staff let youths train other youths in the garden, but if, for example, one youth shows another to cut off the wrong part of a collard plant, Ramsier may well choose to step in.

During the growing season DIG crews work in the morning from Monday to Wednesday, planting, weeding, mulching, watering, building, and harvesting. But they also are given classes in photography and art. Thursdays are a day of rest. On Friday evenings, the crews focus on harvesting for the market. Saturday mornings, beginning at 6:00 a.m., everyone loads the truck, and while two or three then take the produce to sell at the farmers' market, the rest prepare a lunch for everyone to eat together.

Cutting vegetables and breaking bread together does more than feed the family spirit; it also gives the youths a chance to learn more about nutrition and expand their palate in a safe environment. Anthony, a DIG teen leader, says that DIG "helps you to learn to try new things . . . I realized [chard's] not so bad. It teaches me, don't be scared to learn new things, try it."

During their time at DIG, without fail, the youths grow beyond the job in unexpected ways. One way they grow, quite literally, is at their homes. Once they learn how easy it is, some DIG youths start their own food gardens at home. Ramsier says their gardening skills grow into life skills as they learn to help feed their families and improve their family diet. Harris talks about the youths who now love cherry tomatoes and figs – items that might never have previously touched their plate. One teen, she says, no longer wants to eat in restaurants because she wants to know where the food comes from.

DIG youths grow beyond the job by also learning more about their own culture and history. These teens have never heard of black farmers who own their own farms. Yet, Harris explains, many people in the neighborhood are only one generation away from a farm. Many of their parents or grandparents lived on a farm, and now they find themselves living in a place with little or no yard and eating mostly fast food. Many of these folks, she says, don't even know how to prepare the food they grew up eating.

SEEDS had been surprised to learn that African-Americans may attach a certain stigma to the word "farm," which can evoke family history of slavery and oppression. Ramsier explains how this stigma is an important motivator to discuss injustice in agriculture, historically and across cultures, and to help teens understand how their individual choices can affect their food system. DIG is redefining the meaning of farming for these teens, says Ramsier. At least once a year, to broaden their understanding of today's farming world, SEEDS takes the teens to a farm owned and run by African-Americans, to learn firsthand what it means to be black and a farmer. In DIG teens also learn about other ways they can impact their food system – through policy, or joining a CSA, or advocating for farm and migrant labor, or starting a business that supports affordable healthy food for everyone.

DIG isn't pushing its teens to become farmers – it is working to open doors of choice and opportunity.

Tier Four: Internship

The internship program is the latest addition to the Food Project's suite of job opportunities, in part a response to older youths who wish to continue working on food issues in their own communities. In 2006 the Food Project partnered with a historic preservation association and national historic landmark, the Shirley-Eustis House in Roxbury, to create the Urban Learning Farm. Now citizens, neighborhoods, and schools come to the historic mansion and its urban farm to enjoy workshops, which may be led by Food Project interns, who might do everything from building raised-bed gardens for urban gardeners to teaching home gardeners about soil safety and addressing lead and pesticide contamination. These older interns are learning community leadership by also helping to manage farmers' markets and CSAs and giving presentations to community leaders about food issues. Through this work, teens improve their communication and leadership skills and sharpen their "job readiness."[126]

Tier Five: BLAST Internship

In 2003 the Food Project added yet another layer to its program menu. The BLAST — for Building Local Agriculture Systems Today — program is intended to create an international network of youths and adults to prepare young people to become practitioners and find gainful employment in sustainable food system work. Over the years BLAST interns have created a local food guide, built a school garden, conducted a workshop on healthy food, and run the "Eat In Act Out" week.[127]

While the jury is still out as to whether BLAST will have the same enduring impact as the Food Project's core summer and academic-year programs, it seems that BLAST may be a reflection of an increasing pressure being felt nationwide to grow future farmers and food system experts. The Food Project believes a connection with the land can and should be more than a teen "summer job" memory. It wants to help youths stay connected to agriculture into their adulthood and make sustainable agriculture a viable career option.

What It Takes

The Food Project is huge, by urban agriculture non-profit standards, with 35 full-time year-round staff and even more seasonal staff to work with summer youths. In 2009 the Food Project managed a budget of nearly $3.3 million: 40 percent from foundations; 28 percent from individuals; 11 percent from revenues from CSA shares, farmers' markets, and honorariums;

9 percent from corporations; 8 percent from miscellaneous sources; and less than 3 percent from government grants.[128] To keep this multibranched tree healthy and flourishing, the Food Project has five dedicated fund-raising staff; one manages development, two write grants full time, one manages the database, and another manages individual donors.

Running a small-scale farm using only 20 to 25 students, one staff suggests, would require at least three people: one full-time grower and two people to work with the students. But, she adds, to emulate the Food Project, you will need two pieces of land: one in the suburbs and one in the city, so the youths can work at both to experience the diversity of growing and distributing food in the two communities.

The Food Project employs seven full-time farmers. Three manage the Lincoln farm, two work on the North Shore, one is at Beverly, and one in the Roxbury neighborhood gardens. From mid-December to January the farmers plan the crops and field rotations, place seed orders, and prepare for starting seeds in the greenhouse in late February. Every morning in summer the crew leaders walk the farm with the farmers, learning the day's tasks before the youths arrive. Weed this row, stay out of that row, pull out this row and replant it with these trays of seedlings. In this way, the farmers show the crew leaders what their kids should do.

What If . . . ?

The Food Project success makes me wonder. What if *every* community and region had its own Food Project? What would be the long-term impacts on our local food systems, our land, our water, and even our public health? Franklin D. Roosevelt created the Civilian Conservation Corps in 1933 for three million young men who, over nine years, planted nearly three billion trees in soil-conservation projects to counteract the devastating dust bowl and who created with their own hands more than eight hundred parks with trails and recreation areas. The CCC is said to have influenced an entire generation, creating new and lasting interest in the outdoors and in conservation.[129]

I can't help wondering: What if in all communities, large and small, hundreds of thousands of youths — even millions of youths — were to experience their first job on a structured farm like the Food Project, learning skills, responsibility, confidence, and leadership, while also learning about sustainable farming, healthy food, and nutrition? Certainly, this would not be the magic-wand answer to relocalizing food production or reducing our dependence on petroleum-based shipment of food, but it could be a powerful contributor to the answer. At worst teens would earn money while producing healthy food for the community. At best it would be a win-win-win-win: youths earn money while learning skills; the community increases its supply of fresh food; our nation grows new crops of young leaders; and we decrease our long-term public health costs while improving our children's health.

"When hyperactive inner-city kids come to my farm and sink their hands into our rich compost," says Will Allen, of Milwaukee's Growing Power (see page 74), "you can *see* the change. It grounds them, they calm down."[130] Could it be so simple?

So I can't help wondering. The Centers for Disease Control and Prevention cites a 2007 study showing the annual "cost of illness" for ADHD, in 2005 dollars, at between $36 billion and $52 billion.[131] What if we

Managing Volunteers

A typical nonprofit complaint is that volunteers can take more work than they're worth. The Food Project proves that it's all a matter of learning how to manage them. By 2009 volunteers at the Food Project were numbering nearly 3,000 per year and coming from any organization willing to help – churches, schools, corporate groups. Based on its extensive experience working with volunteers, the Food Project has created an invaluable detailed guide for others to learn from their lessons. The guide, which would be useful to food projects working with volunteers in any size community, is available for free online; see the resources section.

didn't have to spend billions every year on *remediating* ADHD but instead spent it on *prevention*? I don't pretend to know how we get there. But what if we used the ADHD billions to create Youth Community Food Farms, where every child in the United States could have the opportunity, even the right, to spend time on a food farm for education and personal health? Youths of all ages could explore nature, land, and food. They could build things, grow things, be creative in the outdoors. They could be restored to health. And as the Food Project shows, youths could be employed in ways that foster transformation and empowerment.

Will Allen argues, "If we're really serious about getting kids off the street, then we'll put money into non-profits that are working with kids. When I talk to kids, they don't want to be out there hustling on the street."

My vision is admittedly a little wild and extreme by intention. But surely Cheney's vision for the Food Project was considered a little "out there" in 1991. Sometimes wild and crazy is what finally makes sense. Health, education, transformation, empowerment. The Food Project is pointing the way.

Lessons Learned
Working with Youth

Bundle skills and learning.

In some of the early youth projects at Growing Power in Milwaukee (page 74), the kids worked in the garden, learning how to use their hands, and also had to write about their experiences, so their writing and reading skills improved. According to founder and CEO Will Allen, this link inspired them to want to learn more. Today, Growing Power youth workshops include gardening, canning, and building, so kids learn how to use tools, how to care for plants and gardens, and how to cook and eat healthy foods.[132]

To attract teens, pay them.

Most teens won't want to spend their entire summer working on a farm as a volunteer. To attract teens, particularly those that come from low-income families, it is important to value their time by structuring their work as a paid job. Both Boston's Food Project (page 136) and Durham, North Carolina's SEEDS Inc. (page 140) have found success with this method. Paying teens for their work in the program fosters a host of valuable lessons, including, for example, responsibility, accountability, managing money, and growth and maturity.

To foster learning and growth, mix youths from different backgrounds.

The Food Project goes about youth empowerment and transformation in a bold way. Unlike most programs that focus on a narrowly defined group – the homeless, or those in public housing, or the residents of a specific low-income neighborhood – the Food Project has proven that intentional mixing of different groups can be a powerful recipe for growth and transformation. The Food Project is not afraid of conflict and structures its program to enable teens to learn how to recognize and deal with their differences. Hard work, combined with shared experiences as volunteers at service agencies and in educational workshops on agriculture, sustainability, diversity, and personal finance, brings a sense of camaraderie to the mixed group.

Youths can be efficient producers of real food for real people.

Often community food projects involving youths consider the program strictly educational and forgo concern for production. The Food Project demonstrates that a concern for production can be instructional for youth and that they in turn can produce large quantities of food to feed their community. During my visit at the Lincoln farm, I witnessed a crew planting new broccoli starts using a system of "leapfrogging" and was stunned with their speed and efficiency. While one teen worked the wheelbarrow with the new seedlings, the lead planter would place the broccoli starts in their right position in the row, a second and third would leapfrog in digging the holes, a fourth and fifth would leapfrog to place the broccoli in the hole and refill the hole, a sixth and seventh would leapfrog in carefully tamping the soil around the seedling and finishing it off, and another was following behind for quality control and cleaning up the row so it looked good. They were doing it naturally and were proud of their speed, efficiency, and quality of work. We can't doubt teens' ability to be excellent production workers. It's all a matter of learning and teaching a system, and the Food Project shows how to teach it.

College students can be powerful change agents.

If your community has a college nearby, count yourself lucky. When it comes to community food projects, students can be a significant asset, starting projects that take on a life of their own. While still an undergraduate student at Evergreen College in Olympia, Washington, Richard Doss was so inspired by Dan Barker's Home Gardening Project (page 13) that he initiated the college-based Kitchen Garden Project in 1992, which has since grown into a much larger community food nonprofit known as GRuB. On the other side of the country, Seth Williams, an undergraduate student at Hampshire College in Massachusetts, wrote a thesis on community gardens and worked with community leaders to initiate La Finquita community garden in south Holyoke. This, too, grew, expanding from one garden to nine, along with a host of other programs that form the basis of the large Nuestras Raíces nonprofit (page 123).

In my own experience, college students often have the energy and optimism needed to take on the task of starting something new. Also, most colleges now encourage or require community service, an opportunity to engage students in empowering communities. In fact, one way to overcome the town-gown tension typical in college towns is to help students connect with the community through volunteer, service, or class projects.

Hire staff who are enthusiastic about the ethics you hope to foster.

In a doctoral study involving the Food Project, researcher Lianne Fisman suggests that the group may not be realizing its full potential. While its workshops teach different components of sustainable agriculture, she found during her time there that most crew workers still treated fieldwork as drudgery rather than a learning opportunity. She observed that most small talk led by crew leaders in the field focused on issues of race and class, not about what they were doing on the farm or the larger context of food systems. And this, she argues, led to missed teachable moments.[133]

Some might dispute her experience. Jake, a 17-year-old on the 2009 summer crew, says, "I'm not sure I'm going to be a farmer, or directly involved with social policy, but it's going to be hard to do anything without having this in the back of my mind – the role food plays in everything, and all facets of social issues. Whether I'm a doctor, a lawyer, whatever – a farmer – I'm going to be thinking about this." Johnny, one of the crew leaders, says that he came to the Food Project in part because his grandmother, a Haitian who felt a strong connection to the land, influenced him to find that same connection. "I needed to do more with my life than just talking about social justice. It needed to be the social piece, the environmental piece, and the agriculture piece."

Still, Fisman's point is important. Staff members will convey their natural passions to the youths in their groups. So if you want to expose youths to the idea of farming as a profitable occupation, then hire people who have had experience working at for-profit farms. If you want to expose youth to the importance of sustainable farming and stewardship of the land, then consider hiring people who have a passion for conservation and sustainability. And if you want to expose youth to even more, say, a sense of spiritual wonder, an opening and connectedness to a larger whole, an understanding of the farm as a place for contemplation and learning, then consider hiring people who will express this natural spiritual connection to place and land.

Keep track of your graduates.

The people who have grown through your program are all potential supporters and teachers in future years. Invite them to come back and speak to the group. Look for places in the community where graduates can build on what they've learned, such as horticultural scholarships or food system projects. SEEDS Inc. reports that some of its graduates want to become chefs and landscapers, and they hope to find restaurants or landscape businesses that would be willing to accept an intern.

5: Education

Food, Nutrition, and Agriculture in Schools

Most parents live in hope that their children will have long, healthy, happy lives. So the news that today's children face the prospect of a *shorter* life expectancy and lower quality of health than their parents, due to the long-term health impacts of childhood obesity and diabetes,[134] can be shocking, and it is galvanizing parents, teachers, and health providers into action.

The front line of attack against childhood obesity might logically be where children spend most of their day: schools. Through a variety of Farm to School programs, the grassroots food movement is tackling this issue with creativity and tenacity, examining every aspect of food and nutrition in schools, seeking ways to give more consistent messages and to make it easy (or at least less painful) for children — and their parents — to change their food buying and eating behavior. Some projects are working through after-school programs, while others are focused on creating curricula for classrooms, and still others are offering field trips to farms and other special events focused on exposing children to local food and agriculture. Some are working to catalyze change by teaching children and their parents about where our food comes from, the principles of nutrition, and how to prepare fresh foods. Others are focused on the school lunchroom, educating cafeteria workers in nutrition and the old-fashioned skill of from-scratch cooking and working to bring more fresh, locally sourced foods into the cafeteria. Some are working to change policies, such as banning sugary drinks and snacks from vending machines, removing fryers from the cafeteria kitchens, and keeping physical education as a requirement for children of all ages.

These grassroots projects work on different fronts and in different venues, but all are trying to turn the tide of the obesity epidemic, even as it washes over us and our children. All are working from a place of conviction and hope — conviction that we must and can change our children's relationship to their food, and hope that we can do it quickly enough to protect their health and longevity. And this aspect of the grassroots food movement has garnered top-down support from the highest levels; consider First Lady Michelle Obama planting a backyard garden at the White House and launching her Let's Move! campaign in February 2010, with much media fanfare.

Because these issues strike the very heart of our nation — our children's health — this arm of the grassroots food movement is likely to be with us for some time, or at least until systemic changes are successfully reflected in more slender and healthy youth. It also may be the arm of the grassroots food movement that is best capable of touching the most Americans the most quickly, through our schools, and therefore it may offer the greatest promise for influencing America's food choices and systems. In fact, we may find that reclaiming our food — rebuilding our connections to what we eat, where it comes from, and the role it plays in the health and vitality of our communities — begins with education. The future of food may belong to our children.

Nourishing Kids by Connecting Farms and Schools

The Farm to School Movement

by Debra Eschmeyer

Farm to School programs build connections between schools and farms, supporting the introduction of fresh, local food into school menus and the development of nutritional and agricultural literacy for schoolchildren, while supporting local farmers and communities. With Farm to School programs in **all 50 states**, the movement is rapidly gaining support from local, regional, and national leaders.

Whether greeting the lunch line with groans or glee, 32 million children clamor in the nation's school cafeterias as they come to quell their grumbling bellies. The cacophony of the cafeteria raises to a fever pitch this question: does today's school food environment fail our kids?

Unfortunately, the average school cafeteria operates on the lowest common denominator of cost, not quality. The USDA currently reimburses schools $2.72 for every free lunch they serve, and lower amounts for reduced-cost and full-price meals. This leaves about $1 to cover actual food costs, once labor and overhead costs are factored in. What can we expect food-service directors to feed our kids on a dollar?

Thankfully, kids, parents, food-service staff, teachers, school administrators, farmers, and other community members with a taste for change have been working to incorporate local products into school menus through what is called "Farm to School."

Where We Stand

School meals are a vital part of our responsibility to ensure the health and well-being of future generations. Over the past 60 years, school meals have helped our nation make impressive strides toward improving childhood nutrition and reducing childhood hunger. Yet in recent years school meals have confronted new challenges. School food services are fighting an uphill battle to provide kids with healthy food. Soaring food and energy costs, the lure of fast food outside the school campus, and financial pressures caused by tight state budgets and diminished tax revenues all stand in the way of food services being able to provide healthy and delicious meals to schoolchildren.

The Childhood Obesity Epidemic

Improving the quality of school meals, and making them accessible to all children, is essential to our nation's future and an important tool for addressing our nation's burgeoning obesity epidemic. Consider the following:

- Obesity rates among children have doubled in the last 10 years and tripled for adolescents.

- One-third of U.S. children are obese or overweight.
- 1 in 3 children born in the year 2000 will develop diabetes – make that 1 in 2 if the child is black or Hispanic.
- For the first time in two hundred years, today's children are likely to have a shorter life expectancy than their parents.

Like school food services, today's farmers are facing numerous challenges. The farmer's share of every food dollar has dropped to 19 cents, down from 41 cents in 1950. As a result, many farmers have a hard time just breaking even. Every year thousands of farmers are forced to give up farming, and the average age of farmers nationally is 57 years. The United States, with fewer than two million farmers, now has more prisoners than farmers.

There is a solution that can help turn around both of these trends: Farm to School. School food is a potentially lucrative market, estimated at more than $12 billion per year. By selling to schools, farmers can augment their income and stay on the land. Yet today's family farmers often don't have access to this market.

Farm to School programs provide that access, bringing local, farm-fresh foods to school breakfast, lunch, and snack programs. They ensure that our children eat the highest-quality food available and cultivate long-term healthy eating habits. They also often include educational initiatives about nutrition, agriculture, cooking, and sustainability. They are a win-win for kids, farmers, communities, educators, parents, and the environment.

Thanks to the efforts of social entrepreneurs, Farm to School programs have blossomed in thousands of schools. The National Farm to School Network (NFSN), a national advocacy group working to support existing programs and promote new ones, describes Farm to School programs as essential components of strong and just local and regional food systems, ensuring the health of schoolchildren, farms, the environment, the economy, and communities. Working to that end, the NFSN supports 50 state leads, eight regional lead agencies, and national staff in facilitating training and technical assistance, information development and dissemination, media and marketing, policy advocacy, and networking at the local, state, regional, and national levels.

Government support for the Farm to School movement has blossomed over the past few years. In September 2009 the USDA announced its Farm to School Team, intended to "support local and regional food systems by facilitating linkages between schools and their local food producers."[135] In May 2010 the White House Childhood Obesity Task Force recommended Farm to School, school gardens, and nutrition education as community-based solutions

What Is Farm to School?

The National Farm to School Network collected input from programs and leaders around the country to define the core principles of Farm to School:

Farm to School is broadly defined as a program that connects schools (K–12) and local farms with the objectives of serving healthy meals in school cafeterias, improving student nutrition, providing agriculture, health, and nutrition education opportunities, and supporting local and regional farmers. Since each Farm to School program is shaped by its unique community and region, the National Farm to School Network does not prescribe or impose a list of practices or products for the Farm to School approach.

At its core, Farm to School is about establishing relationships between local foods and schoolchildren. It includes but is not limited to:

- Local products in school meals (breakfast, lunch, after-school snacks) and in classrooms (snacks, taste tests, educational tools)
- Local-foods-related curriculum development and experiential learning opportunities through school gardens, farm tours, farmer- and chef-in-the-classroom sessions, culinary education, educational sessions for parents and community members, and visits to farmers' markets

Farm to School aims to enable every child to have access to nutritious food while simultaneously benefiting communities and local farmers.

to childhood obesity. Specifically, the task force's report states, "USDA should work to connect school meal programs to local growers and use Farm to School programs, where possible, to incorporate fresher, appealing food in school meals."[136] Michelle Obama's Let's Move! campaign also specifically promotes community gardens, school gardens, and Farm to School programs. In addition, FoodCorps, a national public-service school garden and Farm to School program, has recently launched, with the goal of improving the health and prosperity of vulnerable children while investing in the next generation of farmers (see page 149 for more on FoodCorps).

Better Health on All Fronts

The major aims of Farm to School are healthy children, healthy farms, and healthy communities. Farm to School programs are based on the premise that students will choose healthier foods, including more fruits and vegetables, if products are fresh, locally grown, and picked at the peak of their flavor and if those choices are reinforced with educational activities. These programs provide benefits to the entire community: children, farmers, food-service staff, parents, and teachers.

Farm to School programs influence students on many levels, increasing their knowledge and awareness about food sources, nutrition, and eating behaviors and lifestyles. Eating locally sourced products becomes part of the educational framework that turns kids on to healthier food options. A connection with the source of their food also deepens students' appreciation for food and agriculture. Some examples of the impact of Farm to School programs on children, collated from various programs,[137] are:

- Increased consumption of fruits and vegetables (for example, studies in Portland, Oregon, and Riverside, California, have found that students offered a farm-fresh salad bar consume roughly one additional serving of fruits and vegetables per day)
- Improved knowledge and awareness about gardening, agriculture, healthy eating, local foods,

and seasonality (in Philadelphia, the percentage of kindergarteners who knew where their food came from increased from 33 percent to 88 percent after participation in a Farm to School program)
- Demonstrated willingness to try new foods and healthier options (in one school in Ventura, California, on days in which there was a choice between a farmers'-market salad bar and a hot lunch, students and adults chose the salad bar by a 14-to-1 ratio)
- Reduced consumption of unhealthy foods and sodas
- Reduced television watching time and positive lifestyle modifications such as a daily exercise routine
- Increased social skills and self-esteem

For farmers, Farm to School programs open the door to a multi-billion-dollar market. Historically, small farmers have found it difficult to access local school food markets, given the complexities of the procurement process. Data from Farm to School programs suggests that when schools dedicate a significant percentage of their purchases to local producers, local farmers gain a significant and steady market. For example, the New York City school district signed a $4.2 million contract with farmers in upstate New York to provide apples for New York City schools over a three-year period. The 60 farms providing products to local schools in Massachusetts, meanwhile, are generating more than $700,000 in additional revenue each year.

But farmers and students aren't the only ones to benefit. School food services typically find that Farm to School programs increase participation rates in school meal programs, enhancing their overall financial viability. Food-service staff show increased knowledge about local foods and interest in working with teachers to strengthen classroom–cafeteria connections. Teachers, too, demonstrate a positive attitude and eagerness about integrating Farm to School–related information in curricula.

A Farm to School program incorporating parent education can ensure that messages about health and local foods are carried into homes and

How to Start a Farm to School Program
Adapted from guidelines of the National Farm to School Network

1. RESEARCH. With Farm to School programs active in every state, you can learn from other programs' successes and challenges and begin to identify what you want and what would work best in your school. Visit the website of the National Farm to School Network (see the resources) to acquaint yourself with model Farm to School programs and connect with a network of experts in your area.

2. ORGANIZE. Coordinate a group of cross-sector stakeholders in the community for a meeting to discuss Farm to School, including food-service directors, parents, teachers, farmers, students, school administrators, local nonprofits, and so on. Inspire potential supporters with an activity such as a farm tour or a farm-fresh taste test.

3. ASSESS. Facilitate conversations with various stakeholders to determine the feasibility of the program in your area — discuss where to buy local foods, assess how to serve them at school, develop the budget, and identify staff or volunteers to support the program.

4. PLAN. Create a short description of your ideal program and then list specific first steps. Successful Farm to School programs are based on relationships of mutual respect and trust; taking the time to define your direction will ensure a sustainable program.

5. BEGIN. Take small steps such as working with one or two products that are easy to process and popular among kids. Local apples, potatoes, or strawberries are a good choice when they are in season. Get comfortable with ordering, delivery, invoicing, and food prep before you scale up.

reinforced there by parents and other caregivers. Farm to School education inspires parents to incorporate healthier foods into their family's diet and better equips them to do so through both shopping and cooking tips.

FoodCorps

FoodCorps is a new national service organization that seeks to reverse the childhood obesity trend by expanding vulnerable children's access to, knowledge of, and engagement with healthy food through school gardens and Farm to School programs. As pointed out earlier, one in four U.S. children struggles with hunger and one in three is obese or overweight. FoodCorps addresses the root cause of both: access to healthy food.

Let's focus on the calories. Only 2 percent of children meet the dietary guidelines established by the U.S. Departments of Agriculture and Health and Human Services (and presented in the well-known food guide pyramid). For many kids, more than half of their calories come from school food programs: breakfast, lunch, and snacks.

We need to capitalize on the captive audience we have in schools. If we give children access to healthy food in the cafeteria, nutrition education in the classroom, and hands-on learning through school gardens, a lifetime of healthy eating can take root.

This is the mission of FoodCorps. Service members will work in school districts suffering disproportionate rates of childhood obesity, building and tending school gardens and working to implement and support Farm to School programs. The program will at once serve vulnerable children, improving access to healthy, affordable school meals, while also training a cadre of leaders for careers in food and agriculture.

As a staff member of the NFSN and cofounder of FoodCorps, I often hear from parents and school staff, "Oh, we love Farm to School and we love school gardens, but our budgets are tight. We just don't have the sweat equity and the labor to pull it off." The people power is what FoodCorps is here to provide. In so doing, the program will help transform America's school meals and give our kids a fighting chance for a passing grade in health.

Putting Farm to School to Work
Growing Minds and Abernethy Elementary School

The Farm to School movement is making headway in **all 50 states**, thanks to the enthusiasm and creativity of parents, teachers, school nutrition directors, school chefs, and farmers. The two programs described here demonstrate how this movement is working on the ground, through facilitative partners and direct action from parents, teachers, and school staff. Based in Asheville, North Carolina, Growing Minds is helping schools throughout the region connect children to fresh, healthy foods through the development of food-based curricula, farm field trips, school gardens, and local sources for cafeteria food. In Portland, Oregon, Abernethy Elementary School highlights how a program that begins with one passionate individual can eventually, having gained the support of administrators, school chefs, teachers, and parents, transform kids' relationship with fresh fruits and vegetables.

The Farm to School movement has exploded in the last decade. Schools are a huge community expense, especially in rural areas, where they may be a county's single largest budget item. With all this money pouring into our children's education to bring test scores up, to provide current technology, to leave no child behind, somehow the lunchroom was left behind. In the late 1990s new emerging data about youth obesity, diabetes, and ADHD began to startle advocates into action, leading to the creation of the National Farm to School Program in 2000, the formation of the National Farm to School Network in 2007, the first Farm to Cafeteria conference in 2002, and new policies, programs, and projects. As the data solidified, confirming that the health of our nation's youth was at risk — that obesity and diabetes had become unprecedented epidemics — the Farm to School movement also solidified. Farm to School programs grew from only a few pioneers in the 1990s to four hundred programs in 2004 and then, leaping fivefold in just five years, to two thousand programs in 2009.[138] By 2010, Farm to School programs could be counted in all 50 states. Here are just a couple of examples.

Growing Minds

A Farm to School program can begin with someone who cares — the school's nutrition director, a parent, or a teacher. It can also begin with a community partner like the Appalachian Sustainable Agriculture Project (ASAP) in Asheville, North Carolina (page 243), whose Farm to School program, Growing Minds, is the southeastern regional lead agency for the National Farm to School Network and throughout this region facilitates the integration of fresh food into curricula and school lunchrooms.

Emily Jackson, director of Growing Minds, emphasizes that Farm to School can succeed only when it is fully integrated into the school curriculum. "Integrating Farm to School into the curriculum is a critical piece — so much so that we now are integrating Farm to School instruction into pre-service teacher and dietitian training at Western Carolina University," she says.

The program supports the development and maintenance of school gardens, which are often a first focal point where children can grow the same kinds of fresh food that they eat in their cafeteria, providing a rich educational experience. The garden is an educational space, Jackson emphasizes. "We really don't encourage schools to try to grow food for the cafeteria. The garden should be about the magic of growing food, not having to concentrate on food safety and production." Using the school garden as an educational focus, children can learn everything from math and science to history, politics, art, and music.

Growing Minds also works with schools to develop materials and programs that help educate children about nutrition and diet, expanding their palates and understanding of food through tasting and cooking demonstrations. "This component of Farm to School is really resonating, not only with kids, but with parents too," says Jackson. "Not all families will go to a farm, not all families will grow a garden, but all families have to cook, at least to some extent. We are putting a lot of emphasis on this."

The program also facilitates school field trips to local farms and farmer visits to classrooms. To make field trips as easy as possible for teachers, Growing Minds offers a guidebook with teaching curricula for different age groups, as well as checklists for organizing and managing the field trips. When children visit a farm, they can see, smell, hear, and feel its bounty and learn about their food in ways not possible in a classroom. The experience is as appropriate for children in the region's rural districts as for their urban counterparts, Jackson says. "Many people worry that children in inner cities don't know about farms, yet we often forget that children in rural areas often aren't familiar with farms either."

One particularly popular component of Growing Minds is its Farm to School Cooking Program. Growing Minds recruits chefs from local restaurants and trains them to conduct cooking classes that are tied to the school curriculum. Then it plays matchmaker between schools and chefs, and it provides the funds for local food purchases and cooking kits for the classes.

Changing school menus to incorporate fresh, local food may be the most difficult challenge for any Farm to School program. As Emily Jackson testified to the U.S. Senate's Agriculture, Nutrition, and Forestry Committee in 2007, USDA regulations are a major challenge. "Confusion about USDA's rules related to local procurement," she said, "means that even with competitive prices and a desire to support farmers in their region, schools are hesitant to purchase local products from family farmers in their region."[139]

But Growing Minds has found a way to encourage schools to give local foods a shot: by starting small. Instead of tackling the huge county school system (with 25,000 students and 41 schools), Growing Minds started with just the Asheville schools and smaller rural school systems, working with their established food distributor, Carolina Produce. Located just across the border in South Carolina, the distributor was already trying to supply food that was more local to Asheville, and it was eager to find more sources. Carolina Produce and Growing Minds became partners and decided to use ASAP's Appalachian Grown certification process (see page 243) to identify local food that could be served in school cafeterias.

Slowly, as the number of local farms serving the Asheville and small rural school systems increased, Growing Minds reached a tipping point: there were sufficient quantities of locally grown food reaching local schools to begin talking with the larger county

Bright Idea: Tools to Facilitate Farm Field Trips

Growing Minds has developed two guides that make it easy for teachers to plan and implement field trips to farms. "The Hayride: A Resource for Educational Farm Field Trips" is a booklet that describes exactly what is available at every farm in the region, so teachers know what to expect and can plan accordingly. Every school system should develop its own! The "Farm Field Trip Toolkit," designed for both teachers and farmers, provides field-trip-related activity and curricula suggestions for the subjects of math, language, science, social studies, and healthful living. Both guides are offered at the ASAP website (see the resources).

school system. Today, the county school system's local food program is such a success, says ASAP director Charlie Jackson, that providing local food is a core aspect of the distributor's business model.

Abernethy Elementary School

If anyone is responsible for Abernethy Elementary School's innovative garden program and kitchen, it's the principal, says an Abernethy parent. While the principal may not have a "Cadillac" budget, depending on who you talk to in the City of Portland, the school's real wealth derives from its super-motivated, highly educated parent volunteer base.

As with many school kitchens around the nation, in the 2000s Abernethy's kitchen had been dismantled. It had lost its ability to prepare fresh foods. It didn't have the staff or tools to slice, dice, and chop fresh fruits and vegetables. It would simply reheat items delivered into its freezer storage. One parent, however, went on a mission to change this, working to dig and build a school garden and even putting on a smock in the school kitchen to chop the apples and

---

Lessons Learned
Bringing Local Food to School Lunchrooms

Get the child nutrition director on board.

Without the backing of the school's child nutrition director, transforming school menus is impossible, says Emily Jackson, director of Growing Minds, the Farm to School program of the Appalachian Sustainable Agriculture Project in Asheville, North Carolina (page 243). School food service is entrenched in a web of government subsidies, regulations, and guidelines, which together create real obstacles to change. All of the typical challenges of bringing fresh, local food into the school cafeteria — concerns about farmer liability insurance, finding farmers with products that can be used, coordinating the food with menus and delivery, training cafeteria staff to be able to manage fresh food preparation — become easier to address when the child nutrition director is on board, Jackson says. She can invigorate and excite her staff to support the introduction of fresh, local food into the school food-service program.

Get food-service staff excited about local food.

Growing Minds is working with six Southern states to foster the introduction of more local, fresh food in schools. To help school staff become familiar with, and maybe even fans of, local food, Growing Minds bought CSA shares, not for the students, but for the staff. Growing Minds also took school staff out to the farms, so they could see for themselves what kinds of food were available locally and where it came from, and to help put a face on the local farmers so that the staff would remember them and be more willing to order their products. The tours were also intended to thank the staff for the extra work that will be required in preparing local food. "They loved the farm," Jackson says. "They ate their way across the farm. It was awesome!" Jackson says, "We've found that it's not just kids that need positive experiences with healthy, locally grown food. Adults often don't know where their food comes from, how it's grown, or even how to cook it."

Work within existing systems.

Jackson says this has been one of the biggest lessons for the Appalachian Sustainable Agriculture Project (ASAP). They were able to tap in to the network of a local distributor, Carolina Produce, that was already delivering to schools and had the capacity to expand. A distributor can work with a great number of suppliers, whereas schools are logistically limited to just a few. By tapping into an existing system, ASAP was able to provide many more farmers with access to the school market, without having to sink resources into an alternative distribution system. And during a time of transformation for their food-service operations, schools were able to work with a proven, experienced distributor, which gave them more trust in the process.

vegetables. Eventually the school was able to scrounge an honest-to-goodness operational commercial stove and mixer, cast off from other schools that were still ditching the space-cramping superfluous equipment. Now, fresh food and the garden are integral parts of the school's curriculum, and the school offers fresh food on a regular basis in its lunch line.

Such accomplishments happen slowly. One parent emphasizes the importance of patience and meeting people where they are. It's not uncommon for a key administrator to block or fight a parental push for fresh food. But miracles can happen, the parent says, when someone has the patience to work with the gatekeepers and can coax them along by reminding them of their own food roots.

Still, despite the hands-on contributions of parents, the inherent inequity of this situation was not lost on the school district. Why should Abernethy children benefit from fresh fruits and vegetables when others couldn't afford the same luxury? It was too expensive, and it also was unfair. Abernethy wasn't ready to give up, however, and the matter went to mediation, which led to a temporary resolution designating Abernethy's kitchen as a pilot kitchen for the entire school district. Much is hanging on whether Abernethy can pull off the experiment.

One Abernethy teacher whom I meet in the hallway believes it's a slam-dunk. "These kids here know what vegetables are. And they eat them." The teacher says the school's kitchen chef is the key to the school's success in introducing fresh local foods. The school has a Harvest of the Month program, in which a particular fruit or vegetable grown in Oregon is incorporated into classroom curricula and featured as the primary ingredient in one of the school lunches. "When our chef does Harvest of the Month," the teacher says, "she knows how to prepare the vegetables in a way that's appropriate for kids. This month the Harvest of the Month is pear. So the chef prepared many kinds of pears for sampling and talked about how sometimes pears are 'paired' with other foods — with cheese, desserts, and other foods."

The teacher emphasizes how important it is for the food to *taste* good. "We can try to introduce food to kids, but if it's cruddy food, the kids won't eat it," she says before heading on to class.

The school chef tells me something unexpected: the fresh Oregon fruits and vegetables actually cost *less* than the processed produce ordered through normal channels! These savings are counterbalanced by the additional kitchen labor needed for hand preparation, but a high student participation rate in the school lunch program helps close the gap on labor costs. The school's garden curriculum and passionate parents constantly encourage families to have their children eat in the school cafeteria. According to one parent, this constant messaging, helped by the chef's popularity and ability to make sure the food tastes good, actually works. An unusually high number of Abernethy students do eat school food.

Abernethy is showing how fresh, healthy foods can become integral to the school environment. The children enjoy a salad bar every single day, as well as the Harvest of the Month. Fourth graders get to help serve the food every day. Fourth graders are also the ones who actually plan the gardens, using math to design and plot them out. And, most impressive, all school students are involved with the garden at least one hour each week.

Sarah Sullivan coordinates the school's garden curriculum as well as the AmeriCorps garden teacher and has seen firsthand how children benefit from a curriculum that includes hands-on garden experience. "There is so much to be learned from gardens, and some children actually learn better by *applying* skills — math, science, and art," she says.

The key to success at Abernethy has been involving the teachers from the beginning. "We pull kids out of a teacher's class once a week, for an hour, which is a lot," says Sullivan. "So it's important that we involve the teachers and ask them how we can complement what they're doing, to make sure we're using student time to help meet the standards — whether it's learning about Oregon in grade four, or learning about plant parts in first grade."

Sullivan praises the school curriculum, which has been shaped by years of AmeriCorps work. "I've worked at other school programs," she says, "and this curriculum really impresses me. There are probably six thousand AmeriCorps hours that have gone into this. What makes it so unusual is that it is standard benchmark-targeted, teacher-approved, and aligned precisely with what [the students are] learning in the classroom."

The garden and fresh food also fit perfectly with Abernethy's Wellness Policy, designed to increase access to healthy foods and promote healthy eating, which in 2010 was in its fourth year.[140] As part of its policy to improve the school food environment, instead of seeing physical education as a distraction from learning, Abernethy sends its children out to play every day.

Sullivan's position has been funded for two years entirely by the PTA. She says a group of "super-moms" coordinate all the fund-raising and community events, which enables her to focus on writing grants and trying to effect change system-wide. The garden teacher is a full-time AmeriCorps position that the gung-ho PTA also funds. While some might criticize Abernethy's program for these privileges afforded by its active PTA, others suggest that Abernethy can help pave the way for its less-endowed sister schools by working out the kinks, sharing lessons learned, and creating curricula that others can use.

Sullivan and the garden teacher are trying to bridge the gap between everyday school lessons and food, making the garden and food come alive for children. By all counts, they're succeeding. Sullivan tells me about a risky experiment when the Harvest of the Month was Brussels sprouts. She and the garden teacher covered a lot of ground in the classroom, dissecting Brussels sprouts, pulling them off the stalk, roasting them, mashing them. Most kids, surprisingly, actually liked them. Sullivan says, "It became a funny, cool thing for kids to say, 'We actually like Brussels sprouts, do you?'" When it came time for the Harvest of the Month day in the cafeteria, Sullivan and the garden teacher went to every classroom and gave out samples again, and they even gave students a recipe to bring home to their parents. Sullivan says parents were pulling her aside and asking, "What have you done to my kids? They want Brussels sprouts!" One parent even told her the story of being in the grocery store and allowing her son to pick out one treat; her son dragged her to the produce section and picked out Brussels sprouts. The program works, she says. When given a reason, an opportunity, and encouragement, kids will definitely eat fruits and vegetables.

Lessons Learned
Building a School Gardening Program

Develop a long-term budget and commitment to manage the gardens.

Growing Gardens of Portland, Oregon (page 17), has been working to support school garden programs, inside and outside the classroom, for years. Because a garden may not be seen as immediately relevant to the goal of raising educational standards, the group has found that a school garden program must have total buy-in from the administration, with long-term funding, staffing, and community commitment, to be effective. Otherwise, the best of intentions can turn a productive garden into an eyesore when key staff leave.

"One person might be really enthusiastic and jump in and build the garden but then not be able to carry through with the management," says Debra Lippoldt, Growing Gardens' executive director. If the administration doesn't designate a team member or allocate dollars for the garden management, it can languish.

"Our intention is that whatever we help create, it will be sustained after we leave," says Lippoldt. So Growing Gardens makes a three-year commitment to assist a school in building a sustainable program with the infrastructure and community commitment it will require to endure past the initial launch.

Begin with a school garden committee.

One of the first critical steps in starting a garden program at a school is to establish a garden committee that will provide the vision and ongoing motivation and support for program development. When Growing Gardens begins working with a school, it seeks to form a committee that will include at least the principal and a few teachers and parents. The committee's first task is to develop a vision for the school's garden program and how it fits into the school's overall goals. Before setting up a garden, the committee has much to consider – logistical issues of finding a space that offers access to sun, soil, water, and a place to store garden supplies and tools, and curriculum issues of how garden-based learning fits into after-school and academic learning.

Piggyback on existing resources.

When working with low-resource schools, which are already stressed to the limit, the best strategy for losing support and interest is to create demands that can't be met. When Growing Gardens begins working with a school, it doesn't create new demands. It has learned to start by piggybacking onto existing resources, such as after-school clubs or programs. "This is a really easy and low-cost way to integrate gardening into a school, without having to make inroads into the actual class curriculum," says Nell Tessman, the program's youth educator. If the school doesn't have any garden space, Growing Gardens has developed fun, hands-on curricula for after-school garden clubs that can be done indoors and outdoors in portable containers on a very low budget.

Start small and go slow.

Growing Gardens doesn't encourage schools to jump into a garden quickly; the school will benefit by going slow and carefully thinking through how it will use the garden and who will take responsibility. There can be many challenges, particularly in resource-stressed schools that may need to focus heavily on remedial studies. Children may need to be pulled out of clubs for special attention, creating discontinuities. A teacher may be willing to participate but pessimistic about the garden's viability. A parent may be charged up and highly motivated, but taken away midstream by other demands. "You have to be willing to persevere, and do a little at a time," says Lippoldt.

If the school doesn't currently have a garden space, the school can still start with after-school fun activities that teach students about soils, plants and food. "We make the after-school club a fun learning environment that engages students physically. We work within the system that exists. Over time, this allows trust to be built within the students, teachers, staff and parents," says Caitlin Blethen, manager of the group's school-based youth program.

As trust grows, Growing Gardens may be invited to come in and provide one science class for one day. And from there, the school may discuss how to begin incorporating garden-based learning into its broader curriculum.

Strive for sustainability.

A sustainable program will engage and train both parents and teachers, says Lippoldt. When Growing Gardens agrees to work with a school to build a gardening program, it makes a three-year commitment to the project, and it aims to leave the school with a core group of teachers and parents who will carry on the program. One method of achieving this outcome is to recruit parents who will co-teach the after-school garden club in the program's first year and then assume more responsibility in the second year. Tessman points to one school that is now paying a parent to teach the after-school garden club, an ideal outcome.

At the level of a school system, Growing Gardens is also striving for sustainability and is developing a template for how a school system can grow a system-wide garden program. One goal, suggests Lippoldt, is for a school system to hire a full-time garden program staff who can provide consistency, networking, resources, and support to the schools within the system.

A School Garden Takes Root
Kelly Elementary School

The nonprofit Growing Gardens (page 17) is working with neighborhoods and schools throughout **Portland, Oregon**, to bring gardens and fresh food to urban populations, and especially children. One example of its work is at Kelly Elementary School, which serves a mostly lower-income, culturally diverse population. The school partnered with Growing Gardens' Youth Grow initiative to develop a school gardening program, with a mission to "improve nutrition and decrease the risk of food insecurity by teaching children at risk lessons in where food comes from, the importance of eating fruits and vegetables, and how to grow food through fun hands-on activities." The program encompasses, among other components, a garden, after-school and summer garden club programs, and support for development of a school garden committee.

The garden program at Kelly Elementary extends both outdoors and in. Out in the garden, kids help with planting, weeding, harvesting, and other maintenance tasks, while indoors segments on natural life cycles, plant physiology, insect life, and soil are woven into classroom curricula.

Teaching as a Tool for Building a Healthier Community

Jones Valley Urban Farm

with research by Jessica Ray

The Jones Valley Urban Farm has transformed over 3 acres of downtown **Birmingham, Alabama**, into a neighborhood farm that grows organic produce for farmers' markets, restaurants, grocery stores, and a CSA. As part of its mission to be a "model sustainable urban farm that teaches youth and the Birmingham community about sustainable agriculture and nutrition through outdoor experiential education," JVUF runs numerous educational programs for people of all ages on all aspects of food, including sustainable agriculture, nutrition, and healthy lifestyles.

In Alabama's largest city, Jones Valley Urban Farm is demonstrating how we can change attitudes about food and nutrition with purpose and grace. Once an industrial manufacturing giant of the South, Birmingham is now a banking center that lays claims to more green space per capita than any other city in the nation and a world-class culinary scene.[141] But more green space and haute cuisine don't mean Birmingham is healthier than other cities. The city's obesity rate of 31.2 percent was the second highest in the nation in 2007.[142] It mirrored Alabama's overall obesity rate of 31.4 percent, the second highest in the nation in 2008.[143] And if overweight people are included in this statistic, the figure rises to a whopping 67.9 percent.[144] "We are literally eating ourselves sick," Bob Blalock of the *Birmingham News* writes.[145] In 2008 the state took radical action, announcing that its employees would have until 2010 to get fit, or they would have to pay a $25 monthly "fat tax" for insurance.[146]

It's too soon to know whether a fat tax will change eating habits, but in the meantime the folks at Jones Valley Urban Farm are working to change Birmingham's eating habits in another way: education. The group's executive director, Edwin Marty, and his cofounder, Page Allison, completed an apprenticeship in agroecology at the University of California at Santa Cruz and then traveled the world working on small projects before returning home to Birmingham.[147] Here, in 2001, they founded Jones Valley Urban Farm, converting vacant land in the downtown to organic farmland.

From the beginning, Jones Valley Urban Farm has wanted to use urban farming to teach. The farm has morphed through the years, adding and dropping sites as needed. Now JVUF has consolidated its farming operations to a 25-acre site outside the city and two smaller sites right in the city. At one of the city sites, called the Gardens of Park Place, situated next to a highway and Hope IV housing development, JVUF is working with a local foundation and the community to create a model urban agricultural center. At 3 acres the site is large, and so is their vision. Park Place will eventually offer a variety of uses, including 40 community garden plots, a farm center and kitchen for educational programs, an Early Childhood Learning Center for pre-kindergarten children, and a teaching farm.

The educational programs pioneered by Jones Valley Urban Farm reflect a core belief that change can happen at any age. Some programs are for children and others for adults. Some are targeted to specific populations, like educators, and others are self-selecting. JVUF's package of educational programs continues to evolve as staff learn what works and doesn't. As innovators, they are paving the way for other farms that want to be agents of change for healthier communities.

Accrediting Agriculture

One of JVUF's cornerstone educational programs grew out of necessity. When JVUF was just getting off the ground, the farm desperately needed help. Marty worked with a friend at the Alabama School of Fine Arts (ASFA), a public school for grades 7 through 12 in Birmingham, to develop outdoor activities that students could participate in to earn credit. The result was a state-accredited experiential hands-on science course. Starting as a summer program for credit, the course became so popular that, after two years, JVUF expanded the program to a rigorous semester course offered during the school year.

The name of the course — "The Arts and Science of Agriculture" (ASAP) — reflects its interdisciplinary mix. Both in the classroom and at the farm, students study botany, soil chemistry, entomology, microbiology, the work of water and sunlight, and even weather and air quality. They maintain garden beds, select crops, amend soil, and harvest vegetables for use in the school cafeteria. And students also learn about ecosystems and the regional, national, and global history of food systems. As required by most high school courses, they do homework, listen to lectures, conduct experiments and research, and take tests.[148]

As a science elective, the course popularity continues to grow. By 2009 Rachel Reinhart, the JVUF program director, was teaching two ASAP classes each semester. Bryding Adams, a JVUF board member, says, "ASFA math and science kids like the course. It's applied science, and [it] makes them think differently about their food and what they're eating."

Graduating to Internships

Of course, success often brings new challenges. The popularity of the school-year courses meant that fewer ASFA students were signing up for the summer course — when the farm really needed help. So JVUF designed another approach: the group now hires ASAP graduates as interns to work on the farm during the summer. The interns have proved to be a huge help, because they arrive pre-trained, already knowledgeable about plants, bugs, and fertilizer, and they just need to be pointed in the right direction. The summer interns are more than helping hands; they continue to learn and even conduct research projects. "These students are amazing," Reinhart says. "Some have gone on to study sustainable agriculture, environmental sciences, and nutrition."

Expanding Educational Tools

JVUF is also exploring ways to help the larger Birmingham community develop greater awareness of fresh food and, it hopes, healthier eating habits. It's going about this in several ways, including community gardens, school field trips to the farm, training in basic organic gardening, and even training for other educators.

Community gardens are key, JVUF believes, and it hosts an increasing number of them at its different farm sites. But it couldn't keep up with the demand from the community to help neighborhoods set up their own gardens. So JVUF began a "Growing Together" training series to teach community members how to develop and sustain community gardens themselves. For the first training JVUF thought 20 people might show up, but they had 87 people graduate. By 2009, graduates of the JVUF program had successfully established nine neighborhood gardens all over the city, and some were maintaining a rooftop garden at a women's shelter.

In connection with its efforts to expand community gardening, JVUF believes that it is important for people to learn to garden organically. So, during its "Weekend Workshops" series, the general public can learn about basic organic gardening methods, why local food is important, and, through a self-guided tour of the demonstration farm, sustainable farming.[149]

For community youth, JVUF is trying to connect food in the field to food on the table. School classes can schedule a half-day "Seed 2 Plate" workshop at the farm, where they experience a hands-on agriculture lesson in the field, a nutrition lesson in the classroom, and a culinary lesson in the kitchen, in which they make their own chef-supervised snack. A class can schedule as many as five of these field trips to the urban farm, each one focused on a different farm-to-fork theme. As if that weren't enough, JVUF is also offering short programs for preschool gardeners and a one-week "Foodie Camp" in the summer for children to learn about growing and preparing food.

Not surprisingly, other educators have expressed interest in learning from JVUF's success. So JVUF now offers a series of half-day workshops for educators — teachers, community leaders, master gardeners, parents — to learn how to use a garden as a teaching tool and how to develop an effective experiential curriculum for people of different ages.

For Real Change, Change School Food

"If we want to change the way that students and youth are eating in Birmingham," Marty says, "the low-hanging fruit is to change what's actually served at the schools." The quality of school food is shocking, he says, full of fat and fried foods. And if anyone doubted this, Jamie Oliver, of *The Naked Chef* fame, went on a crusade with his spring 2010 show, *The Food Revolution*, to show people just how unhealthy most school food is — and also how difficult it is to change it.

One of Birmingham's biggest triumphs, Marty says, was to achieve the seemingly impossible: removing all the fryers from all the schools. JVUF is a core partner in the local citizen-led Obesity Task Force, which the Jefferson County Department of Health has authorized to lead and coordinate county efforts against childhood obesity. The Obesity Task Force identified school fryers as a major issue that could be easily remediated. So the task force developed a policy paper and submitted it to the school board, which balked. But the task force continued to push, giving the school board ongoing information about alternative options and what other schools were doing. During the summer of 2009, the fryers were pulled out.

The next-lowest-hanging fruit, Marty says, is to train cafeteria staff to work with fresh food — to order it, to plan the menu so that the food is affordable and meets state dietary guidelines, and to prepare it. The idea that cafeteria staff are trained to work only with commodity food (for example, prepared foods that are ready to heat and serve, canned goods, and industrially grown fruits and vegetables, bred to be hardy and have a long shelf life), rather than fresh food, seems incredible, but others have the same story to tell. At one Virginia university, cafeteria staff reportedly couldn't deal with fresh, locally grown tomatoes because, unlike the tomatoes they usually bought, which were bred to sustain the abuses of transportation and handling, these heirloom tomatoes were too ripe and soft to put through the automatic slicer. The supervisor needed to teach his staff how to slice the tomatoes efficiently by hand. Though it was more time-consuming, the cafeteria staff ultimately agreed that students deserved this extra effort because the tomatoes were so different — so delicious.[150]

In Birmingham, Marty explains, the Obesity Task Force strategy is to work with cafeteria staff in August, before school starts. In 2009 they conducted a pilot nutrition training program to get cafeteria staff excited about serving fresh foods and decreasing the amount of fat and salt they use. (If cafeteria staff are required to take food *safety* training, he asks, why shouldn't they also be required to take *nutrition* training?) One year later, Marty reports, the county school system established a new requirement that *all* school cafeteria workers participate in JVUF's training program, to learn about preparing delicious, nutritious food.

As part of its Farm to School focus, JVUF has hired a new staffperson who will work at both the policy and practical levels to bring healthy, local, fresh food into school cafeterias. This person will also work to fold nutrition into classroom learning and to expand school activities outside the classroom to include farm visits, gardening, and healthy cooking. "We have high hopes for changing the food system!" says Marty.

Making Sustainability a Campus Issue
Farm to College Movements

with research by Benjamin Chrisinger

As centers for experimentation and transformation, college campuses **across the country** are experiencing an awakening of student and faculty interest in food-related issues. Some campuses are finding ways to increase the amount of locally grown food offered in cafeterias; some are creating school gardens and farms where students can connect with the land and grow food for themselves, the school, or local food pantries; some are offering food-related courses, certificates, or even degrees in food planning, food policy, food production, food business, and food justice. Food is becoming a part of the academic vernacular on campuses, and these farm-to-college efforts may bear continuing fruit for generations to come by inspiring students to take an active interest in what they eat and where it comes from.

The typical all-you-can-eat college meal plans allow students to eat unregulated diets that are influenced by a host of factors, from lack of sleep to stress, hectic schedules, and irregular meals. These factors, coupled with alcohol consumption, which can block nutrient absorption, are thought to explain why college students can experience the perverse coupling of weight gain and malnourishment. When I began to research trends in campus food, I assumed the burgeoning obesity epidemic alone would have sounded the alarm for a change in dining-hall offerings. Yet this particular public health alarm seems strangely silent.

There are significant forces pushing for change in college dining, but I found that they originate largely from a desire by students, faculty, and administrations to increase campus sustainability. In the dining hall, concerns for reducing carbon emissions, waste, pollution, environmental degradation, globalization, and dependence on foreign oil are translating into a desire to support local farmers, the local economy, and buying fair trade and organic or sustainably grown foods.

The good news is that, after 20-some years of talking and teaching about it, sustainability is now an imperative for campus action. Colleges are intensified

microcosms of the world around them — hotbeds of diverse ideas, innovative thinking, and cutting-edge research — but they are still large and clunky. And, like large carrier ships, they turn very slowly. Though slow in starting the turn, campuses are, indeed, turning toward sustainability in every aspect of their operations. As they become practitioners of sustainability, they are looking inward, owning responsibility for their environmental impacts, and seeking creative, cost-effective ways to live with integrity on the land and in their communities. They also are taking leadership in sustainability research and fostering a generation of leaders who will have sustainability literacy.

Sustainability Gains Ground

The college sustainability movement began to gain steam in 1990. At an international conference in France, university leaders developed the Talloires Declaration, a 10-point action plan for integrating sustainability into all aspects of campus life — operations, teaching, and research. By 2010 over 400 university leaders from more than 50 countries, including 163 from the United States, had signed the declaration.[151] You might think, rightfully, that this is a pitiful percentage of the tens of thousands of academic institutions worldwide. Yet the Talloires Declaration set standards for campus sustainability that, if nothing else, forced academic institutions to think about what they could, should, and would do on their own campuses.

Ten years later, in January 2010, at the World Economic Forum in Switzerland, the International Sustainable Campus Network and Global University Leaders Forum articulated three core principles for a sustainable campus in the Sustainable Campus Charter. A handful of Ivy League universities, including Yale, Harvard, Brown, and Columbia, were among the charter signatories. These early adopters were, again, setting a standard and pointing the way for others to follow.[152]

During this decade, too, some organizations wanted to measure campus progress and sought ways to hold academia's feet to the proverbial fire. The Sustainable Endowments Institute, founded in 2005 by the Rockefeller Philanthropy Advisors, has now become the go-to source for information on campus sustainability with its independent annual "College Sustainability Report Card." In some ways the 2010 Sustainability Report Card was disappointing. It shows how much campuses have achieved — and how much further there is to go. Only 26 public and private campuses, or less than 10 percent of the 300 assessed, received an A– or better in all assessment categories, earning them the honored title of Overall College Sustainability Leaders.[153]

An assessment of campus food, however, tells a different and far more impressive story. The Sustainability Report Card found that campuses were progressing most rapidly in the areas of administration and food and recycling. It awarded 40 percent of the 300 schools an overall A for sustainable administration. And in food and recycling, it awarded 36 percent, or 119 schools, an overall A.

I'm intrigued by the possible connection between campus progress in administration and in food. Why would sustainability in administration and food move forward almost in lock-step, while leaving other factors measured by the Sustainability Report Card, like shareholder engagement and endowment transparency, behind in the dust? The answer may be obvious: of the 52 indicators the Sustainability Report Card evaluates, administration and food have highest visibility on campus, and therefore they are logical first targets for greater student concern.

In my years at the University of Virginia, and in conversations with faculty at other schools, grapevine wisdom has it that "students rule." While faculty and public pressures can be temporized, students are more difficult to ignore. So when students pressure the administration for a cause, change is more likely to happen. Students, seen in this light, are hardly empty vessels to be filled with knowledge but important and powerful change agents on campus.

When it comes to food, this has certainly proved true. Students are often the impetus for a campus to take baby steps with pilot programs — establishing a café that purchases only local or sustainably grown food, or convincing the dining hall to buy only cageless eggs, or bringing a farmers' market to campus. And these baby steps, again with student prodding, often lead to deeper institutionalized commitment to supporting local farmers and the local economy.

Universities Take a Bite into Local

One of the first campus campaigns for local food was initiated by a concerned parent — who happened to be chef Alice Waters of San Francisco's award-winning Chez Panisse restaurant. Waters had founded the Edible Schoolyard garden at a middle school in Berkeley, California, in 1995 and was a recognized champion and veteran at introducing fresh food into schools. When she approached Yale University president Richard Levin in 2001, he agreed to offer locally grown food at the Berkeley undergraduate residential college where her daughter was enrolled. Accomplishing the unimaginable, Berkeley transitioned to a menu of entirely sustainable and, when possible, organic ingredients. Berkeley College's website notes that its dining hall is "perhaps the most envied on campus" and had a role "in founding and developing the Yale Sustainable Food Project, which revolutionized food service at Yale with national acclaim."[154] The food at Berkeley College became so popular that students reportedly used fake IDs to get into its dining hall,[155] leading the university to extend its menu to other Yale dining halls. Yale's Sustainable Food Project has since expanded to host a farm, internships on the farm, and an academic concentration in sustainable food for environmental studies majors.

Brown University's acclaimed university dining program began in 2002 with a student request for fair-trade coffee. Virginia Dunleavy, associate director of dining services, decided that, while Brown couldn't afford fair trade, the university could begin to buy more local produce.[156] Dunleavy launched Community Harvest, a "commitment to socially and environmentally sustainable purchasing practices," as its mission statement says, which has expanded over the years and today does guarantee that all campus coffees and teas are fair-trade certified.

In 2004 Brown hired Louella Hill to champion the cause of bringing more local food into its dining hall. As is the case for universities today, Hill and Dunleavy faced the logistical challenges of serving an incredibly large population (at the time amounting to 1.5 million meals a year) while finding ways to incorporate food from small, local farms.[157] But Hill was stubborn and creative. When she wanted to serve local milk in Brown's dining hall, she faced the daunting barrier of stainless-steel milk dispensers that required 5-gallon bags, not the ½-gallon bags provided by the newly formed local dairy cooperative, Rhody Fresh. So the Brown dining hall agreed to put the Rhody Fresh milk cartons out on ice until the co-op could fix the packaging issue.[158] The university dining contract was a huge boost for the fledgling dairy cooperative. Most dairy farmers in the state had already sold out, cashing in on the high value of their land. "As a Rhode Island dairy farmer, I'm an endangered species," said Louis Escobar, president of the Rhody Fresh co-op, in an interview with the *Chicago Tribune*. "[The Brown contract has] been like a shot of antibiotics for a sick animal."[159]

Hill also helped open a direct line of communication between growers and chefs by cofounding the Farm Fresh Rhode Island collaborative. Today, much like North Carolina's Appalachian Sustainable Agriculture Project (page 243), this collaborative is an active nonprofit that helps connect growers with buyers through conferences, meetings, and one-on-one relationship building, and it supports growers through a host of market-growing services.

Brown and Yale are just two of the more well-known examples of colleges aiming for food sustainability. Others abound; Middlebury College in Vermont, for example, reports that *all* its dairy products are purchased locally, and it also purchases local maple syrup, local eggs, and Vermont cheddar cheese. Bates College in Maine spends 30 percent of its dining dollars on local food and also recycles, composts, or diverts to a local pig farmer a remarkable 82 percent of its dining waste. California State University at Chico, about 90 miles north of Sacramento, boasts an organic farm and organic dairy and passed a resolution in 2007 mandating that 25 percent of its food be both certified organic and produced in California.[160]

Oberlin College in Ohio may be the nation's leading pioneer in buying local food, purchasing an incredible 45 percent of its food locally. The healthy cooking initiative that the college launched in 1998 has revolutionized its dining hall food. In 2001 it began preparing its own stocks and dressings and purchasing sustainable seafood; in 2002 it switched to rBGH-free milk and in 2003 to antibiotic-free chicken; in 2004 it eliminated all trans fats; in 2005 it switched to antibiotic-free turkey, cage-free eggs, and grass-fed natural beef chuck; and in 2006 it eliminated all seafood with anything above minimal levels of mercury. And its list goes on.[161]

These examples are interesting because of what they represent: institutional change. It's true that institutions could one day say, "We can't afford to do this anymore." But the forces that are pressing campus institutions in these directions are not likely to reverse. It's hard to make these changes; they require perseverance, building relationships, negotiating new kinds of deals, and above all lots of patience. These changes are not a result of snap or faddish decisions. They are changes to an entire *system* of food purchasing, with positive implications for local economies, frayed town-gown community relations, public health, and sustainability.

In *Deep Economy*, Bill McKibben argues that we too often forget that oil is a critical factor in conventionally grown and transported food. "Our food arrives at the table marinated in oil — crude oil," he writes.[162] When oil prices climbed to new heights in 2008 and 2009, an interesting thing happened: the cost of local food relative to that of conventional oil-based food reached a tipping point. As Phil Petrilli of the Chipotle restaurant chain, who helped bring meat from Virginia's Polyface Farm to the local Chipotle restaurant (see page 220), says, the cost preferential shifted and local meat suddenly became relatively more attractive and cost-effective.

As the world depletes its oil reserves, the economics of supply and demand suggest that it is only a matter of time before oil prices once again begin to climb. And when they do, campuses that have already put into place a system for purchasing local food will be well positioned to weather the transition away from an oil-based economy. This is one reason why purchasing local food is considered so integral to the push for campus sustainability. And this, too, is why the forces driving campuses toward local food are likely not to fade away but to grow stronger and more urgent.

Using Food to Educate

The purchase of local food by campus dining halls can provide a major boost for local farmers, economies, and farmland preservation. Equally important are efforts to integrate the food movement into higher-education curricula. The subject of local food offers a foundation for educatinge students about nutrition, agriculture, and the social and economic values of buying locally, and colleges are finding ways to broaden and enrich the student experience through courses, internships, and work that foster a deeper understanding of and relationship with the natural world.

We have often looked to our land-grant institutions to handle all things agricultural, but today liberal arts schools too are gaining ground in food planning, food marketing, food history and anthropology, food-based business development, plant medicine, edible landscaping, and even food production. It's not so unusual anymore for a nonagricultural liberal arts graduate to pursue a food-related career in public policy, community nonprofits, social justice, marketing, or even farming itself.

Recently, the Harvard Graduate School of Design introduced a course called "Food Forms: Agriculture in Urban Systems," which was reportedly enthusiastically received by the student body. The Massachusetts Institute of Technology School of Architecture and Planning has added "Food Systems and the Environment" to its course offerings for graduate students. And at the University of Virginia, students can take my own graduate-level urban planning course on food systems as well as numerous other courses exploring the politics of food, food justice, edible landscape design, and more.

Though food has always served as a focal point for student social events, only recently has food become a hot topic for academic discussion, training the future leaders of our world to think critically about what, when, where, why, and how we eat. Whatever the goals may be for these varied programs, they share a common motivation: fresh food is something students want the opportunity to enjoy. And as a subject for academic inquiry, the study of our food system is coming into its own, enabling students to understand core community issues through the lens of food. In this way the farm-to-college food movement is changing not only what we eat, but how we think about and understand the impacts of what we eat.

Cultivating Confident, Competent Leaders
Warren Wilson College Farm

with research by Benjamin Chrisinger

For more than a hundred years, Warren Wilson College in **western North Carolina** has operated a farm of about 340 acres to serve its students and community. Today this student-run farm produces a variety of rotated crops along with grass-fed beef, pastured pork and poultry, and eggs, and a 5-acre student-run garden produces vegetables and herbs. Some of the meats, eggs, and vegetables make their way into student dining halls, while some are sold through a farmers' market and weekly community-supported agriculture (CSA) basket. In addition to producing food, the farm and garden are part of the college's mission to grow students into confident leaders.

One interesting model for integrating agriculture with academics can be found in the southern Blue Ridge mountains at Warren Wilson College. Founded in 1894 as a school to educate and feed Appalachian boys who came from farms and would return to work at those farms, Warren Wilson put its 340-acre farm into immediate production.

Most unusual about Warren Wilson is its founding holistic ethic, known as "The Triad," an imperative for three paths to educating the whole person: academics for the mind, service-learning for the heart, and work for the hands. As a designated work college, Warren Wilson requires all residential students, nearing 900 strong in 2010, to work at least 15 hours each week and complete at least 100 hours of volunteer service-learning before graduation. "Academics alone doesn't create people who are balanced," says Chase Hubbard, Warren Wilson's farm manager. "All these other things help round out a person."

Warren Wilson produces a distinctively different graduate. I can't help but notice that every Warren Wilson graduate I've met has exhibited a quiet confidence, competence, and maturity, coupled with an unusual spiritual connection to land and community. Ian Robertson, the dean of work, says, "We graduate a student who knows what it takes to get something done, knows how to do it with others, knows the meaning of community in terms of interdependence, who is engaged very quickly in their community, who is very confident."

Of course, it could be argued that the Triad attracts this kind of person to the college in the first place. But Warren Wilson doesn't bungle the job. "All three aspects of our Triad come together and really educate the whole person," says Robertson. Students wrestle with an issue in the classroom and then see that same issue play out in a real work or service-learning situation. That's what brings it all together, creating what Robertson likes to call a "distinctive experience."

The campus sits up on a hillside. Just below, the college farm and garden stretch out across the sun-drenched valley. A short trip up the road and around the bend are more fields for pasture. The farm averages from 130 to 200 cows and 60 to 100 hogs at any given time. Roughly 250 acres of the farm are devoted to crops of corn, barley, oats, and alfalfa-based hay in a five-year crop rotation. Mixed into this are rotations of a grazing mixture or buckwheat, depending on the needs of the animals.

The two full-time campus farm managers are both Warren Wilson graduates, and 25 to 30 students during the school year and five or six students during the summer provide eager working hands. A crew of three students works with the pair of huge Belgian draft horses that help plow and disk the fields, pull a manure spreader, and "snig" (pull) logs from the managed woodlands to the campus sawmill. These beauties also get to show off their snigging abilities during an annual Plow Day event.

Nearby, at the college garden, one full-time garden manager works with 25 students during the school year and seven students in the summer. Situated on 5 acres, with one heated greenhouse and two unheated hoop houses, the garden produces vegetables, fruits, flowers, and herbs that are sold to the college dining service, weekly campus markets, and a small CSA.

Altogether, Robertson estimates, 60 to 70 college students work during the course of the year in some aspect of food production. Of those who graduate each year, Robertson guesses that as many as 15 to 20 enter some kind of agricultural production and often choose to work in small-scale agriculture, where they can personally interact with buyers.

Put simply, this small liberal arts college is successfully graduating a small but steady stream of people who are helping to grow local food systems. When they graduate, the students are successful at small-scale farming, says Robertson, "because that's what we do here."

Growing a College Market Garden

Prior to Robertson's tenure, the Warren Wilson college garden was a tiny plot run by a volunteer faculty member who had a passion for strawberries and sunflowers. In 1981 Robertson was invited to create a larger garden, along with classes that would provide students with gardening skills. "Students at that time wanted to control their food," Robertson says. "And that meant growing it yourself." He also wanted to see if the garden could provide fresh produce to the college's dining service.

Robertson started small, creating a 1-acre garden from a former pig field. But it was an intensive, difficult birth. "There was an incredible amount of compaction," Robertson recalls, "because pigs are like 300-pound ballerinas on point all the time, walking

across your land." So Robertson decided to prepare the entire acre with Alan Chadwick's biodynamic method of double-digging — an arduous, long process executed by hand. Robertson felt this work would help students understand the soil at an intimate level, and, because he knew the ways of college administrations, he wanted the students to become so invested in the garden that the administration couldn't possibly do away with the project. His plan worked. The garden has since expanded to 5 acres where, to this day, students continue to double-dig beds, on a rotating basis. "It's a great winter project," says Karen Joslin, the garden manager. "I use it as a teaching tool."

Of the 5 garden acres, 3 acres are always in active production. One of Joslin's summer students works full-time on the medicinal herb garden and creating teas and hand salves. The other six tend the acres of garden fields, the Belgian draft horses, and the chickens. As with most decisions about the farm and garden, the choice to raise chickens was all about the students. Chickens are a common element of diversified small farms, providing a small income stream, some bug control, and manure fertilizer, so the school thought it would be important for students to learn how to integrate chickens into a small operation.

Warren Wilson grows its food organically but has not pursued the USDA label of organic certification. Robertson believes that the college's reputation is a more worthy stamp than government certification, because people know the college and trust that it will do the right thing.

Robertson's vision of providing food to the college cafeteria has come true. The garden sells about 60 percent of its produce, by weight, to the two college cafeterias. Meanwhile, the garden is working through several kinks, says Joslin. Crops that grow well in the garden may not always be the crops wanted in the cafeteria. Seasonality is another issue. Every school garden and farm faces this challenge: the easiest time to grow is during the summer, precisely when most students are gone. So the program has begun dabbling in food processing, mainly freezing, to enable students to enjoy the garden's produce in winter. They blanch, cool, and then freeze the vegetables. But Joslin says it's hard to produce a quality frozen product without expensive flash-freezing equipment, which is an investment of thousands of dollars. Still, they're working on ways to improve the quality of their frozen vegetables.

The biggest challenge, however, lies in the very nature of a college garden: turnover. Joslin says that students who want to work at the garden must commit to stay for two years. There still are times when she can't always fill the summer slots with experienced students, in which case she has to train new students at the busiest time of the year, a less than ideal situation. And students don't always get it right — they might plant all cucumbers instead of a mix with zucchini and squash. But that is also part of what it means to be a college garden: it's a laboratory where students are always learning, from both real mistakes and real accomplishments.

In addition to selling food to the campus dining service, the garden sells food at a farmers' market on campus twice a week and also at a weekend market at a nearby fire department. Students operate the table, introducing the food to customers, explaining the different varieties, suggesting how they can cook tasty dishes, and explaining why they should buy local food. "They know the chitchat," Robertson says. He says the ability to chat up a customer is critical for the students' later success as farmers, because what makes farmers' markets successful is the relationship between farmers and buyers.

Growing Competence and Confidence

On the Warren Wilson farm, just a short walk from the garden, students work with cattle, pigs, and field crops. The farm has stayed strong for more than a hundred years, largely because it was managed by just one family for more than half that time, says Chase Hubbard, farm manager. The Larson family grew and prepared food for the college students from the 1930s to 1990s. It was only late in these years that people began to realize how unusual it was to have a college farm, says Hubbard.

Hubbard, a Wilson graduate himself, says that students vie for farm jobs. Still, students know it's a lot of really hard work, so there is never a flood of applicants. "The ones who do apply are very ambitious," says Hubbard. "They're the kind of student who want to suck the marrow out of their farm experience. They're drawn to the work because it's very compelling — they see how it affects the landscape and people. It's also dangerous, and college kids like the fact that it's adventurous and strenuous."

During the summer, Hubbard's five student interns are paid a straight salary, though that doesn't stop them from averaging between 60 and 100 hours a week. Students start by learning the basics: running a string trimmer, distinguishing between a Phillips and a flathead screwdriver, coming to work on time, and so on. After mastering the basics, they move up to working on different animal crews: swine, cattle, and more recently, because they're easy to get into, a few hundred broilers.

"No freshman gets placed on the farm crew," says Julius Stuart, a rising junior in the summer of 2009. "You have to come down and volunteer first." Stuart volunteered a few times and was successful in getting on the farm crew in his second year. "During the summer you're doing everything," he says. "It's a really great learning experience."

Hubbard says his summer crew members are all above average, responsible, and communicative. After all, the crew needs to run itself, evaluate itself, and then be evaluated on its evaluations. I believe him, too. Hubbard leaves me with Stuart, who drives me over to the pasture fields, talking confidently about his work, explaining what he's looking for to determine whether to move the cattle down the field — if the paddocks look about the same, if manure piles are steaming, if the field looks patchy. Stuart tells me about the time they decided to mob-stock a field (bringing in stock at an ultra-high density to graze), instead of cutting the field for hay, a technique they had read about in an article written by Joel Salatin of Polyface Farm (page 220). The mob stocking got off to "a rough start," says Stuart, "because we didn't know how to gauge the paddock sizes. We went from too big to too small. It was good to determine what our limits were; that way we could find a good medium." I can also see how this experience would give students confidence to innovate and experiment on their own, later in life.

Stuart decides it's time to move the cows and tells me where to stand and what to do. Any concern about being trampled fades when he grabs a fence post, reels up the wire, and begins calling out, keeping them bunched as he moves them through. The cows know they're going to a fresh pasture, and they almost start to stampede, but Stuart somehow slows them down to a reasonable trot. Not yet in his third year of college, Stuart doesn't think he'll be a farmer, but he loves the work and wants to raise just enough to support himself and his family. He'd like a couple of cattle to "just raise and slaughter in the fall."

Colleen Vetti, another rising junior, says the farm has the hardest-working crew on campus. "I've probably learned more here than I have in my classes, in a different sort of way," she says. "It's the only work crew where you do everything yourself." Vetti is on the swine crew and says that her typical workday begins with hosing and scraping out the pens, followed by feeding. The night before was a "hog cook," and Vetti took part in killing, slaughtering, and butchering — a pretty intense process, she says. She volunteers that she'd seen it before, but this was her first time participating. "More than being upset about it, you're stressed about getting it right," says Vetti. She tells me, "You can taste the freshness. You know the taste after catching a fish? That's what it tasted like to me. When we cooked this pork, it tasted awesome."

We walk over to the alfalfa field to check on the sows. Vetti looks each one over, moving carefully around them so they don't get riled up. She inspects one that was born with an infection on her neck and is receiving antibiotics. The others are getting close to their market weight of 250 to 300 pounds and will be sold in quarters to the community through the local market. "But we can't sell her," says Vetti, gesturing to the one she just examined, "because we sell only antibiotic-free pork."

Vetti says that she had wanted to become a veterinarian when she first started college, and then she had thought she might be an obstetrician. Now she's not so sure. "I don't want to be a farmer," she says, "but it's kind of worth it. It feels fulfilling, like going for a really long hike. You feel like you've accomplished something. The farm is as good as you want it to be." During the school year, when she gets frustrated by her school work, Vetti comes down to the farm. "How can you not love it?" she asks, gesturing to the pigs and fields. "When your blood, sweat, and tears go into it every single day, how can you not love what you're doing?"

Starting from Scratch

If a college wants an operation like the one at Warren Wilson, starting from scratch with bare land, Hubbard says that he thinks it would take at least ten years to establish. Acquiring the land, building fences, hiring staff — it all takes time. For only ten head of steer, some chickens, and some vegetables, Hubbard guesses it might take only five years to become fully operational. He suggests that a new college farm start small, get something in place to show success, and begin building from there.

Not including the start-up costs, which are significant, Hubbard thinks a college farm can be run at a fairly low cost. The Warren Wilson farm grosses about $200,000 a year on 275 working acres from sales of its broilers, beef cows, and hogs, which covers all operating expenses and a portion of student and management labor. The additional annual cost outlay, he says, is about $25,000. To my ears, that sounds like a very small amount for a flagship college program with so much to offer. "We are a long way from turning a true profit," says Hubbard, "but as a nonprofit our goal is to offer real-life experience for students for education."

If students want to make lots of money, Hubbard wouldn't encourage them to take up farming. After all, he explains, Warren Wilson isn't trying to make farmers; it's creating students who are better prepared for whatever they want to become, whether that's a raft guide or a doctor. "Education is way more profitable than farming," he says. "We do not turn out tradespeople, we turn out broad thinkers who are exposed to a broad curriculum."

Lessons Learned
Establishing a College Farming Program

Determine how a farm or garden will complement the institutional mission.

Before plowing up a new farm or digging in a new garden, a college should first determine how it would complement the institutional mission. Ian Robertson, dean of work at Warren Wilson College in Asheville, North Carolina, says that the farm or garden, no matter how small, is a part of the academic environment. The ideal, says Robertson, is for a college to give the farm or garden serious time and consideration, determine how it will be integrated into the overall mission, and then find a competent farm manager who will be responsible for growing food as well as cultivating student skills and competence.

Starting a farm or garden may be easiest at colleges that offer programs in agriculture, environmental sciences, or sustainability, says Karen Joslin, Warren Wilson's garden manager. These programs will provide a healthy contingent of students who are interested in working on the farm or garden and can help train others.

Joslin also advises colleges to start small. As the farm or garden becomes established, infrastructure is put into place, and systems are worked out, then the operation can be expanded.

Hire knowledgeable managers.

The ideal farm and garden managers are passionate about the work, passionate about talking to others about the work, patient, and willing to be nurturing mentors to students on a daily basis, says Robertson. They should be treated and paid as faculty, says Robertson, because they do far more than helping students learn about food production; they help students build competency in life skills, reflect on their life, and sort through their own priorities and goals.

Candidates should have practical hands-on farming experience and technical experience with tractors, plumbing, and irrigation, says Joslin. She has two agricultural degrees and lots of classroom experience but feels she could have avoided mistakes and difficulties if she'd been more seasoned. If possible, for a first manager find someone who's been farming for 10 to 15 years, who will be able to help build the foundation and infrastructure for the long term, suggests Joslin. After the foundation is built, subsequent managers may not need the same level of experience.

Consider the farm or garden an outdoor laboratory.

Robertson draws a distinction between liberal arts colleges that prepare students for jobs and colleges that prepare students for *work*. He says that Warren Wilson College has a long history of using the outdoors as a living outdoor laboratory to prepare students for a life of work. Here, students are faced with making real-life decisions, with real-life consequences. "If a student makes a mistake working on a pig," Robertson says, "the pig may or may not get better. There is real feedback and consequences here." One key to a successful education, Robertson believes, is to provide students with opportunities to fail, and to have people who can help them reflect on and learn from what they're doing. This, says Robertson, is why Warren Wilson graduates such confident individuals.

Even campuses without farms or gardens can help students learn and grow through the real-life work, because most offer students the opportunity for federally subsidized work-study. Robertson suggests it is a matter of shifting the academic culture and expectations of professors. Many aren't thinking about how to help students draw connections between their work and their study. Robertson gives the example of a student who might work at a fast-food restaurant. Robertson suggests professors could ask, "What are the challenges there? What's it like to work collectively?" Sociology professors could ask, "What type of socio-economic groups are you working with?" Creative writing professors could ask, "What's the life story of the person you're working next to?"

Focus on production.

For a campus farm or garden to succeed in teaching students about real life, responsibility, and consequences, Robertson emphasizes that it must be concerned with production. Warren Wilson strives, and succeeds, in having its farm and garden sell enough food to at least cover their costs, not including the farm and garden managers' salaries. Because the college places a premium on the education and mentorship that the managers provide, it classifies and pays them as faculty. Robertson knows that running a campus farm and garden requires far more than just teaching students how to move the cows, double-dig a raised bed, or wean piglets from the mother sow. The managers also help students sort through important life issues.

But production is what makes all these other lessons possible, says Chase Hubbard, farm manager. "If it was only about education, then the farm would be more like a watered-down petting zoo. By focusing on production, we create really intense learning opportunities."

Include all stakeholders in decisions about serving campus products on campus.

For a college to consider serving food from its campus garden or farm in its cafeterias, one of the first key steps is to create an inclusive campus stakeholder committee to discuss the issues, concerns, and goals. The food-service manager, contracted food vendor, and business manager are all obvious key players who must be at the negotiation table, says Robertson. The real issue is who else needs to be invited to the table. At Warren Wilson College, Robertson says it was important to include the local food committee (which is setting standards for what is meant by "local"), faculty who teach sustainable agriculture, the dean responsible for the student work program, and the farm and garden managers. Each institution will be different, with different players and interested parties.

A key role for this group is to develop specific contractual "gatekeeping" targets for the percentages of food sourced locally and from campus production, says Robertson. A first goal of the group should be to reach agreement on the campus definition of "local food." Does it mean food grown and processed within the same state, or within a four-hour drive, or within 100 miles? A second important goal is for the group to negotiate specific targets for local food purchasing, perhaps starting with a modest goal of 15 percent and ratcheting up over time to 40 percent or higher. At Warren Wilson, where the farm and garden are integral to campus life, Robertson's goals would be to have the college food vendor source 40 to 50 percent of its food locally, with 30 percent coming from the college farm and garden initially and increasing over time. These amounts are averaged over the year, he explains, so that the garden may provide 100 percent of the lettuce eaten during the spring but nothing during the hottest summer months.

Use local distributor prices as guides.

A campus food service that uses food grown by the campus garden or farm should be paying the fair market price, as this establishes a real-life business relationship that enables the garden or farm to pay for itself. To determine a fair market price, Warren Wilson College and its food service use a small distributor of local food as a price guideline. If the distributor is selling a case of peppers for $12.50, then that is what the food service will pay for a case of peppers from the campus garden, says Robertson.

Consider hiring out the food processing.

When a vegetable is ready for harvest in abundance, like zucchini in July, neither the garden staff nor the food-service staff has the time to process it for later consumption during the school year. For this reason, Robertson advises that it's helpful to have a third party that is focused on managing just the food processing.

An outside, contracted food processor may have access to its own equipment. If the college desires to process food on-site, as part of its student education and work program, then it must set up the necessary infrastructure and equipment. The Warren Wilson program has access to a commercial kitchen, but it's 14 miles away, and transporting produce there and back is a burden. Ideally a campus processing center would be closer to the farm and garden, Robertson says.

6: Food Heritage
Preserving Cultural Identities

It shouldn't come as a surprise that more and more communities are focusing on regional resilience, as multiple stresses across the globe have tested communities to their limits — tsunamis, hurricanes, earthquakes, volcanic eruptions, flooding, rising oil prices, food shortages and riots, the economic crisis of 2008, not to mention threats and acts of terrorism.

"Resilience" is a twenty-first-century buzzword for good reason, as communities now desire sufficient flexibility with sufficient resources to withstand and recover fully from a blow, whether economic, physical, or social. Most are approaching resilience in a fairly straightforward manner by planning environmental and physical infrastructure, developing emergency plans and services, and building diversified economies.

Food is arguably at the very core of community resilience; everyone needs to eat. And a community's depth and breadth of food security is an indicator of how quickly it might be destabilized. Few, however, have realized the potential power of heritage foods. This chapter is an attempt to connect the dots — to demonstrate why support for biodiversity and place-based heritage foods suited to local climate, soils, and culture is important for achieving greater food security and, therefore, community resilience.

By definition resilience excludes practices that weaken a community's adaptability and survival. Supporting biodiversity and heritage seeds and breeds that have withstood the test of time, that are not dependent on costly synthesized chemicals and mechanized applicators, is a commonsense strategy for enhancing local resilience. Importing foods that cannot be produced locally is another age-old strategy. Salt, for example, was being traded as long as ten thousand years ago. There's no inherent contradiction between efforts to support biodiversity and native foods and importing essential foods that round out the diet.

But two strategies would weaken resilience. Importing foods to *replace* local heritage foods, our trajectory for the past 50 years, threatens resilience in the form of public health epidemics of obesity and diabetes; higher costs for packaging, refrigeration, and transportation; and the loss of seeds and breeds that sustained us for hundreds (if not thousands) of years. And genetically engineering crops threatens resilience by infecting and destroying native and heritage seed races. In fact, in recognition of this threat, at least 40 countries around the world have now mandated that genetically engineered crops and foods be labeled, while a number of nations and regions have banned genetically modified organisms in agriculture altogether.[163]

In this chapter you'll learn how Native Americans are seeking to preserve their cultural food and agriculture knowledge, to reverse their diet-induced obesity and diabetes epidemic, and to enhance community resilience. Though their efforts mirror the grassroots food movements throughout America, their emphasis on preserving biodiversity and food heritage is unique. In fact, the trajectory of First Nation people offers a sobering lesson in food security. Gary Nabhan, an ethnobotanist who chronicled the life of Nikolay Vavilov, a Russian pioneer in seed and plant banking, in *Where Our Food Comes From*, explains how the Hopi and Navajo survived the extreme Dust Bowl drought and famine:

> The Hopi and Navajo of the Tuba City area continued to be 90 percent food self-sufficient up through the onset of World War II. How was that possible for people who lived in a stretch of the Painted Desert that received less than 10 inches of rainfall in the average year and far less during the Dust Bowl era? What buffered the Hopi and Navajo from famine during drought was a mixed subsistence strategy that drew on a diverse set of crops adapted to different agricultural habitats.[164]

Yet this amazing cultural resilience began slipping away, as they, like everyone else in America, shifted toward commercial food. Historically, nearly every Hopi family had grown its own full set of vegetable and fruit crops. By 2002 the majority of Hopi farmers surveyed no longer grew enough corn to meet their family needs. Nabhan reports that the varieties grown during their era of food self-sufficiency had declined by over half, with only "47 percent of the varieties still grown and shared among them."

Again, we might ask whether this really matters. Isn't the food that Hopi or Navajo buy from the supermarket today just as good as the crops they once grew? The answer for all of us — regardless of race or culture — is a clear and resounding "No."

Nabhan writes, "Several studies have now verified that the imported foods that have replaced the native crops in their contemporary diet are poorer in protein, minerals, and dietary fiber but flush in fats and sugars . . . Belatedly, they are rediscovering that the best cures for those maladies may be the very crops that once grew outside their own back doors."[165]

So the First Nation people who are rediscovering and trying to revitalize their native foods and agriculture — for physical health and community resilience — offer a powerful lesson and inspiration for the rest of America.

Why Preserve Heritage Crops?

You have noticed that everything an Indian does is in a circle, and that is because the Power of the World always works in circles, and everything tries to be round . . . . The Sky is round, and I have heard that the earth is round like a ball, and so are all the stars. The wind, in its greatest power, whirls. Birds make their nest in circles, for theirs is the same religion as ours . . . .

Even the seasons form a great circle in their changing, and always come back again to where they were. The life of a man is a circle from childhood to childhood, and so it is in everything where power moves.

— Black Elk (1863–1950), Oglala Sioux holy man, in Black Elk Speaks, *by John G. Neihardt*

Coming full circle, seeking wisdom from our past carries power for our future.

Numbers can be cited about our staggering food heritage losses – the loss of fruits, vegetables, and animal breeds – but some may wonder why we should care. The United Nations Food and Agriculture Organization estimated in 2006 that over the previous five years the world lost one animal breed per month,[166] and about three-quarters of the original varieties of agricultural crops have been lost from farm fields since 1900.[167]

The UN's 2005 Millennium Ecosystem Assessment, initiated in 2001 and representing a consensus view of the largest body of social and natural scientists ever assembled to assess knowledge in this area, concluded that human actions often lead to irreversible losses in diversity of life on Earth, and these losses have been more rapid in the past 50 years than ever before in human history.[168] And five years later a 2010 UN-supported study reported that the world's pressures leading to biodiversity loss are increasing (resource consumption, invasive alien species, nitrogen pollution, overexploitation, and climate-change impacts) and that, while there are some local successes and increasing responses, the rate of biodiversity loss does not appear to be slowing.[169]

Still, the question remains: why should you and I care?

Cary Fowler, a self-proclaimed biodiversity warrior, says in his Technology Entertainment and Design (TED) lecture that his interest in crop diversity was sparked when he read scientist Jack Harlan's claim that crop diversity "stands between us and catastrophic starvation on a scale we cannot imagine."[170] Fowler soon found that Harlan wasn't a nut case, as he had feared, but the most respected scientist in the field.

Fowler learned from Harlan that crop diversity is the "biological foundation of agriculture, the raw material, the stuff of evolution in our agricultural crops."[171] And he learned that "a mass extinction was under way in our fields, and that this extinction was taking place with very few people noticing and even fewer caring."[172]

Biodiversity Enhances Resilience

The UN's Millennium Ecosystem Assessment explains that we should care because biodiversity contributes over the long term to human well-being in four ways:

- **PROVISIONING SERVICES** (food, fiber, fuel, genetic resources, biochemicals, fresh water)
- **CULTURAL SERVICES** (spiritual and religious values, knowledge system, education and inspiration, recreation and aesthetic value, sense of place)
- **SUPPORTING SERVICES** (primary production, provision of habitat, nutrient cycling, soil formation and retention, production of atmospheric oxygen, water cycling)
- **REGULATING SERVICES** (invasion resistance, herbivory, pollination, seed dispersal, climate regulation, pest and disease regulation, natural hazard protection, erosion regulation, water purification)

Translated into plain English, this report seems to be pulling out all stops to make us sit up and pay attention to the fact that biodiversity may hold the technical and cultural keys for resilience – resilience that will enable us to feed ourselves, to heal ourselves, and, most importantly, as our planet is increasingly resource-stressed, to survive.

At a practical and local level, the UN report suggests that biodiversity can offer the seed and animal diversity needed for times and areas of great drought, great flooding, and other extreme conditions. Biodiversity also contributes to options for medicines – these are among what the UN calls "biochemicals." Today not only do naturopathic, Chinese, and other indigenous medicines rely heavily on natural herbs and plants, but so do manufactured pharmaceutical medicines. The UN FAO reported in 1993 that in the United States, 25 percent of all prescriptions dispensed by pharmacies were substances extracted from plants, while another 13 percent came from microorganisms and 3 percent from animals. Worldwide, at least 119 important drugs in current use are derived from plant species.[173]

At its simplest level Cary Fowler suggests that the answer to the question of "Who cares?" is one word: *options*.

Biodiversity creates options for growing food that will sustain our race through whatever conditions may come, good or bad. Different breeds of animals and cultivars of plants have evolved through millennia to survive specific conditions of temperature, moisture, soils, and vegetation. Fowler explains that genetic diversity is valuable for breeding strains with particular traits. So even if a heritage apple cultivar doesn't have a particularly memorable taste, it can have other valuable traits such as resistance to a particular pest or disease.

Whatever cultural, climatic, and resource challenges lie ahead, biodiversity is a winning card for one simple reason: options increase our resilience.

Restoring Balance through Traditional Agriculture

North Leupp Family Farm

The North Leupp Family Farm in **northwest Arizona**, on the Navajo reservation west of Flagstaff, is a cooperative farm focused on reestablishing traditional Navajo agricultural and culinary traditions, with the goal of promoting food security, agricultural sustainability, self-sufficiency, and health for Navajo communities. Among other initiatives, the farm offers plots for community gardeners, educational programs, and community-building events.

The North Leupp Family Farm is one of many efforts throughout the nation working to restore American Indian farming traditions. A family-based agricultural co-op dating back to the 1980s, NLFF transitioned in 2006 to its current name and community farm status when 30 families signed on to farm plots of land ranging in size from 1/4 to 1 full acre. Situated in the flood plains of the Little Colorado River, the soil here is fertile and loamy, in contrast to the sandy soils that cover much of this region. Each family pays just a small annual fee (only $20 in 2009), which helps pay for the farm's irrigation. NLFF is, in a sense, a 100-acre farm version of a community garden.

Bringing People Back to the Land

At the North Leupp Family Farm, Stacey Jensen, the farm's manager in 2009 and a landscape architect by training, explains to me that it is important to keep the gardening project simple. "Don't get too complicated," he suggests. "It's mainly a matter of bringing the elders and young children back out to the fields. It's a very easy sell to the elders. With the children it's a little harder because a lot of them don't know what they're doing. You have to go and teach them step by step."

Jensen explains that the cooperative farm promotes growing your own food and knowing where your food comes from, and he believes it is making a difference in the community. He explains that the people who farm have different motivations. For some it's a hobby. For others, who may be closer to the edge financially and need the extra cash from selling crops, it is a matter of sustenance. Some are serious farmers, extending their season into winter, freezing large quantities of produce, and saving their seeds. The farm also includes a market garden, which sells produce at the local market to support the farm operations. The farm irrigates from a well fed by the Coconino Aquifer, pumping 200 gallons per minute and enabling them to irrigate entire fields through a network of drainage pipes that feed into surface drip-line irrigation down each row.[175]

"There's so much potential here," says Jensen. "We had several youth groups that came out this year, and a lot of them went away with the great idea of having a farm."

Each year the farm has workshops and a seasonal kickoff, to which the entire community is invited. Games and activities are mixed with teaching community members how to use the fields. Jensen remembers when he first came to the farm. "It was so desolate. I wondered what would grow. Time and effort have made it nice." As manager the only guidelines Jensen gives people are no pesticides and no wild, exotic seeds, to ensure that invasive species are not unwittingly introduced.

Restoring Health

"We've always wanted to do this," says an elderly Navajo woman, who is hoeing her wheat field when I

Connecting Culture to Land and Food Traditions

The NLFF mission is to "re-engage the local communities in time-honored farming practices and culinary traditions; and to establish healthy communities through healthy eating and dynamic lifestyles. In essence, to re-establish the symbiotic relationship with Mother Earth, that is so vital to the People's existence and survival."[174]

NLFF's goals are openly and naturally spiritual, giving vivid voice to the twin forces of despair and hope that have inspired its work. Elsewhere in our nation these same themes of despair and hope are echoed by many community food projects. Yet the history of oppression of indigenous peoples fuels this fire in the Navajo Nation with greater intensity and clarity. NLFF explains on its website:

> The Diné (the People) culture, traditionally self-sufficient, had been farmers, hunters and gatherers until the mid-1800s when the US government instituted its "scorched earth" campaign against the people to annihilate them into submission and eventual assimilation. The result was four years of forced internment at Fort Sumner in eastern New Mexico and eventual signing of the Treaty of 1868, which secured their release to return to their homelands. Without their farmlands and hunting grounds, the Diné were forced to rely on government food handouts to sustain them. Many of the children were placed in schools against their will, where they were malevolently indoctrinated into

a foreign religion and culture. The consequences of the extreme deviation from a traditional diet is diabetes and other diseases (circulatory ailments, stroke, kidney failure, obesity, etc.) associated with an apathetic lifestyle and an unhealthy dietary intake. It has been apparent for some time that when indigenous peoples stray away from their traditional diet to a "modern" diet, they become exposed to a range of degenerative diseases.

Navajo concerns about diet and connectivity to food and land will resonate with many if not all of the food projects mentioned in this book. But the stakes are even higher for the Navajo and other indigenous nations, who are gaining a sense of urgency about saving an entire culture whose very existence is threatened by disconnection between the people and their land and food.

There are lessons here for those who care about public health and the quality of community life, for the Navajo are able to draw a clear thread that inextricably links connectedness to land and food with our health and spiritual wholeness. Some might wish to dismiss this, saying the spiritual nature of these concerns is peculiar to indigenous cultures and not common in mainstream culture. Yet in all corners of the nation, whether expressed in completely secular ways or more openly in the spiritual domain, people are motivated to initiate or assist community food projects by a desire of the spirit to connect land and food and community.

visit in August 2009. She grew up with her parents and grandparents farming in New Mexico. Now at Leupp she and her husband are farming a 1-acre plot, usually working two or three hours in the morning and returning later in the day.

Nearby the woman's daughter is hoeing the squash field, while her husband hoes the corn. Before I leave she is eager to tell me something more. "We really enjoy this," she says. "We used to be on the heavy side, and last year we worked out here so much we started

losing weight. People were asking if we were sick!" Just by working in her community farm plot, and eating the food they grew, she lost 30 pounds in one season. And her daughter enjoys the farming so much that next year she will ask for her own ¼-acre plot.

Jensen suggests that farming is therapeutic for many of the farmers. "A lot of farmers come out here to get away from the hustle and bustle, the stresses. Many have had triple bypass surgery, some are diabetic." One woman was on insulin for the longest time, says Jensen, but by working at the farm and eating the food she was growing each day, she eventually got off the insulin.

When a community member loses weight and gets off insulin, word gets around. Jensen speculates that is why demand for farm plots has soared. He also thinks that's why children are starting to show interest in farming. "They used to all be into rock and roll, and mainstream America, but they're starting to come back out here and volunteer. In that sense [the farm] has had that ripple effect. I think [the farm] is also advancing traditional crops."

The farm encourages its farmers to use and save heirloom seeds, growing traditional squash that can't be found in their local supermarkets. But Jensen says that each year it's a bit of a struggle to obtain sources of traditional heirloom seed. "Eventually, I'd like to get us a seed bank so we really don't have to worry about it."

Jensen also thinks it's possible the community could become more self-sufficient through its community farm. "It's not too radical. That's within our grasp. What we need is more support, more resources. Farming on the reservation has always been a family affair. Families would always help other families. We're trying to bring that back. It's starting to come around again. I know when I first came here everyone kept to their own field. We're really seeing a change. I see good things out of this farm, a lot of potential."

A New Generation of Navajo Farmers
Navajo Nation Traditional Agricultural Outreach

The Navajo Nation Traditional Agricultural Outreach program, based in **Cameron, Arizona**, is intended to increase knowledge and practice of traditional dryland agriculture, boost food self-reliance, and address food-related health concerns in the Navajo Nation. Among other things, the program helps communities establish community gardens, arranges mentoring relationships for beginning ranchers, and sponsors educational workshops on all aspects of traditional and modern food systems.

As lead coordinator for the Navajo Nation Traditional Agricultural Outreach (NNTAO), a program of the nonprofit Developing Innovations In Navajo Education (DINÉ), Inc., Jamescita Peshlakai is leading an effort to restore traditional Navajo agriculture. The program, Peshlakai says, aims to increase self-sufficiency and combat the epidemics of obesity and diabetes on the reservation.

Peshlakai is a U.S. Army combat veteran of the Persian Gulf War and has seen much of the world. DINÉ, Inc., hired her in 2009, charging her with the daunting task of covering 10 million reservation acres. In just one year, Peshlakai and her staff of seven have successfully helped nine communities launch their own community gardens. The group has also initiated a training program for 50 beginning Navajo ranchers who will have four mentor farmers, in concert with other groups and agencies, to teach them traditional Navajo farming and ranching practices.

In addition to community gardens and mentoring relationships, a key program tool for effecting change has been community workshops. In its first year the project team hosted workshops on all parts of the food system with a specific focus on building skills in Navajo heritage food and agriculture — selecting and growing crops, soil fertility, crop insurance, traditional Navajo butchery, cooking a variety of traditional foods, canning, community gardens, gathering piñon nuts, and nutrition and health — as well as workshops devoted to teaching young people about their broader heritage.

And now Peshlakai is facilitating the formation of the Navajo Nation Food Policy Council, an independent ad hoc organization that will work to improve the Navajo food system. The effort is well under way; on August 26, 2010, the council held its foundational meeting and elected its officers.

These accomplishments are not coming easily. Peshlakai has learned a lot along the way, sometimes the hard way. "Our main challenges are two — water and motivation," she says. And from her stories it's hard to tell which might be the more difficult challenge. She describes two opposing yet simultaneous trends among the Navajo — one moving inward to keep mainstream culture out and another moving outward to adopt the American way of life.

"Over the past 100 to 200 years," says Peshlakai, "people have come to the reservation as curiosity seekers. People just walk into our homes without asking — to take pictures, to see how the Navajo live. There has been much cultural exploitation and curiosity, so many of our people have gotten tired of this exploitation. This is the reason why some tribal nations such as Taos Pueblo are now gated, and access is controlled."

At the same time, Peshlakai explains, chapter houses, the local governance entities, are reluctant to give permission for community gardens because they've become accustomed to a different way of life. "We're Americanized," she says. "People drive to stores for food and fast food and mail-order food. Now people are into the fast-food drive-through, the five-minute meal, doing as little as possible, while making money as easily as possible and creating time for pleasure."

Planting the Idea

Just as many other efforts throughout the United States have discovered that community gardens catalyze interest in building a local food system, Peshlakai has found that one way to rekindle an interest in heritage Navajo food and agriculture is through establishing community gardens. The difference is that Peshlakai and her team try to use every opportunity in the community gardening effort to educate and foster interest in Navajo heritage.

There isn't one single magic bullet that motivates people to come together to build a community garden, Peshlakai says. But the most important factor has been helping communities take ownership, so that the garden is truly their own, built for and by themselves. "If you go into a community and tell them, 'You're going to start a community garden,' nothing will happen," says Peshlakai.

So the first step, she says, is simply to plant the *idea* of a community garden. She will meet with the community, tell them about the program's goals and its work in other parts of the reservation, and how this has resulted in community gardens. Learning what is happening elsewhere inevitably sparks interest, she says. But at this point, Peshlakai says, people usually assume that she and her staff will not only build the garden for them but also maintain it. "They don't yet understand that it needs to be their garden."

The next step, says Peshlakai, is to facilitate the community toward taking ownership of the idea. If they want to have a garden, then it needs to be theirs, and they will need to start it. She says the way to facilitate this critical step is to articulate your role as a *supporting* one, and to ask questions about their vision — what their goal would be, what their garden would look like, what kinds of tools or support they would need. Slowly, as the idea of a community garden takes shape, Peshlakai says, it becomes the community's own.

To create interest and excitement around a community garden, Peshlakai says they have learned that first and foremost any garden-planning event should include food. A potluck is especially effective, as it offers

a way for people to show off their signature Navajo dishes. Events often gain traction by hosting special speakers, such as a worm farmer from Wisconsin to spark interest in learning how to compost. "People showed up just to see what a worm farmer *looks* like!" says Peshlakai. Or they might offer a potting workshop where children are able to take plants home for their backyards. Sometimes they record the events. "When people see themselves on YouTube, they feel like movie stars," says Peshlakai.

Peshlakai emphasizes the importance of working with community elders, who are often lonely because families are not able to visit often. "The elders want to share and are not fully appreciated for their knowledge," says Peshlakai. She suggests that senior centers are effective resources for starting a community garden.

An important part of restoring traditional agriculture is fostering interest in traditional foods. Cooking demonstrations with the elderly can be made into popular community social events by demonstrating heritage foods and making a big deal out of the fact that everyone of all ages should be able to do traditional home butchery and also make heritage Navajo foods, such as kneel-down bread, earth cake, salsa, and steamed-corn stew.

There is no one way that works for everyone, says Peshlakai. She recommends taking a multipronged approach to ensure that the community garden effort has something for everyone. Some people are most attracted to the idea of being a part of a team. Others, she says, are motivated by their natural altruism to do something to help their community.

All of Peshlakai's efforts are aimed at one goal: sustaining her rich cultural heritage. She recalls a quote from J. Cedric Woods of the Institute for New England Native American Studies at the University of Massachusetts in Boston: "We all have a shared communal responsibility to keep our cultural heritage going." Peshlakai is working hard to spread this word and to help others see the need to take responsibility for a successful tribal continuum. "We have obligations to our people to remember — and this brings the past forward into our future."

One Farmer's Work to Restore Traditional Farming

Water is perhaps the single most precious resource in the high desert plateau of the Southwest. This stark landscape, marked by washes, gullies, canyons, and mesas, offers little free moisture, with an average annual rainfall of less than 10 inches. Working under these harsh conditions, through the centuries Navajo and Hopi perfected farming techniques – dry farming, flood-field farming, and others – that enabled them to grow enough food for their own family and ceremonial needs.

Today, however, these tried and tested farming methods are at risk because of changing water patterns and water withdrawals. When I visit the reservation in August 2009, Jonathan Yazzie, a Navajo farmer near Tolani Lake, and one of DINÉ Inc.'s mentor farmers, tells me that changing weather patterns forced his family farm to close down. Without the heavy winter snows, he explains, the spring melt doesn't provide enough water down the wash to divert into the fields. The availability of groundwater is also changing, as major water pipelines extract 1.3 billion gallons of water annually to move coal slurry from Arizona all the way to Nevada.

Yazzie takes me to see his multigenerational family farm in Sand Springs, about 20 miles off-road, where his family once grew corn, melons, potatoes, tomatoes, chile peppers, cabbage, lettuce. "We even had grapes, peaches, and pears," says Yazzie. His father had 15 children to feed, and he was able to do this successfully with his farm and also by raising cattle, horses, goats, and sheep.

When Yazzie shows me his family farm, the fields are empty. The irrigation equipment is silent, and the black irrigation spaghetti lines snaking across the parched earth are dry. Looking down into the Dinnebito Wash, which serves the family farm, though trees dot the streambed, it is dry, with not a drop of water to pull up into the fields. While he was growing up, the wash always had water, Yazzie says. Always. Now the wash has been dry for at least three years.

"Pretty much everyone stopped farming because of water," says Yazzie. "We have all the equipment for farming, but there's just no water. Sometimes I shed tears. Too much emotion. Sometimes I just sit out here."

Today, living about an hour south in Tolani Lake with his wife and three sons, Yazzie is focusing on helping the community reclaim its farming tradition by gaining rights to use water for crop fields. "My main goal is to get people involved," he says. He wants to help people who used to farm return to farming. He ticks off the things that need to happen: get a farm committee restarted, get a water line, get the land back into farming. When he meets me in 2009, he has been working as a DINÉ Inc. mentor farmer and has already helped 30 different families start growing their own food again, some with backyard gardens, some with larger plots of land.

Yazzie is sure that one of the keys to ensuring a future for traditional Navajo farming is its youth. He develops questionnaires and conducts surveys in his community, which show that Navajo youth are interested. Yazzie has been invited to six or seven schools with greenhouses to help their growing programs. He says his own three sons like to farm, and so do their friends. Many youth even volunteer to help plant for the elderly. "I think if we could push that, they'll be more interested in farming, and it can go on."

To help his community reconnect with agriculture, Yazzie plans to conduct workshops on traditional activities, such as corn grinding and making traditional kneel-down bread, which is usually made in batches of 20 breads baked together in one fire pit.

"If there's no farming, there's nothing," he says. "That's the way I see it. You've got to have farming."

I am deeply moved by the desolation at Sand Springs, the emptiness, the loss it represents at so many levels. I wonder at the tenacity of people like Jonathan Yazzie, who continue to persevere in the face of such loss. And in the end I learn something important from Yazzie. He proves to me that success – with commitment and hard work – is possible. In 2010, nearly one year after my visit, he calls with good news: He now has water! The Dinnebito wash is still dry, but they have found a way to bring water to Sand Springs. And now his family's farm – the heritage that he inherited and can now teach his children – is back in full swing.

Learning and Teaching Heritage Farming
Rose Marie Williams, Navajo Farmer and Rancher

Through workshops and talks, Rose Marie Williams is an active force in her home of **Tuba City, Arizona**, for teaching others about Navajo heritage food and agriculture. She operates DanRose Farms with her husband and children and hopes that her farm will inspire others to support traditional Navajo agriculture.

In Tuba City, not far from the eastern reach of the Grand Canyon, Rose Marie Williams, a Navajo farmer and rancher, meets me for breakfast. She eats her bacon and eggs slowly as she describes her childhood, stories pouring forth. She and all her siblings had been daily field hands for her grandfather's farm. They planted seed, pulled weeds, chased away crows, and harvested corn, squash, melons, peppers, peaches, apricots, and more. Yes, she assures me, hundreds of peach trees — in the desert! She clipped sheep and milked cows. For a feisty young girl with ambitions, the work was hot and tedious. She hated it. One day, fed up with it all, she announced loudly to her grandfather that she would never — not in a *million years* — own one single cow. They were disgusting, she ranted. They smelled. She was going to leave the farm as soon as she could. You wouldn't catch her doing this dirty work!

After her grandfather died, the family became too busy to tend the farm. His beloved peach trees all died. His fields reverted to desert. Sitting across from me, Rose Marie tilts her head of long, wavy black hair and erupts into a big laugh. The joke is on her, and she revels in it. Now, with her husband and seven children, she works her grandfather's farm, where they grow white, blue, and yellow corn, watermelons, crosan melon, casaba melon, honeydew, and cushaw squash. On the ranchland to the north, she raises 45 Lincoln sheep and manages a herd of over two hundred Beefmaster and Black Angus cattle for the family. Now she can't get enough of farming. Now farming is her life.

After breakfast we drive north to her home, where she shows me pages and pages of detailed planting records carefully bound in a spiral notebook. She brings out tall jars of native corn seed she carefully selects and saves every year to begin the next year's crop. Then we drive south of town to walk her fields. We climb a rock outcropping from where we can view the length of the valley. She gestures north and south. When she was young, the entire high valley was one vast field of lush green. Every Navajo family cultivated their land and grew their own food. Now, in all directions, the valley is golden-red desert, and Rose Marie's fields of corn, melons, and squash — some 10 acres — seem just a small splotch of green. She points to an empty field where she will, one day, restore her grandfather's orchard. There are no signs it ever existed. A windstorm is sweeping up the plateau, driving the whirling sand at us in thick blankets, and we are beaten off our perch. In the distance her husband, Daniel, and two of her sons continue working, not deterred by the clouds of stinging sand.

The Art of Dryland Farming

Williams grew up learning from and listening to her father and grandfather, and she is still learning from and listening to her land. She learned dry farming, natural springwater farming, and flood-irrigation farming, three distinct traditional methods for farming in the desert, each appropriate for different conditions. She keeps meticulous records of everything, and she uses these records to make continual improvements. In 2009 she is actively farming about 10 acres of the farm, but she hopes to someday replant the 4 acres of orchards where her grandfather grew apples, pears, purple and green grapes, freestone and cling peaches, and apricots.

The work is hard and begins early. Come January, the family needs to start cleaning the fields and digging out the irrigation ditches to be ready for the one-time late-April irrigation. Once a year, with the spring melt, water flows down the wash and the local Hopi and Navajo farmers divert it to their fields in a tightly governed allocation and rotational system. Williams says the system can be difficult, and it requires being there when it's your turn, or you're out of luck.

Completely dependent on the wash, farmers here can irrigate their fields only once a year. The remainder of the year, Williams is grateful for whatever drops of rain Mother Nature brings to the desert, which is never much, as the average rainfall is less than 10 inches a year. Yet over the centuries the farmers here have successfully fed their families and communities by devising unusual farming methods and by careful selection of seeds suited to the climate.

Irrigating the fields is not as simple as it may sound. Done right, it is its own special art form. The fields are laid out in a kind of grid pattern, with each field edged by high berms that will contain the water. When irrigating, you must gauge how fast the water is running, how quickly it will flood the fields, and how quickly the water soaks in. Some of the fields take 4 hours to fill, some 6 or 10 hours, says Williams. "If we feel the water sank too fast, we'll rerun it," she says.

Ideally, the water should take one and a half to two days to soak in. "If I filled the field with water last night and this morning find the water's gone, I reirrigate right away," she explains. "I don't wait a week. I don't wait three days." The timing is important, and if you wait too long to reirrigate, you can ruin the field. "You really have to know what you're doing," she says. "You can't just go out here and plant."

The next step is to let the fields sit for at least three or four days. "The fields are so drenched, you can't even get in there for a couple of days," she says. She describes the perfect condition for the next phase of planting: the land curls up a little and gets white on top, and a tiny bit of grass will be starting. "That's when you plow, then disk." Sometimes, however, if they think it is going to be a particularly difficult year,

they may do their single plowing prior to irrigating, as it will allow the water to soak in deeper.

Planting is another part of the art. Each crop is different, and Williams pulls out her map to show me how she keeps track of everything. She is careful to rotate her crops, as her dad and grandfather did. Though they didn't keep a map, like she does, her father always told her to "tattoo to your brain," meaning remember this. Or "use your upstairs," meaning use your brain. Now she tells me of an older farmer she visited who complained of not getting any corn from his fields. She visited him to see what was wrong and discovered he had not been rotating his crops. "The corn is spongy," she told him, "because you have killed your land."

Williams does not use any commercial fertilizer (or Miracle-Gro, as someone once suggested). "I say, if my grandfather didn't do that, what makes you think I should be doing that?" She attends workshops, where she continues to learn about new methods of composting, and she reads as much as she can. In her own fields they finely chop the cornstalks and hand-plow them back into the soil, to feed the soil. She might also allow her cattle into the field, but never horse or sheep, as their manure is too acidic. (Though she does use the sheep manure for filling sinkholes, so long as it can be kept more than 18 inches underground and is covered by sand, so it won't affect the crops.)

To prevent the corn from blowing over in the intense desert winds and bursts of hard rain, they plant their corn unusually deep. In one hole four to six seeds of corn are planted at least 8 to 10 inches deep. Each clump is spaced 4 to 6 feet apart, creating an effect of tiny bouquets of corn speckling the landscape. If they doubt the seed's viability they might plant as many as 10 kernels, and if a particularly harsh summer is predicted, they might plant the seed as deep as 12 inches. Williams uses a hoe or shovel for the task, while her daughter has devised a T-shaped instrument that sinks into the ground.

The seed saving itself is yet another art form, passed down from generation to generation. To ensure continual renewal and adaptation to her climate, Williams mixes seed for planting on a large tarp, combining 5 gallons of last year's corn and 1 gallon from each previous year up to seven years back. It is *not* necessary to collect the best corn, she insists. In fact, her grandmother always said that seeds from the plant that struggled the hardest, the runt, will yield the most. So she looks for the runts, and she also has learned to seek out seeds from other dry farmers, to mix with her seed.

"If you use the same seed over and over, it will weaken the strain," she explains. "So you need to keep mixing the corn with strains from other dry areas. It can be next door, or a mile away, from whoever has corn that yielded well and was planted in dry farming." A few times she bought cheap corn seed from other areas, such as Shiprock and the San Juan River Valley, and none of it was worth it. The corn grew and died; three times she planted, with the same result each time. It was not suited for her climate, just as her corn seed would likely not be suitable for other regions. One year, at the edge of a nearby canyon where people would dump things, she found a jar of very old, black, shriveled corn seeds; she mixed these in with her planting seed, and that year her corn yielded beautiful huge corn kernels.

Williams is eager to share other lessons she's learned over the years. Never plant squash near melons, as the squash bugs move to the melons and steal their flavor. And the best way to eradicate grasshoppers is to place horse or cow manure in metal cans and burn it to smoke them out.

Still, all of these mechanics of planting don't begin to capture the real meaning of what Williams is doing, or why she loves to farm. For her people the very act of planting is sacred. Every time a seed is placed in the ground, it is a form of prayer. "Being out in the field, it seems like the holy people are with you," she says. "They tend to heal you. It's peace and quiet there. You are there to be holy and humble."

The Power of Agriculture

Through her farming and ranching, Williams is a leader in the Navajo grassroots food movement who not only provides her people with traditional foods but also teaches them how their forefathers farmed and encourages them to reclaim these traditions. As testimony to the growing community interest in traditional Navajo foods, her farm stand in Tuba City routinely sells out, and her sheep are usually completely presold for ceremonial purposes well in advance.

When I ask Williams for her advice to others, she is quick to respond. "I wish everybody knew how to live that way. Everything would be peaceful. It seems like now everything has been corrupted. If we could all do it, talk to the holy people, and plant and utilize our corn and melons as our food and for ceremonial purposes, I think everything would fall the right way. If we were to harmonize with the holy people, their spirits can lead us the right way. We wouldn't have hunger or food poisoning. The food is off the land — straight to you, not through any other process. If we were to make it that way, a lot of people would be stronger."

Her words are deep, and I realize they are intended not just for mainstream culture but also for her own people. She believes deeply in the power of traditional farming, which is at the core of Navajo culture, ceremonies, and beliefs. And she is working hard to teach others through workshops and consulting at chapter houses through the Navajo Nation, at Hopi and Navajo K–12 schools, and at an annual agricultural outreach conference for Native farmers and ranchers.

She is also spreading the word through her children. In the Navajo, land is owned by and passed down through the women. And two of her girls are eager to take over the farm when they're older, says Williams proudly. They see, from her example, that they can make good money at farming, as well as learn the knowledge of farming and ranching that will keep the family culture and tradition alive. "This is who we are," says Williams.

When I leave her I think of how Rose Marie Williams's experience is different from mainstream farming in so many ways. Yet too, she gives voice to familiar themes that echo through many of the projects I've visited — themes of harmony, peace, and healing.

When her children help plant the corn, Williams tells them, "You plant it, but make sure it's in the ground, or it will be crying." Surely others from all cultures would listen, understand, and nod in agreement.

Land, Corn, People: Preserving Cultural Identity

Carl Honeyestewa, Hopi Farmer

Outside the small Hopi village of **Moenkopi**, located on the eastern edge of **Tuba City, Arizona**, Carl Honeyestewa and his father, Luther, practice traditional Hopi farming of corn, melon, squash, and even some fruits such as grapes and apples. They are among the dwindling numbers of Hopi farmers and serve as a startling warning that food and agriculture heritage is a precarious asset that can be lost all too quickly. Without a grassroots food movement to reinvigorate and reclaim traditional agriculture, they say, the future of Hopi agriculture is threatened.

The word "culture" evokes a host of complex images to describe a particular cultural identity, from the simple physical manifestations of dress and language to the more abstract manifestations of beliefs, morals, law, customs, and religion. To use an old eighteenth-century word attributed to Jonathan Herder, the *Volksgeist* (folk or national spirit) of a people implies that there is more to a culture than its clearly defined, discrete parts. Something else is there, the indefinable *je ne sais quoi* that makes a culture distinct and unique. Historian Woodruff Smith describes this as "irreducible, irrational spiritual forces that lie close to the foundations of perception and behavior and explain why people of one nation must differ radically from those of another."[176] Culture, according to those who believe in the *Volksgeist* concept, is what confers identity, a sense of place, a foundation for knowing your place among your people and in the larger cosmos. In short, it imbues life with meaning.

Agriculture and food are so fundamental to our survival as a race that it is almost trite to say they are necessarily core elements of every distinct culture. For some cultures, however, agriculture and food are far more than a manifestation of a way of life: they are the life itself. In northwest Arizona, in the small village of Moenkopi, a Hopi outpost set in the midst of Navajo land, this is a concept that Carl Honeyestewa and his father and mother, Marie and Luther, try to help me understand.

Honeyestewa tells me that his mother is a nurse, retired after serving more than 30 years with the Indian Health Services, and his father is retired from the labor union. All the men in his family are proud marines, their photos on the wall in uniform, and both Carl and his father saw combat — Luther in Korea and Carl in Iraq. When Honeyestewa invites me to meet his parents, I find them eating a breakfast of eggs, bacon, toast, and orange juice. They are Americans, proud to serve, and they share the same concerns as Americans everywhere — jobs, taxes, politics, schools.

Yet everywhere, too, are signs that the Honeyestewas live and breathe another, far older culture imbued with different values and meaning. The kitchen windowsill holds large jars of sacred white corn (*ho-mah*) for ceremony and prayer, which Marie brings over to show me. I nod in admiration of their stash; this corn is highly valued and expensive, a sacred commodity. Carl Honeyestewa explains that corn is a part of their daily life. In the morning it is an offering that may be sprinkled on the ground for the morning sun to take. A prayer made at any time of the day is always accompanied by an offering of corn. Corn is also used in ceremonies, to feed the spirits, to give them sustenance.

Several miles to the east, in a long stretch of desert extending toward the mesa, the Honeyestewas have staked out a 10-acre plot of land and tend a large field of corn. This field is the epitome of true dry farming. With absolutely no source of water for irrigation, only the moisture in the soil and occasional summer rains, they are somehow coaxing the land to grow traditional Hopi corn. Planted deep in the sandy soil, the beautiful clumps of corn dot the landscape like individual green bouquets. Father and son work together, erecting chicken wire around the entire field to keep out rodents and squirrels. Their tracks are fresh, showing the damage they've done to the corn since the harsh windstorm of the day before.

Luther is a village elder, edging toward 80, but he continues to work hard in the fields. "It's the way of the Hopi," he says. "If you don't plant, you're no good."

"This year is a good crop," says Carl. "Last year I don't think we got any [corn] from up here. This low-lying area was all farmland, back when the old men were committed to farming. Back then you could live out here for a week just on what you could bring: dry corn, piki flat bread, watermelon. But now, we've got to have a car, we've got to have satellite, washer-dryer; you have to work and have a job."

Things are changing all around them — the culture, the climate, the crops. "Our growing season seems to get a little bit longer and longer," says Carl. "The seasons are messed up. It doesn't snow anymore. You know, we used to have fog, too — more moisture, any type of moisture would help. We've also seen a difference in wildlife. Lately, I've been told other farmers are having trouble with a black bird that has an orange spot. The black birds around here that I know of don't have orange spots. These have really, really hurt other farmers. I hunt quite a bit around here and back that way. I don't see the snakes anymore. I hunt only rabbits. We had a diet on jackrabbit. They're not out anymore. They're not alive."

Corn Is Life

Corn is at the center of Hopi life. "Corn is life," explains Marie. "Up here especially, us Hopi people, everything revolves around that. Weddings and dances, food, everything we use corn for. That's life for us."

Carl explains how the life of corn and his people are intertwined and metaphorically the same. "We consider the corn are our children. We plant them with love and care. Then urge them on to become fruitful. In each stage of a corn's life — I don't know all this but — they're just babies and considered to be getting older and older. Once they've fulfilled their role, they bend down in old age like these. When they're strong they're up. When they're older, older, they're on the ground — almost bent over like an old man. Then when they die the stalks are called *ka-tunqwa*, the shell of what was once there. That's what you also call your body when you die — it's just a shell. It's the same, the life, of man and corn. Your life is out of you, your spirit is gone, so what's left of you — *ka-tunqwa*."

Near the Honeyestewas' home is a building where they clean corn, hang it to dry, grind it into meal, store dried kernels in huge glass jars, and store yucca for weaving into baskets, and where Marie engages in the difficult art of making piki bread. The piki sandstone slab is sanded to a fine flat finish and is heated over the open fire pit. When it's hot, she spreads a thin blue corn batter over the greased griddle with the palm of her hand. It is tricky to get the batter just the right thickness, not too thick, not too thin, and all too easy to burn the hand. The batter cooks in no time and must be lifted quickly and rolled lengthwise like a jelly roll. Piki bread keeps for long periods of time and is a handy, dry lunch for men in the field.

All of this, say the Honeyestewas, is in peril. Their way of life in growing corn is in peril from changes in water, climate, wildlife. Along with the corn, the ceremonies. And along with the ceremonies, their identity and meaning as a people. Carl tells me that he is the last in his family who will farm and that other families are facing much the same fate. "I would hate to think [farming] would die out with my father, as old as he is. He's still strong. He keeps coming. I'd hate to see it die with him. I'd rather see it die with me than him." He's definitely found the positive: as long as Carl lives, he will keep the farming heritage alive.

His mother explains with quiet acceptance that they are living a prophecy she never thought would come true. "In Hopi it is the prophecy that everything we once had will go away. Great societies like Aztec, Mayans, they had their calendars, everything was set in motion. Now we're down to our last intricate society, the Kachina Society. Once we're done with that, we'll just be dancing there, tapping our feet with no words coming out of our mouths." Marie taps her fingers on the palm of her hand to illustrate. "Right now that is how some kids participate. They don't know the songs because they don't know Hopi words. They don't participate. They're just standing there — no words coming out of the mouths. That is prophecy."

She speaks as someone who has come to accept a hard truth, without anger or bitterness, but with matter-of-fact resignation. "We've forgotten our way, how we were taught, because it was a long time ago. In his time," she says, pointing to her husband, "and my time, they used to talk about these things. They all talk together way into the night. They say this will happen. You're going to disregard everything. You're going to go to White Man's way. You're going to forget your Indian way, and you will forget to pray. Then you'll be nothing. It won't rain anymore."

To prove her point her husband chimes in. "Yesterday, they had a snake dance. Years back when they had a snake dance and the home dance, it usually rained. Now it doesn't."

Marie nods in agreement. "Clouds don't come. You have to believe. You need heart."

The land and corn and people, together, form an inseparable, coherent whole. Without one or the other, the culture cannot persist. Marie Honeyestewa continues to share more of the prophecy. "They say that pretty soon we will not have any crops. It won't rain. Pretty soon the crops will be about this small"; she puts her hand out at about waist height. "And they will freeze. It will be cold. You can't plant when it's time to plant. You will be using your gloves and big coat to plant. That's when you'll starve. You won't have any corn. Nothing will grow. Either we'll have starvation, earthquakes, flooding."

As a child she never believed these prophecies and says she thought her uncle, who told her them, was lying. "Unbelievable, but I'm beginning to see the things that are happening. I never thought this would happen. I never thought the water in that one wash would dry. It's dry now. There's no water. There used to be lots of water."

So many people and food projects around the nation are focusing on the power of our youth — teaching, influencing, and inspiring them to understand their world in new ways by learning through gardens, nutrition, and food. Here the stakes are even higher. And there's so much more to teach — an entire culture's traditional knowledge about seeds, cultivation, traditional foods, prayers, ceremonies — in essence, the meaning of life within a cultural identity.

The ability to preserve, reclaim, and reinvigorate Hopi food and agricultural heritage may depend in large part on the Hopi community's ability to interest and inspire its youth. Carl says their schools have gardening programs with greenhouses, and the children are excited about doing their own planting, especially chile plants. At the end of the school year, children bring their plants home to their parents and grandparents to plant in their own gardens.

Carl knows that not everyone shares his love for farming, but he is passionate about what it means to him. He got hooked early. While his brothers were involved in sports, his father would bring him down to work on the farm. "I find peace there, I really do. Even though I'm alone there, I'm not figuratively alone. I'm with the plants that are there. It just seems they're there to provide me with the comfort I need that I won't get with my friends. I feel I can give them the service they need to sustain life."

This is the kind of passion that can help inspire a new generation of Hopi farmers. As Native American grassroots food movements take hold, it will be up to people like Carl, who are keepers of their tribe's agriculture traditions, to provide energy and strategies for preserving and reinvigorating food and agriculture traditions.

Marie is still hopeful. "If we all come together and do it — everybody — then maybe the good spirit will find a way. We're just renting his house — Earth."

Reviving Hope for Tradition and Heritage

Farmers of the Navajo and Hopi

In the high desert plateau of **northern Arizona**, the Navajo and Hopi people tended gardens and livestock for hundreds of years, using inventive dryland farming techniques. But over the past century, as bureaucratic policies have redistributed land and water, and as young people have left the reservation and Western culture moved in, the once thriving farmlands have dried up, and with them have gone not just age-old agricultural knowledge but an entire cultural tradition. Here, though, in a place where — perhaps as a microcosm of larger society — the growing disconnect between people and land threatens the basic foundations of community, we find oases of growth and hope.

One spark of the grassroots movement to reclaim tribal food and agricultural traditions is Navajo rancher Rose Marie Williams, who is working to resurrect her grandfather's farm, running cattle and sheep and tending fields of corn, melon, squash, and other crops, with the help of her husband and children. She is an outspoken advocate of traditional Navajo farming techniques, and she passes along her enthusiasm and knowledge at local schools, conferences, and community meetings.

Williams is not alone in her mission to preserve tribal knowledge and build community resilience on the reservation. Jonathan Yazzie, another Navajo farmer, whose family's long farming heritage goes back for generations, is now working with DINÉ, Inc., a nonprofit using education and community outreach to support Navajo communities and revive tribal heritage, including the techniques of dryland farming. Like Williams and Yazzie, Carl Honeyestewa, a retired U.S. marine and one of the dwindling number of Hopi farmers, practices traditional dryland farming methods, keeping his tribe's agricultural legacy alive in a time when the entire Hopi culture is endangered. And the North Leupp Family Farm, a Navajo community farming project not unlike many others around the country, works to rebuild connections between people and land, engaging communities in growing food to supplement the family table, finding respite in gardening, and rediscovering a sense of place.

As is the case in the wider world, the Navajo and Hopi are experiencing a loss of interest in farming among youths, coupled with loss of farmland, loss of farming knowledge, and loss of agriculture-based traditions. In both worlds, people are increasingly distanced — if not alienated — from a physical connection to the land and understanding of where food comes from. These pioneers of the Navajo and Hopi, like the leaders of grassroots food projects across the country, are leading the way in shoring up, rebuilding, and building anew the practices and systems that make up a strong, healthy local food culture.

Jonathan Yazzie sets up fencing in the arid landscape of his farm. When his father, a well-known medicine man, died a few years earlier, in the Navajo tradition Yazzie rested his father's fields — literally leaving plants, gear, and tools where they lay — for two years. His farm, in the back country of the Navajo Reservation, came back to life in the spring of 2011.

Rose Marie Williams (at right) visits with her aunt, Tsosie Hyden, in Pasture Canyon, an oasis-like wash several miles outside of Tuba City, where year-round springs flow from caves. Access is difficult, but Rose Marie and others are testing crops here, eager to take advantage of the fertile soil and natural irrigation.

Top: Rose Marie's husband, Dan, sells corn from their truck just off the main road into Tuba City. *Bottom:* As part its effort to support local farmers, DINÉ, Inc., helps with marketing and infrastructure. For DanRose Farms, that help included replacing small signs fixed to cardboard boxes (at left) with more visible, permanent billboards (at right).

Tyrone Thompson is the farm manager at the North Leupp Family Farm, a community program providing land, education, and basic infrastructure, like irrigation, for members who want to grow their own food.

The Summer Harvest Festival brings together people from within the Navajo communities, as well as from the outside. Most of the food comes from the farm members' plots, and the focus is on traditional offerings like kneel-down bread (at top), which is cooked throughout the day over hot coals buried in the ground.

Top: A photo on the wall of Carl Honeyestewa's home shows his great-grandfather Yukioma, one of the founding elders of Hotevilla, site of the famous Hopi terraced gardens. *Bottom left:* Carl and his father, Luther, tend several plots of land between Tuba City and Hotevilla, growing corn and a variety of melons. *Bottom right:* Traditional Hopi corn comes in diverse varieties and colors.

Top: In Moenkopi, just outside Tuba City, Carl points out the seasonal run of water through a nearby wash, which provides some irrigation. *Bottom left and right:* Further out from town, where there is no available ground water, the Honeyestewas use traditional dry-farming techniques, conserving water from the occasional rains in ways that will sustain the plants for the entire season.

Preserving Culture through Native Seeds
Mvskoke Food Sovereignty Initiative

with research by Jessica Ray

The Mvskoke Food Sovereignty Initiative is a grassroots Native American food move-ment centered in **Okmulgee, Oklahoma**, the Mvskoke (Creek) Nation's capital. It seeks to preserve Mvskoke food and agriculture traditions and enhance the health and well-being of its six tribes: the Muscogee, Seminole, Chickasaw, Choctaw, Cherokee, and Yuchi. It operates numerous programs — community gardening, cooking and garden-ing workshops, written manuals, a heritage seed bank, a farmers' market, and special events — all to enable Muscogee people and their neighbors to provide for their food and health needs now and in the future through sustainable agriculture, economic development, community involvement, and cultural and educational programs.

Just as community food programs are taking hold around the nation, Native American communities are taking the initiative to launch their own food projects to revive their agriculture and food traditions. And because Native American agriculture and food tend to be tightly interwoven with cultural identity, these projects are teaching much more.

In Okmulgee County, Oklahoma, the Mvskoke Food Sovereignty Initiative, founded in 2005, is seek-ing to preserve cultural traditions of the Muscogee, Seminole, Chickasaw, Choctaw, Cherokee, and Yuchi, for whom growing, preserving, and using traditional foods plays an important role in cultural activities. Starting with a community food assessment of 17 com-munities, MFSI learned that a major community inter-est and concern was in traditional foods. This led to one of their activities, sponsored by the Taos Economic Development Corporation: buying fishing licenses for a fishing expedition, followed by a fish fry. Since then MFSI has expanded to numerous activities and hands-on workshops, including installing gardens at people's homes to allow them to begin growing their own food.

One of MFSI's most difficult but important chal-lenges is to find, acquire, and begin saving traditional seeds, as traditional crops are a key part of their cultural heritage. Vicky Karhu, executive director of MFSI, herself a longtime organic vegetable gardener and commercial grower, describes how someone happened to find a native Indian pumpkin at a roadside stand, and starting with just that one, the project was able to grow eight hundred more pumpkins. It continues to save the seed from this rediscovered heirloom variety in a seed bank for preserving and restoring endangered seeds that are culturally linked to Native American gardens. Another success has been the project's ability to rescue from near extinction a particular corn known as *sofkee*, used by the Mvskoke in a traditional dish of thick cracked-corn soup, often served with bits of meat or fish.

Karhu also says that, with the help of the Jessie Smith Noyes Foundation, which supports social change movements around toxics and environmen-tal justice, reproductive rights, and sustainable agri-cultural and food systems, MFSI has connected to minority farmers who have helped mentor their efforts. To assist the regeneration of Native American farmers, the project is developing a resource manual for farmers and ranchers and has established a local outlet for their produce: the first local farmers' market in Okmulgee since the 1930s. MFSI says the market is providing fresh, affordable, locally produced fruits and vegetables to the community.[177]

Running to Raise Awareness

Youths are also getting involved in MFSI's efforts to raise awareness about the link between agriculture, food, health, and fitness literally by running. In an annual 420-mile relay, Muscogee National Walk/Run, youth cover the Muscogee (Creek) Nation boundaries, tracing the Opothle Yahola Trail (Trail of Tears). In other events the Muscogee youth participate in Spirit Walk/Runs and have joined with First Nations United, a nonprofit advocating prosperity and unity for First Nations people through redefining identity and connecting with the past to run another 290-mile relay along the Opothle Yahola Trail.

Just as runs are used throughout the United States to raise awareness about specific issues, running is a powerful way for Mvskoke youth to bring their communities together and to raise awareness about widespread diet-related diabetes, obesity, and addiction issues among Native peoples. Through running, youth have inspired other communities to participate, sometimes on the spur of the moment. For the Muscogee, Cassandra Thompson, staff for MFSI, writes that youth run to encourage Native Americans to reinvigorate their traditional agricultural practices, to provide for themselves locally grown fresh foods, and to live without diabetes and alcohol and drug addictions. To inspire their First Nation communities to remember an earlier time when they were healthier, physically fit, and growing and eating their traditional foods, the act of running the Opothle Yahola Trail connects youth with their heritage and builds understanding of how their historical relationship with the federal government continues to affect their community conditions and health today.[178]

Changing the Food System

In 2009 MFSI began to work on establishing a tribal food policy council to make changes at a more system-wide level. Karhu envisions that the council will help set policies to preserve heritage seeds and food, such as banning genetically modified crops, which could infect and destroy the tribe's heirloom seed diversity, and work to help more young people choose farming as a career.

One year later, on September 25, 2010, with a unanimous vote, the Muscogee Nation Food and Fitness Policy Council became the first tribal food policy council in the United States to be established by tribal law. The tribal resolution establishing the council affirms that the Muscogee were "traditionally excellent farmers, fishermen, and hunters and maintained a sustainable food system and were physically fit" and that the council is intended to address the "diet-related health problems and excessive dependency on outside sources of food."[179]

"It is unprecedented to have this level of collaboration between the tribal government and an independent community-based organization," says Karhu, noting that the resolution establishing the council gives the Mvskoke Food Sovereignty Initiative two designated seats on the council, along with seats for the Indian Health Service; Muscogee Nation Division of Health; legislative, executive, and judicial branches of government; farmers and ranchers; and other stakeholders. "This is big news and promises to be an avenue for real and lasting change to benefit all people living within the Muscogee (Creek) Nation," says Karhu.

The efforts of the council and the MFSI are about more than diet, says Rhonda Beaver, chief performance officer of the Muscogee Nation Division of Health. The programs that encourage people to revive heritage food gardening and cooking are also about "improving relationships with ourselves, and improving relationships with our food, so that we are overall healthier people."[180]

Native Foods Define a People
White Earth Land Recovery Project

with research by Regine Kennedy and Ben Chrisinger

The White Earth Land Recovery Project is a grassroots Native American food movement that seeks to protect and encourage the food and agriculture heritage of the Anishinabeg (Ojibwe and Chippewa). Located in **northern Minnesota**, west of the Chippewa National Forest, the WELRP's mission is to facilitate recovery of the original land base of the White Earth Indian Reservation while preserving and restoring traditional practices of sound land stewardship, language fluency, and community development and strengthening spiritual and cultural heritage. As part of its effort to protect the Anishinabeg culture, WELRP is protecting and reviving native seeds, heritage crops, naturally grown fruits, animals, and wild plants, as well as traditions and knowledge of indigenous and land-based communities.

In Minnesota the White Earth Land Recovery Project, founded in 1989 by Winona LaDuke, has gone to court to protect the agriculture and food heritage of the Anishinabeg, also known as the Ojibwe or Chippewa. Wild rice, or *manoomin*, is not only a key part of the ecosystem of the northern Minnesota lakes region and considered a grain unique to northern North America, it is central to Anishinabeg culture and tradition, through its migration stories and prophecies.[181] As a sacred food, "wild rice not only feeds the body, it feeds the soul," writes LaDuke.[182]

LaDuke suggests that the designation of wild rice as Minnesota's official state grain in 1977 eventually led, ironically, to its endangerment. By the early 1980s, she writes, the University of Minnesota's efforts to create domesticated versions of wild rice were so successful that domesticated rice grown in paddies had outstripped the native wild rice lake harvest. By the late 1980s more than 95 percent of the wild rice grown in the United States was domesticated; most was grown in California and harvested by machines — though marketed as authentic Minnesota lake wild rice, with pictures of Native Americans harvesting the rice in canoes with paddles.[183]

LaDuke writes that lake wild rice and paddy rice are completely different, as illustrated by "a favorite joke among most connoisseurs of real wild rice," which goes like this: "How to Cook Paddy Rice: Put rice and water in a pot with a stone. When the stone is soft the rice is almost done." By contrast, lake-harvested wild rice has an "amazing aroma" and cooks in only 20 to 30 minutes.

In 1988 the Ojibwe filed suit (*Wabizii v. Busch Agricultural Resources*) against the agricultural research and manufacturing company that was marketing the paddy rice, claiming false and misleading advertising. They eventually settled out of court with an agreement that the nonnative rice would have to be labeled "paddy rice," in lettering no less than half the size of the words "wild rice." LaDuke writes that this was a small but important victory on a "slippery slope in the age of globalization."[184]

With the turn of the millennium, the tribe viewed the advent of genetically engineered rice and University of Minnesota mapping of wild rice DNA as a major threat, because of the possibility of contamination of native Minnesotan wild rice stock. Again the Ojibwe sought legal protection, this time from the Minnesota state legislature, which in May 2007 passed a bill (H2410/S2096) that protected native lake-harvested wild rice with a two-year moratorium on genetic engineering of wild rice and also required the state to conduct a study of the environmental threats to natural wild rice stands and present recommendations to the legislature.[185] One year later the Minnesota Department of Natural Resources submitted its report, along with six major recommendations, including that the state "increase intensive natural wild rice lake management efforts and accelerate the restoration of wild rice stands within its historic range."[186]

Reclaiming Culture

The White Earth Land Recovery Project is deeply engaged in other ways of preserving Anishinabeg food heritage and culture. LaDuke says that Native American populations suffer from the highest rates of type 2 diabetes in the world, and 30 percent of Minnesotan Native American adults have been diagnosed with diabetes.[187] To combat this epidemic WELRP is providing food packages to Native American families with traditional foods such as buffalo meat, hominy corn, chokecherry or plum jelly (made with honey), maple syrup, and wild rice. The vision of WELRP is to reclaim the Anishinabeg land granted by an 1867 treaty but lost through the same process that caused many Native American and Hawaiian peoples to lose their land — a process of dividing the communal land into individual allotments for taxation and leading to forfeiture, without the people's understanding or consent.[188] Slowly, through purchase or donation of one parcel at a time, WELRP has reclaimed more than 1,000 acres of its tribal land. Much of this land is sugar bush, which enables WELRP to collect and make another one of its traditional foods, maple syrup. Sturgeon are one more important player in the native ecosystem, and LaDuke describes how WELRP has collaborated with fisheries biologists to release more than 50,000 sturgeon into reservation lakes and streams.

"They are returning to the White Earth Reservation," writes LaDuke. "As they return, they teach us a lesson of connectivity and our own relationships with each other. That lesson, we believe, is . . . that we can begin undoing some of what we have done to each other and with the realization that we are all ultimately connected."[189]

Coming full circle, seeking wisdom from our past carries power for our future.

Lessons Learned
Preserving Cultural Food Heritage

Cultural heritage is a powerful tool for positive change.

A community farm can be an important link in the chain of restoring community health. As the website for the North Leupp Family Farm in northwest Arizona (page 174) states, the farm is intended to "offer the people an opportunity to engage in agrarian activities that offer the benefits of a return to consumption of traditional local grown, nutrient-dense foods. The common sense approach to lifestyle changes that would help reverse the diet-related health epidemic trends are traditional foods, increased exercise, and avoidance of unhealthy foods and habits. The mainstream approach of nutrition education offered through federal and tribal health programs doesn't always work. What [is] needed [is] a more radical behavioral modification type approach through practical hands-on application that is consistent with indigenous people's commitment to helping one another. Regular farmers meetings and workshops offer information about the traditional agrarian culture, and sourcing traditional slow foods and recipes."[190]

In other words education alone does not work. This lesson is echoed by the national Farm to School program (page 146), which suggests that the best approach to teaching a new generation about better nutrition is multifaceted, hands-on, practical, and fun. An additional lesson we may learn from the Navajo projects, however, is that framing this education around cultural heritage can be a powerful motivator to help people begin to live a healthier lifestyle.

Document your cultural food and agricultural heritage.

The national organization Renewing America's Food Traditions (RAFT) has established good models for documenting heritage agriculture and foods. One way to collect this information is through community workshops where people are asked to answer several questions, such as:

- What foods have unique traditions in your landscapes, seascapes, and cultures?
- Which of them offer flavors, textures, and pleasures that are cherished in your food shed but can't be found anywhere else on the continent?
- How many of these foods—traditionally foraged, fished, hunted, or grown—might now be at risk in their home place?
- What can we do to ensure their survival and to support their original stewards in their struggles toward food sovereignty?

Another important aspect of documenting food heritage is to collect individual stories about the meaning and context of the food, to keep these traditions alive for future generations.

While much information about your agriculture heritage may be elicited by the questions above, it may also be important to document agriculture traditions and methods specific to your region or people. Rose Marie Williams, a Navajo farmer and rancher in Tuba City, Arizona (page 180), knows how to save her seed and plant according to the traditions she learned from her father and grandfather – looking for runt seed, how to mix seed from previous years, how to know when the soil is right for plowing, how to know how much to irrigate, how to know how deep to plant the seed, and much more. This kind of traditional wisdom is central to keeping a specific heritage agriculture alive and useful for future generations.

Establish a seed-saving bank to preserve agricultural heritage.

Collecting and saving the seed of heritage fruits, vegetables, nuts, and herbs allows the continuation of agricultural and food traditions that are inextricably linked to particular cultures or regions. The practice also preserves for future generations an important source of genetic information about crops that thrived in the local soils and climate conditions. Possible spin-offs of this effort could include the development of heritage food enterprises that also support the local economy.

Some regions already have seed-saving efforts, such as Southern Seed Exposure Exchange in Virginia, which specializes in saving heirloom seeds suitable for the Mid-Atlantic and Southeast. If you're interested in starting a seed bank, you might begin your efforts by contacting the national nonprofit Seed Savers Exchange to find out who in your area may already be saving seed and to learn how you can support that local effort or, if there is none, how you can begin your own. You might also contact your local Cooperative Extension Service to see if they are aware of a seed-saving effort in your area and to enlist their support.

Food heritage can be protected by legal and/or legislative action.

The White Earth Land Recovery Project in northern Minnesota (page 196) offers lessons from its experience with native wild rice – unique to the northern lakes of North America and a food central to the Anishinabeg past and future survival. When a native food that is central to a culture is threatened, and when private negotiations and collaboration fail, more formal action may be needed, such as legal or legislative action. The tribe first filed lawsuit in 1988 to prevent misrepresentation of domesticated paddy-grown wild rice as the real lake-sharvested wild rice and won a small but important victory. In 2007 the tribe pushed for a state moratorium on genetically engineered rice, which led to a state study that recommended native wild rice production be protected and expanded. To determine what legal precedents may already exist in different states, or to seek assistance in thinking through viable options, consider contacting organizations such as Genetic Policy Alliance, which includes representation from the agriculture, consumer, health, faith, labor, environmental, social justice, and business sectors.

7: Sustainability
Food for the Long Term

The degree to which a system is sustainable over time depends on multiple elements and factors, the pushes and pulls within the system as well as outside forces, both the known and the entirely unpredictable, that will influence the system. A system that depends on a nonrenewable material to produce a valuable widget, for example, needs to figure out how it will continue to operate after it has depleted that material. It can find a substitute material for the same process, invent a new process altogether, or transition to another paradigm where that widget is simply no longer needed.

Our food system is facing these difficult choices. Through the nineteenth and twentieth centuries, the Industrial Revolution and its younger sister, the Green Revolution, constructed a food system that became increasingly dependent on nonrenewable energy resources; namely, petroleum and natural gas. This dependence is now completely integrated into all parts of the food system — in fertilizers and pesticides; in machines that till, drill, harvest, bale, process, flash-freeze; in transport of live animals halfway across the country to central megaprocessing plants; in transport of foods halfway around the world; in refrigeration and packaging. This dependence is so great that some calculate that it now takes an average of 10 petroleum calories to produce just 1 food calorie (by comparison, in 1940, 1 petroleum calorie produced 2.3 food calories).[191]

The motivations behind the grassroots food movement are many, as the stories in this book prove, and not least among them is the desire to transition to a food system that is less dependent on, or perhaps even completely independent of, our waning and increasingly costly and precious petroleum and natural gas reserves. A complementary motivation is to transition to a food system that is more sustainable environmentally, one that doesn't degrade our soils, water, and air.

To make this transition is no simple matter. Though our food system is so enormously complex that only the uninformed pretend to understand it in its entirety, these complexities melt away in the face of some very simple facts. A sustainable food system that is less (or not) dependent on nonrenewable energy and that supports healthy soils, waters, air, and ecosystems, requires one additional element: farmers who can stay in business.

These factors can be discussed separately, but in the end, they are inextricably linked. A farmer aiming for environmental and economic sustainability must work in a completely different paradigm. Here, the farmer builds and sustains soil health with methods that require less or no reliance on oil and natural gas to produce reliable yields of good food. And this farmer must also sell this food profitably to keep the farm in business. Without a good income to stay in business, there is no farmer.

This chapter takes a look at innovative ways that the grassroots food movement is exploring and creating a new paradigm of sustainability, including ways to sustain the farm. As you'll see, some established farmers are working with innovative models for supporting new farmers in ways that don't require them to assume mountains of debt, as well as creative approaches to making the farm a successful and sustainable business. Others are showing that it is possible to farm profitably and successfully on renewable energy and that farming can actively regenerate and heal the environment. Fisheries, too, are proving that they can be sustainable, and the grassroots food movement is contributing to new ways of thinking about our waters and oceans. While these are still early reports from the grassroots food movement, the goal is clear: connecting with neighbors in new ways and shaping a community that is more environmentally and economically sustainable.

The Green Revolution: Billions of Dollars in Hidden Costs

In some ways the so-called Green Revolution, which brought hope to so many in the last half of the twentieth century, was just one more step in humanity's continuing march toward ever greater efficiency and productivity. Pesticides, herbicides, fertilizers, and hybrids were apparent answers to the world's prayers for more food to feed its growing, and often starving, numbers. Finally, we could beat back the mosquitoes that spread malaria, the potato bugs, the cotton boll weevils. We could grow good, clean food in greater quantities, faster, better.

With the exception of a few skeptics who refused to adopt the wonder seeds, drugs, and machines that were quite literally transforming the shape and quality of our farm landscape, most farmers were pressed into this technological revolution. It was scientific. It was proven to work. We could count on it. The agricultural treadmill knew only one speed: faster.

Undoubtedly, the many farm families who started down this path believed they were doing the right thing, the only sensible thing, for their land, their animals, their community. If the ghost of Agricultural Future could have materialized to show the agronomists, agriculture and animal scientists, and farmers what future impacts their choices might have, we suspect many would have been puzzled, if not frightened. It was unthinkable to many that science might lead us astray. If we can't trust our farmland doctors, who can we trust?

On the plus side the Green Revolution did make certain foods a lot cheaper and more abundant. Farmers also learned to become more scientific in their approach to farming. With the advent of soil testing and regulatory programs adopted in the late 1990s and early 2000s to set maximum application rates for nitrogen and phosphorus, specific to each field and crop, farmers today are using an amazing 40 to 50 percent *less* fertilizer than they did during the pre–Green Revolution days and are achieving the same high yields. And too, with GPS satellite equipment, farmers can seed, fertilize, and spray with unprecedented precision.

Yet the consequences of our Green Revolution are still in the process of revealing themselves – and unfortunately, they are still accumulating. The new systems designed to cure such environmental ills as plagues of pests and diseases eventually caused new, largely unforeseen environmental ills of their own – soil erosion, water pollution, an annual dead zone in the Gulf of Mexico, increased insect resistance to pesticides, increased bacterial resistance to antibiotics, and the loss of wildlife and habitat. The Union of Concerned Scientists reported in 2008 that "many of the costs of industrial agriculture have been hidden and ignored in short-term calculations of profit and productivity, as practices have been developed with a narrow focus on increased production."[192]

But that's not all. The Centers for Disease Control points out that "American society has become 'obesogenic,' characterized by environments that promote increased food intake, nonhealthful foods, and physical inactivity."[193] While the factors behind America's newly obesogenic lifestyle are complex, our intensified, technological farming system is most certainly one of them. The concept, admittedly oversimplified here but carefully laid out by others, such as Michael Pollan, goes something like this: The federal system of incentives and supports leads farmers to grow and sell huge quantities of corn and mass-produceed commodities at artificially low prices. Food-manufacturing corporations can make huge quantities of highly processed food from these taxpayer-subsidized commodities and sell them for artificially low prices.[194] In turn, consumers can buy two to four times as many calories for their dollar in unhealthy, energy-dense, and nutrient-poor subsidized foods – chips, sodas, cookies, crackers – as they can in healthy, nutrient-dense, unsubsidized fruits and vegetables.[195] "Given that," writes Bryan Walsh for *Time,* "it's no surprise we're so fat: it simply costs too much to be thin."[196] It's also no surprise that farmers have gotten bigger or gotten out. America's unwritten commitment to a "National Cheap Food Policy" ironically imposes a cost/price squeeze on food producers that has forced out thousands of small farmers.

When we include the related costs of environmental degradation, the obesity epidemic, the foodborne illnesses, the cancers and neurodegenerative disorders linked to pesticide exposure, and other by-products of industrial agriculture, the real costs of industrial food soar. Consider obesity alone: the Surgeon General's "call to action" as long ago as 2001 reported that the public health costs of obesity stood at $117 billion per year – more than the $108 billion per year spent on the war in Iraq.[197] Now add the cost of cleaning our 40,000 nutrient-polluted waterways. Just to create the individual cleanup *plans* – not including the actual cleanup itself – the EPA in 2001 estimated the cost at between $900 million and $4.3 billion – *annually* – for 15 years.[198]

Today these costs are externalized to society. These costs are not paid for directly by those who have created the environmental and health impacts, and therefore are not included in the price of our food. Instead, these costs are paid indirectly by each of us, through increasing health-care costs, days lost on the job due to illness, state and federal cleanup of our waterways, and taxpayer-funded cost-share programs to encourage farmers to adopt more sustainable practices, such as planting trees along streams, fencing cows out of the streams, no-till farming, and integrated pest management practices. Some costs, of course, such as permanent loss of habitat, may never be paid.

Is the only solution a full-scale return to small family farms selling to their local communities? Probably not, for a host of reasons, not the least of which are the historical difficulty in changing large, complex systems quickly; the political clout of conventional agriculture; and the economic difficulty of large operations scaling down.

Perhaps the wisest solution is a diversified approach incorporating farms of varied sizes united by a universal commitment to employ the most environmentally sustainable production techniques for environmental health, produce the most nutrient-dense food for human health, and support healthy regional economies that in turn support farmers and their families by passing on to consumers the full long-term costs of the food we eat.

Perhaps it was best said by Sir James Michael Goldsmith, who warned in 1994 on the *Charlie Rose* show, "We have to change priorities. . . . Instead of just trying to produce the maximum amount for the cheapest direct cost, let us try to take into account the other costs. Our purpose should not be just the one-dimensional cost of producing food. We want the right amount of food, of the right quality for health, of right quality for the environment and employing enough people so as to maintain social stability in the rural areas. . . . We are worshiping the wrong God: economic index. . . . The economy, like everything else, is a tool which should be . . . subject to the true and fundamental requirements of society: [serving the people]."[199]

– coauthored by Michael J. Ellerbrock

Permaculture: Regenerating Place
Radical Roots, Three Sisters, and Innisfree Village Farms

by Christine Muehlman Gyovai

mid-atlantic

Permaculture, or permanent culture, is a system of ecological design that creates more sustainable (and regenerative) homes, neighborhoods, and communities. Through understanding patterns in nature, permaculture practitioners learn how to grow food, manage water catchment and storage, utilize renewable energy, and build community. Three farms on the front edge of the wave of regenerative farming, demonstrating innovative permaculture design strategies, are Radical Roots Farm, Three Sisters Farm, and Innisfree Village Farm.

Farming may soon look significantly different from how it looks today. With the increased pressures of peak oil, peak water, loss of topsoil and arable land, and increasing population, farmers will have many challenges to face in the future. Several farms are pioneering ways to regenerate the very soil under their feet, building the health of the land and the health of the people they support through innovative permaculture design strategies.

Permaculture, or permanent culture, is a system of ecological design that focuses not only on sustainability but also on regeneration. Developed in the late 1970s in Australia by Bill Mollison and David Holmgren, permaculture is based on several disciplines, including ecology, sustainable agriculture, natural building, and community building. At its core are three permaculture ethics: earth care, people care, and fair share. Out of these ethics emerge the principles that guide permaculture design. Among others, these include:

- **DIVERSITY.** Polycultures (systems composed of diverse elements) are nature's most productive and efficient systems. For example, a farm might include bees, chickens, an orchard, kitchen gardens, a wildflower garden, and a pond to create a dynamic edible landscape.

- **MATERIALS CYCLING.** There is no waste or pollution in the system — the output from one natural process is utilized by another natural process (there is no *garbage* in nature). Local resources are used and reused as often as possible. For example, all organic materials are composted to return their energy and nutrients to the farm soil.

- **ENERGY CONSERVATION.** System components are sited and designed to minimize their use of energy (both human and fuel). From a functional perspective, this means things that are used together are placed together. For example, the composting area would be located near the primary source of its materials (kitchen or barn) and close to (but uphill from) the gardens.

- **REDUNDANCY.** Every function of the system can be performed by more than one element. For example, water on a farm may be provided by a well, a rain barrel, a cistern, and a pond.

- **STACKED FUNCTIONS.** Every element in the system has many uses and functions. For example, nitrogen-fixing plants in raised beds not only improve the fertility of the soil but also help hold the soil in place; when chopped down in place, they serve as mulch; and as they break down they contribute to the soil tilth.

Permaculture design strategies may be applied in both rural and urban settings at the home, neighborhood, community, region, or national scale.

Radical Roots: Establishing Sustainable Systems

Radical Roots is a 5-acre farm located in Virginia's Shenandoah Valley. Farmers Dave and Lee O'Neill have been farming at Radical Roots for seven years and currently sell their produce through a one-hundred-member CSA, two farmers' markets, and restaurant and wholesale outlets. When the O'Neills moved to their farm in 2003, it was one large hayfield. Their first step was to create a permaculture design for the entire farm, complete with clearly articulated short- and long-term goals. As a whole, according to Lee, they try not to be overwhelmed by their goals and move forward one calculated step at a time.

One significant improvement the O'Neills made was to install a mile of swales along the west and south sides of their land. Swales, an iconic element of permaculture-oriented agriculture, are drainage ditches dug on contour with the land to capture rainwater, allowing it to percolate slowly into the soil. They created berms — mounded soil beds — below the swales, planting them with different perennial plants, including fruit trees and nitrogen-fixing plants and shrubs such as goumis, rhubarb, and raspberries. They mulched heavily and installed drip irrigation for some of the plants, but for the most part the swales are extremely effective at capturing rainwater for the plants.

As part of their permaculture design, the O'Neills have been experimenting with no-till farming — instead of tilling under the previous year's crop residues and turning the soil, they simply seed crops through the crop residues. They have dedicated approximately 3/4 acre to permanent, raised, no-till beds. Now three years old, these beds are producing some of the farm's best yields. The O'Neills plant these beds with cover crops such as winter rye or vetch and allow them to winter-kill. They then plant directly into the cover crop, or they cut the cover crop down and sheet-mulch over it with layers of cardboard, compost, newspaper, and mulch. To protect the beds they are careful to not

leave any soil bare, even planting parsley on the edges of some of the raised beds.

In the future, Dave and Lee hope to make the farm more sustainable. Some of their next steps include building a water cistern and installing solar electric and hot-water systems. They are working to put in more backup systems, such as devising a way to water the greenhouse when the power goes out. Working less is also a major goal, as is trying to have more family time. Regarding their next steps, Lee says they have reached a point where they want more time to enjoy what they have created, while still refining and tweaking their existing systems and practices.

When I ask Lee what lessons she and Dave would pass on to other farmers, she says that building a community of peer farmers is key. They have developed a support network of other farmers they turn to when questions arise. "Learn from other farmers, and develop a community of resources. Go to other farms, and learn what they are doing. Learning one thing, like a different way to move your chickens, can save you hours, and then you can spend time doing what you really want to be doing," Lee says.

Three Sisters: Regenerating Biodiversity

Three Sisters Farm, located in Mercer County in western Pennsylvania, is an example of an older, more mature permaculture system. With 5 acres of gardens, the farm was started in 1989 with permaculture as its core design principle. "Part of the goal of the farm is an experiment and living laboratory to practice permaculture. We met Bill Mollison in 1981 and were inspired to develop a farm that was a model of permaculture design and to be a teaching center for the region," says farm owner Darrell Frey.

Three Sisters Farm is working to steward the land with the goal of increasing biodiversity, primarily in four ways: First, expand the existing tree and shrub line. Second, expand habitat for beneficial birds and insects with species such as black locust, honey locust, highbush cranberry, mulberry, anise hyssop, and native wildflowers. Third, add ponds. Finally, manage the landscape consciously by paying attention to patterns and minimizing disturbance of animal habitat in farming practices.

One key permaculture concept is that diversity builds diversity, and indeed, Frey says it is amazing how quickly biodiversity returns to the landscape as they establish more perennials, trees, and shrubs. In his book *Bioshelter Market Garden: A Permaculture Farm*, Frey writes:

> The best way to describe the changes permaculture development has brought to our farm is to compare it to what came before. A small field that was previously plowed each year and planted in corn and soybeans now contains many dozens of species of plants.
> It is home to a thriving natural diversity of animal life. Bats, moles, mice voles, and weasels live between our gardens and along the edges. Swallows, orioles, bluebirds, catbirds, wrens, redwing blackbirds and many other songbirds nest here. Herons, ducks, geese, owls, and hawks forage our pond and fields. Turtles, snakes, frogs, toads, spiders, bees, wasps, and other native creatures live here as well. Wildflowers and new insects constantly amaze us as they join our evolving system. Together the gardeners, gardens and nature manifest the essence that is greater than the sum of its parts.[200]

Three Sisters has been selling to some of the same customers, mainly restaurants, for 20 years, and most of the produce is preordered. They rarely harvest anything that is not ordered. The farm's primary staple crop has been a bioregional salad mix, which includes several self-seeding wild edibles such as dandelion and chickweed in addition to other greens such as chard, spinach, and Asian greens. As their nut and fruit trees and shrubs and perennial plants mature, Three Sisters Farm will offer more fruits and berries to their customers. Other crops include lettuces and greens, root and other vegetables, herbs, and edible and cut flowers. Chickens are an integral component of the farming operations as well.

Frey and his companion farmers at Three Sisters Farm have harvested many lessons over their years of farming. Frey says they could do much more to fully realize their permaculture design — they have space for a thousand meat chickens a year, or they could have dairy goats or value-added products such as herbal vinegars and oils. However, they are intent on not becoming significantly overwhelmed with activities. At the same time Frey and his colleagues are working to develop a teaching center and on-farm education program, and they are considering expanding the farm's internship program. "We are looking at more cottage gardens and integrating local food systems into existing gardens, customer needs and demands, and community infrastructure," says Frey, as he explains future plans for Three Sisters Farm and for helping to build the regional permaculture movement. He adds, "The food future has to be delicious and bountiful and beautiful."

Innisfree Village: Regenerating the Land

Innisfree Village, located near Crozet, Virginia, at the base of the Blue Ridge Mountains, is a residential, life-sharing community for adults who have an intellectual disability. Innisfree encompasses 550 acres, with about 280 acres of open fields, and supports a CSA program with vegetables, herbs, flowers, eggs, and fruit, as well as a wood shop, a weavery, and a bakery. Beef is raised on the larger pasture areas, and a large-scale land restoration project using permaculture techinques is currently under way.

Peter and Debra Traverse, who both grew up on family farms in New England, manage the Innisfree Village Farm, which has embarked upon a fascinating and ambitious experiment to regenerate land that lost fertility over decades of continuous grazing. Their project, initiated in 2010 and dubbed the Innisfree Village Farm Environmental Services Project, aims to explore the possibilities of developing a farm business model that uses land husbandry to provide environmental services, all while sequestering carbon in the soil. (Carbon sequestration, considered vital for mitigating climate change, is the process by which carbon is removed from the atmosphere and stored in natural "reservoirs," such as forests, soils, subsurface aquifers, or even the ocean.)

One of the project goals is to gauge how the degraded land responds to different land-management practices. "Initially, we are trying to document rapid regeneration of land," says Peter Traverse. "We don't want to focus simply on land conservation. We feel strongly that land can be restored and regenerated under certain practices. Our projects at Innisfree are about exploration and observation of resulting changes. Exploration leads to new insights and provides new opportunities for scientific research. We are going to innovate like crazy and do our best to measure what happens."

The project's basic premise is that the average farmer (or any land manager, for that matter) can use many of the farming tools and techniques developed over the past century to address degradation of soil. However, there is a twofold challenge for achieving landscape regeneration at a larger watershed scale. The first challenge is social: how is innovation for landscape regeneration communicated to and adopted by people throughout a watershed? The second challenge is practical: how does innovation for landscape regeneration find time, scale, and traction in communities? These questions are guideposts as the project develops.

Traverse says that other key questions include the following: How does innovation get rewarded so it results in problem solving? How do we keep farmland productive and economically viable for farmers while improving the environmental quality of our watersheds?

With regard to the project's goal of sequestering carbon in the soil, increased carbon in the soil translates to increased soil porosity, increased water infiltration, and a greater capacity for the land to capture and hold water. Carbon can be used as an indicator of soil health and is measured in both the short term and long term, according to Traverse. However, there is no agreed-upon method for measuring organic carbon in the soil. This challenge will be addressed over time as the Innisfree project develops.

With various consultants, including a microbiologist and two environmental consulting firms, Innisfree Village Farm is working to gather the best data possible, through 3-D soil mapping and other testing and data-gathering tools, to measure the effect of selected landscape regeneration techniques. These run the gamut from keyline subsoiling (tilling the soil without turning it to facilitate the infiltration of oxygen and water, which encourages the formation of topsoil and controls water runoff and soil erosion) to "cocktail" seeding with leguminous and other deep-rooted plants; inoculating the root-zone layer with mycorrhizal fungi; applying to the soil compost tea, a remineralizing formula, and/or raw milk (which has significant beneficial microbial activity); and employing holistic management methods (intensive rotations of grazing livestock).

Peter Traverse says that the future plans of Innisfree Farm are to "continue the marriage of exploration in land regeneration and research. We hope to have more and more data that encourages even more exploration and, eventually, a market that makes investing in land regeneration practical." He adds that their management goals will be a balance between maximizing production of grass-fed beef and increasing soil fertility. Whatever the balance Innisfree Farm does strike, it will be leading the way for others to learn about landscape restoration for years to come.

Energy for Farm, Family, and Community
Seeds of Solidarity

with research by Coogan Brennan

Seeds of Solidarity, in **western Massachusetts**, has two arms: a for-profit farm and a nonprofit educational organization. Both are committed to environmental health and renewable energy, sharing a mission to "bring to life the connection between environmental and social justice issues through practice and education."

The connection between food and fuel is not always apparent. We can see, smell, and taste our food, whereas the energy needed to cultivate, package, process, transport, and store the food is usually out of sight and therefore out of mind. But at the 2009 Bioneers annual conference, guest speaker Michael Pollan rendered the invisible visible. He asked the audience, "How much oil are we eating?" Then, next to a double quarter-pounder burger with cheese, he poured out the amount of oil needed to create the burger: a full 26 ounces.[201] In his 2008 letter to President-elect Obama, Pollan explained that "chemical fertilizers (made from natural gas), pesticides (made from petroleum), farm machinery, modern food processing and packaging and transportation have together transformed a system that in 1940 produced 2.3 calories of food energy for every calorie of fossil-fuel energy it used into one that now takes 10 calories of fossil-fuel energy to produce a single calorie of modern supermarket food. Put another way, when we eat from the industrial-food system, we are eating oil and spewing greenhouse gases."[202]

At Seeds of Solidarity, a 30-acre farm in the small town of Orange, Massachusetts, near the Quabbin Reservoir, owners Deborah Habib and Ricky Baruc demonstrate how to grow food that is *not* reliant on fossil fuels. In addition to their family farm, they run a nonprofit organization, Seeds of Solidarity Education Center, through which they run Seeds of Leadership (SOL) Garden, a program that inspires activism and leadership among low-income teenagers as they cultivate food. They also work with schools locally and regionally to implement school gardens and related curricula to promote health and connection to food,

especially in low-income communities. And they also host educational workshops and retreats for adults and youth. The nonprofit is supported through grants and contributions and is fiscally distinct from the farm, though many of its educational programs are held at the farm.

Farming Off the Grid

The land that Habib and Baruc bought to start a farm was, at best, marginal, with almost no topsoil in the pockets of open space among woodlands and wetlands. But this was the best land they could afford. They began building the farm, one piece at a time. It now successfully operates completely off the grid and provides them with both food and income. While living off the grid might seem a difficult proposition, Habib believes that it is within everybody's reach. People worry that going off the grid means their quality of life will suffer, she says, but she attributes this attitude to lack of information or even misinformation.

Habib and Baruc's farm demonstrates that living on renewable energy does not involve sacrifice. Their farm is run on solar power, and like most families in the United States, they enjoy lights, running water (from a well), a refrigerator, computers, power tools, a washing machine, and more. People often don't try something new because they think they need to know it all themselves to get it right, says Habib. "But we don't ask these same questions about fossil fuel. There's a trust we put into the system because we've been so used to it." So why, she wonders, is solar energy held to a different standard? "We don't have to be solar experts," she insists. "We can rely on friends and experts!"

Seeds of Solidarity operates with three solar electric systems: one for their house and the greenhouse water irrigation system; another for the apprentice housing building; and a third for the battery bank in their "SOL Patrol" van, a mobile source of electricity for irrigation pumps, power tools, and even the fan that dries the garlic for the farm's annual garlic festival.

Habib says the farm's independent solar system is more secure than a connection to the nation's electrical grid. Over the years only once has lightning struck, taking out their inverter and leaving them without electricity for a day or two. That one time they felt truly on their own, she says. But then Habib points to ice storms that have taken out all their neighbors' electricity, while their independent system continued to work fine, even allowing them to pump emergency water to neighbors.

Reducing Reliance on Fossil Fuels

Seeds of Solidarity grows intensively and by hand, without use of machinery. They practice no-till methods — seeding through the previous year's crop residues, instead of removing all residues and tilling and turning the soil. This not only reduces reliance on fossil-fuel-powered machinery but also, some scientists suggest, may increase carbon sequestration.

The family car and farm truck run primarily on biodiesel (it's blended with petroleum diesel in winter to prevent flow and crystallization issues in the colder temperatures). The SOL Patrol van is outfitted with a Greasecar system that allows it to run on both biodiesel and filtered waste grease, collected from local restaurants and institutions. In addition, in thinking about how far to travel to market their produce, Habib and Baruc adopted a "closer to home" policy, meaning that they sell only within a 30- to 40-mile distance, primarily to restaurants and co-ops.

Drawing a Crowd

How does the farm manage to turn a profit? For one thing they grow a lot of garlic, and they sell most — if not all — of it at the North Quabbin Garlic and Arts Festival. This annual festival was first held in 1999, and it has grown to be an important venue for Seeds of Solidarity and a hundred regional farmers, food producers, and artists. The idea began with a conversation between Ricky Baruc and a neighbor, solidified over a potluck dinner, and has blossomed into an event that attracts 10,000 people each year. The festival is organized and staffed entirely by volunteers, and it is a family-friendly celebration of art and agriculture. While bringing people together over food and music, Habib says, the festival is revitalizing the community while also providing multiple ways of educating people about renewable energy. Although all vendor and admission fees go into the festival fund, the festival does provide a significant income stream for SOL and the other producers who market their goods there. Communities need and want to celebrate through the seasons, says Habib. And a farm is a natural site and sponsor for a ritual of celebration. "A great lesson is that neighbors really *can* come together to put together a massive event!" says Habib enthusiastically.

Habib recommends that celebration events should be affordable and accessible for the whole neighborhood or community, offer lots of activities for children, ensure that people from diverse walks of life feel welcome, encourage a spirit of celebration, and contribute to the local economy. At the Garlic and Arts Festival, about $500,000 is exchanged locally, and its impacts ripple through the community. Over the years the festival has earned enough to provide small grants to community groups for projects relating to the festival's goals of supporting the agricultural, artistic, and cultural bounty of the region.

Sustainability is a primary characteristic of the festival. Because of extensive composting and recycling, Habib says that the festival produces remarkably little trash — in one year only three bags! The music stage is powered by solar-electric panels, drinking water is available for free to discourage the use of bottled water, all tables and benches are built of locally harvested lumber, and at food stands all plates and utensils are biodegradable. Educational booths abound, with information about energy conservation, solar and other renewable power sources, gardening, and more. Making a playful dig at the name of her hometown, Habib says that the many environmental aspects of the festival show that Orange is the New Green.

Modeling a Community-Supported Fishery
Walking Fish

southeast

In the coastal community of **Beaufort, North Carolina**, the Walking Fish community-supported fishery (CSF) is building connections between fishermen and consumers. Spun off from the community-supported agriculture (CSA) model, the CSF has subscribers who purchase a share of a season's harvest up front, and Walking Fish delivers fish and shellfish to them on a weekly basis. Through this system, local small-scale fishermen are finding a way to survive, and consumers are finding a way to engage directly with the people who provide their food.

Walking Fish got its start in 2008, when Josh Stoll and a small group of Duke graduate students studying for their master's degrees in environmental management met Bill Rice, owner of Beaufort's Fishtowne fish market. Bill was looking for ways to support small-scale, high-quality, local fishermen, and Josh was looking for ways to combine his interests in ocean ecology with his academics. They hit it off, and the Walking Fish CSF soon followed.

As small-scale fisheries down the coast have slowly disappeared, they've left behind a patchwork of communities struggling to make ends meet. Beaufort, historically a fishing town, has taken on new life as a tourist destination and site for second homes for the wealthy. Other neighboring communities have not fared so well, sinking into depression.

Fishing in the region had its boom in the 1970s and '80s, says local Clark Callaway, who for many years ran one of the most profitable seafood processing companies in the area. But the industry began to decline due to regulations (a necessary evil, he admits), pollution (such as freshwater runoff from development), and imports driving down prices even as expenses like fuel rose. Clark was forced to throw in the towel, and he now works as an insurance agent. But his wife, Debra,

may have a hand in industry's resurrection. She is the coordinator of and jack-of-all-trades running operations for Walking Fish, which has undertaken to reverse the fortunes of the regional fishing trade.

Like other community food projects, Walking Fish aims to support not just its producers (the fishermen) but also its community, with a focus on the triple bottom line: economic, social, and ecological sustainability. For this endeavor, establishing resilience and security are key. The region's economy has shifted from being resource-driven to being amenity-driven, the group says, and preserving the cultural character, economic vitality, and self-reliance of the region necessitates reestablishing its natural resources as a base of capital. Fishing, for Walking Fish, is about building communities.

The CSF is newly minted as a cooperative, with a board of directors assuming leadership. Bill Rice, Walking Fish cofounder, whose Fishtowne market processes the co-op's seafood, is the new president. The CSF may grow over time, Bill says, but only within reason. And as Walking Fish evolves and matures, perhaps it can serve as a model for similar initiatives in coastal communities around the nation.

Above: Paul Russell, whose family has lived in the Beaufort area for more than three hundred years, farms clams on water first leased by his father. Paul's clams supply shareholders in Walking Fish's community-supported fisheries as well as restaurants and fish markets in Beaufort and Durham.

Following page: Like communities up and down the Eastern Seaboard, Beaufort once depended primarily on fishing for its local economy. Cheap imports, rising costs of living and doing business, and pollution from development have put many fishermen out of business. Walking Fish is attempting to build an alternative business model to better support those who remain.

Top: Lyn Chestnut gigs (hand-harpoons) flounder for Walking Fish. This method yields fewer fish but is exceptionally sustainable because it allows him to harvest only fish of legal size, with no by-catch. *Bottom:* At Fishtowne, a local fish market and Walking Fish's center of operations, the day's catch is on display.

Top left: Henry Coppola, a graduate student at Duke University and the Durham coordinator for Walking Fish, explains to a shareholder how to shuck an oyster. *Top right:* Debbie Callaway is the only full-time employee at Walking Fish. Her husband, Clark, used to own the largest fish-processing business in Beaufort. *Bottom:* Walking Fish suppliers Paul Russell and his brother meet with Amy Tournquist, owner and chef at Watt's Grocery, a farm-to-table restaurant in Durham, and one of Walking Fish's regular wholesale customers.

Sustaining Food from the Blue

We live in a blue world. And much of our global food supply relies on oceans and marine environments. By one estimate more than a billion people rely primarily on seafood for their protein. Yet many of our global fisheries are badly managed and severely overfished, and the connections between seafood consumers and fishermen are even more remote and distant, it seems, than in land-based agriculture.

In an effort to rekindle these connections and to find ways to better protect the health of our oceans, many new and creative ideas are being tried. The community-supported fishery (CSF), based on the CSA model, is one potential new model. There are now operating CSFs from Maine to North Carolina to Alaska, and the number is growing. As in a CSA, subscribers buy a share in what is usually a cooperative of fishers, which entitles them to a selection of locally caught and harvested seafood.

The advantages are clear enough: fishers receive a higher price per pound for their harvest, money recirculates in the community, and there is more transparency about the methods used to harvest the fish and a shift away from the highly destructive industrial fishing methods by which most seafood is harvested. A CSF in Canada, called Off the Hook, boasts support for low-impact fishing methods – fishing methods that are much less harmful than conventional bottom-dragging methods. In Sitka, Alaska, proceeds from the CSF operated by Alaskans Own support the Fisheries Conservation Network, a group dedicated to reducing the impact of fisheries on the marine environment. And some operations offer both a CSA and a CSF, such as Maple Ridge Farm and Fishery in Sabattus, Maine, from whom consumers can purchase not only Maine lobster and scallops but also fresh vegetables, cut flowers, and raw honey.

In Beaufort, North Carolina, a CSF called Walking Fish (see page 208 [this chapter]) offers a diverse array of locally harvested seafood, including southern flounder, sea mullet, triggerfish, clams, and shrimp, all helpfully pictured and described on its web page (how many of us know what a sea mullet looks like?). Shareholders can buy a full or half share, with a full share (which cost $420 in 2010) delivering 4 pounds of seafood a week for 12 weeks. Walking Fish has made arrangements to have the fish processed at a local retailer and delivered via refrigerated truck to the Duke University campus for pickup. Shareholders come prepared with coolers to pick up their seafood; there is a festive atmosphere on pickup day and sometimes cooking demonstrations.

The experience of Walking Fish has been very positive so far, according to Josh Stoll, one of the founders and a Duke University graduate student who saw the CSF as a way of engaging urbanites more directly in the harvesting of seafood they sometimes take for granted. Whether the fishers who participate in the CSF will always use the most sustainable fishing methods is left up to them, as Walking Fish does not mandate particular techniques or fishing methods ("clean gear," as it is sometimes referred to). Most of the flounder harvested for the CSF comes from "gigging," which involves spearing the fish, with virtually no bycatch (marine animals caught unintentionally), but some fishing with gill nets, more controversial because of the potential harm to sea turtles, also occurs. Walking Fish adopts a pragmatic approach, with the knowledge that fishing methods change over time and that connecting consumers and fishers will help push these changes toward sustainability.

Much of the emphasis at this CSF is clearly on providing more income for these small-scale, community-embedded fishers, and Stoll estimates that they are able to pay fishers on average a 30 percent premium over what they might be able to get through conventional markets. Walking Fish now explicitly encourages the fishers to set their own price ("what do you need to make a living"), and Stoll says the CSF is mostly able to provide that.

The comparison of a CSF with a CSA is common, but Stoll believes that at this point the two types of organizations have quite different goals. Stoll characterizes CSAs as a useful tool in an already established local and organic foods movement. CSFs, on the other hand, he says, have a larger goal: they aim to change an entire industry.

For now most consumers of seafood will not find a CSF nearby, though that may change as the idea catches on. In the meantime for most consumers seafood comes from quite a distance, and it is often a daunting task to make an informed choice about which seafood has been harvested sustainably. To help consumers and to steer the seafood industry in a more sustainable direction, a number of educational efforts are under way. One of the best known is the Seafood Watch program run by the Monterey Bay Aquarium in California. Science-based and peer-reviewed, the program rates different types of seafood as "Best Choices," "Good Alternatives," and "Avoid" and presents this information in handy regional pocket guides (downloadable from the aquarium's website; see the resources section).

Seafood Watch also runs a restaurant program in which participating restaurants pledge not to serve seafood on the "Avoid" list, and their staff undergo seafood awareness training (and the aquarium has produced downloadable materials and resources to help with this). It also sponsors cooking events, such as the annual "Cooking for Solutions," which features celebrity chefs, and even a seafood challenge cook-off, all to raise awareness about sustainable seafood and also to raise funds for the program.

Many other efforts around the country also support the development of more sustainable seafood options. Regional organizations such as the Northwest Atlantic Marine Alliance (NAMA), based in Gloucester, Massachusetts, have been working to support the creation of more CSFs and other means to support economically struggling fishing communities in the Northeast. They argue rightly that making management decisions on a fish-by-fish basis makes little sense, and support for more holistic, ecosystem approaches to ocean and fisheries management is needed. That said, they advise also about how consumers of seafood might steer a greener course. They offer the following advice:

- Buy from a local fisherman when possible
- Get involved in a community-supported fishery (CSF)
- If you don't live near the coast, consider eating something else that is local
- Eat fish that looks like fish (rather than being highly processed and reformulated, such as in fish sticks)
- Avoid fake or imitation seafood products
- Eat wild seafood whenever possible
- Ask how, where, and when your fish was caught[203]

On a more global level are sustainable fisheries certification systems, notably the Marine Stewardship Council (MSC), which evaluates and certifies fisheries around the world that meet its principles and criteria for sustainable fishing (notable among them is the requirement that a fishery be managed so it is not depleted or overfished, as assessed by a third-party evaluator). Modeled after the Forest Stewardship Council, which certifies wood from sustainably managed forests worldwide, the MSC works to create financial incentives for sustainable fisheries management. And it seems to be working; as of October 2010, some 95 fisheries globally had received the MSC certification.

– by Tim Beatley

Modeling Sustainability for Freshwater Fish Farms

Sunburst Trout Farm

Sunburst Trout Farm is a third-generation family-owned fish farm in Canton, North Carolina. The farm is a model of aquaculture sustainability, with innovative systems for raising fish ecologically and humanely, to benefit the fish, the consumers, the environment, and the local economy.

For more than 50 years, a small trout farm nestled in the heart of western North Carolina's Pisgah National Forest has repeatedly — and successfully — overcome daunting hurdles, including a devastating arson and theft in 2006 that nearly destroyed everything. Sally Eason runs Sunburst Trout Farm, and she exudes an almost raw energy and entrepreneurial spirit that is reflected in the farm staff's professionalism and enthusiasm. With Eason at the helm, Sunburst Trout has developed an array of mouthwatering products — trout sausage, trout jerky, trout cakes, trout mousse, trout ribs, and even golden-orange trout caviar that has been featured in *Gourmet* and *Bon Appetit* and brought national TV shows such as *Chefs A' Field* and *Food Nation* and the Food Network to shoot at the farm.

Eason inherited the trout farm from her father, who, as an honors engineering student at Yale, bucked his father's advice and dropped out in 1948 to start the "first commercial trout farm in the South."[204] Sunburst is truly a family farm. Eason's husband, along with their two sons and their sons' wives, all work side by side as part of the farm's dozen or so staff. "It's a family-owned business, and we don't ever give on that," Eason says. "We want our product represented on that seafood counter to be as good as we can get it."

Parameters for Sustainability

A plentiful supply of clean, cool water is the critical factor for anyone thinking about starting a trout farm, and Sunburst is proud of maintaining a high water-to-fish ratio. The trout swim in currents of pristine 50- to 60-degree mountain water that is filtered through the watershed of Pisgah's Shining Rock Wilderness Area. Water tumbles out from a nearby dam at a controlled 12,000 gallons a minute, incorporating large quantities of oxygen that rainbow and golden trout need to survive. The trout farm raceways (canals) are layered in three tiers, so that water tumbling downhill into the lower raceways continues to incorporate oxygen for the fish. And because the fish are swimming in a virtual stream, their muscle structure is excellent and flesh firm. "They're doing what they would in their natural environment," Eason says.

Sunburst prides itself on raising fish in an environmentally sustainable manner. Sales manager Lila Eason says, "Sally focused on [sustainability] way before it was a cool thing to do. It was just how she thought business needed to be run." The farm doesn't use antibiotics or pesticides, relying instead on cool water temperatures to maintain healthy fish. In the summer, if water flow drops and, as a result, temperatures creep upward, they move the fish to two cooler sister farms.

Sunburst also works closely with its feed companies to create "the most sustainable feed possible." The farm's fish feed has no animal by-products, no antibiotics, and no growth hormones, and as of 2009 they were working to replace the protein in the feed with insects, trout's natural food.[205] Sunburst trout flesh is bright red as a result of the yeast in the feed, which, unlike synthetic yeast used by other trout farms, is a natural yeast high in antioxidants. "Our fish are loaded with as much or more omega-3 fatty acids than salmon," Sally Eason says.

The farm approaches killing the fish in the most humane way they can. Most harvested trout are about one year old, weighing in at 8 to 10 ounces, a perfect size for a filet. When fish are ready for harvest, Sunburst moves them from the pond to a cooler, where they are chilled in an ice slurry until they slow down, fall sleep, and die peacefully.[206] "It's more humane," Sally Eason says.

Building Community Connections

Sunburst Farm is creating ways to bring the community out to the farm. Two community organizations bring children to fish, and Sunburst hands out awards for the biggest, most, and smallest fish caught and for golden trout, the hardest of all to catch. Following the idea of pick-your-own farms, Sunburst opened its ponds in 2009 to people who want to catch their own fish. It charges $3.50 per pound for the whole fish, uncleaned. It will dress the fish for only 50 cents per pound and filet it for another $1 per pound. People love it, says Lila Eason, because they have a good time and get a beautiful fish for less than half the store retail price.

Sunburst Trout's chef, Charles Hudson, is also working with the community to increase its interest in fresh, local food. He does more than create innovative products from rainbow and golden trout. He volunteers with the Appalachian Sustainable Agriculture Project (ASAP; see page 243) as a chef in schools, and he enjoys exposing children to fresh trout. In recognition of his important work, Chef Hudson was invited to the White House as part of Michelle Obama's Move Chefs to Schools initiative.

Fish Farming in the Long Term

Sally Eason is well aware of the environmental concerns associated with fish farms along the coast or in sea waters — pollution, stressed and infected fish requiring the use of antibiotics, infection of nearby wild species, habitat destruction, unintended release of invasive or genetically modified species into the wild. "We often get people saying, oh no, it's farm raised — it's bad," she says. But she enjoys educating people about Sunburst trout and seeing them change their minds. "This is a different type of fish farming," Eason says.

Eason notes that her farm can raise a quarter of a million pounds of fish on 7 acres, a staggering production of protein on less land and with fewer resources than required by beef or pork.[207] She is selling her prized trout all over the country, retail as well as wholesale, and her customers include more than 185 high-end restaurants.

Yet the long-term future of aquaculture, according to Eason, is uncertain. "We're either going to be catapulted forward into the next two decades and be able to make a difference in world sustenance or we're all going through the tubes," Eason said in an interview with a local magazine.[208] She guesses that a new trout farm would require an up-front investment of at least half a million dollars, probably more. Unlike such commodities as wheat, corn, soy, and milk, fish farming doesn't have a safety net of subsidies provided by the federal government. And if drought or other weather causes a bad year, the fish farmer isn't eligible for federal relief through traditional crop-insurance programs. So starting and running a fish farm takes money, creativity, and the ability to weather difficult years.

Sally Eason has vision, gumption, and entrepreneurial know-how. Lila Eason says the fish are so beautiful, they "look like the sunrise." Sally Eason has proven that a fish farm can be beautiful, too — growing food in a way that is healthy for people, healthy for the fish and the environment, and healthy for the local economy.

The Future of Sustainable Fish Farming
Australis Aquaculture

Founded in 2004 in **Turners Falls, Massachusetts**, Australis Aquaculture has gained an international reputation for its inventive fish-farming technology, its extreme adherence to environmentally friendly methodologies, and, above all, the culinary appeal of its fish: the barramundi.

Joshua Goldman, founder and CEO of Australis Aquaculture, has been interested in sustainable fisheries for decades. He studied biology at Hampshire College in Massachusetts, and later, contemplating starting his own fish farm, he'd wandered the world in search of the "perfect" fish, one with which he could address the pernicious issues that face the fish-farming industry: escaped fish becoming invasive species, farmed fish requiring more protein for their own consumption than they provide to their consumers, the release of polluting effluent into natural waterways, and so on. He came back to Massachusetts with the barramundi, little known outside of Australia at the time, but in that country prized for its mild flavor and meaty texture. In the tiny town of Turners Falls, he set about building a barramundi fish farm, which has since become an unparalleled model of sustainable practices.

To begin, Goldman located his farm next to a manufacturer of styrofoam cups, so he could use the factory's waste heat to warm the water for the tropical barramundi. He designed the farm as a closed-containment system, meaning that the fish are grown in tanks on land, not in water, which limits water waste and pollution from effluent reaching natural waterways. Australis purifies and reuses 99 percent of the water it requires and releases only a tiny fraction of waste compared to traditional fish farms. (According to the Worldwatch Institute, a salmon farm with 200,000 fish can release the waste equivalent of 10,000 to 20,000 households, though fish waste is not as dangerous in terms of pathogens as human or livestock waste.)[209] The setup, Goldman likes to say, is basically a large water filtration system with fish in it. All the waste manure from the fish is given to local farms for fertilizer.

Since the founding of the Massachusetts facility, Australis has expanded to sites in central Vietnam and Indonesia. In these tropical environs the fish are started in land-based tanks, and when they're big enough they're put out in modern ocean pens — at a very low stocking rate, to protect water quality — as they grow to maturity. Barramundi are native here, so if they should happen to escape their pens, they cannot become invasive species, as has been known to happen when fish are farmed outside their natural ranges.

Goldman believes barramundi is the perfect fish for farming. It is durable and can live in freshwater, saltwater, or a mix, so it can be raised in ocean pens or land tanks, depending on what works best for the particular system. It spawns on a lunar cycle, rather than an annual one, making it easier to maintain a consistent supply. It converts feed to protein well and has the rare ability to produce high levels of omega-3 fatty acids when fed a largely vegetarian diet. By eating low on the food chain, barramundi bears a low risk for accumulating mercury, PCBs, and other toxins in its flesh. All of this makes for good business as well as some of the healthiest and most sustainably farmed fish out there.

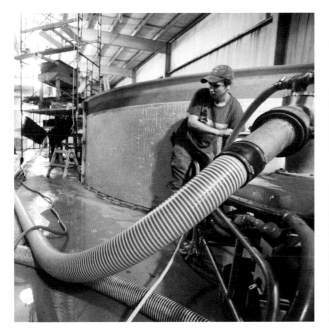

Josh Goldman, cofounder and CEO of Australis Aquaculture (pictured at bottom right), traveled the world looking for a fish that could be truly sustainably farmed. It had to meet his criteria of mild flavor, favorable protein conversion, avoidance of disease and toxin buildups, and the ability to thrive in systems that negated issues like escapes and pollution. He discovered barramundi and built an innovative "water filtration system with fish in it" to raise it sustainably.

Sustainability through Innovation
Polyface Farm

with research by Jessica Ray

Polyface Farm is a self-described "family-owned, multi-generational, pasture-based, beyond organic, local-market farm and informational outreach in **Virginia's Shenandoah Valley**." The family raises beef and pork, along with chickens for eggs and meat, and they market their products successfully through on-farm sales, a buying club, and restaurants, including the Chipotle Mexican Grill fast-food chain. The Polyface mission is "the redemption business: healing the land, healing the food, healing the economy, and healing the culture."

Joel Salatin is Polyface Farm's firebrand innovator, who loves nothing more than to challenge the status quo, shake things up, make people *think*. You may not agree with his self-proclaimed "Christian-libertarian-environmentalist-capitalist-lunatic" ideas. And you may not like his intensely pointed, verbally rich rants that can raise a roar of laughter or thundering applause or stunned silence — especially if they're aimed at you, your beliefs, or people you know. But you do have to admire Salatin for his transparency and integrity. While some people these days are increasingly cautious about riling political, spiritual, cultural, educational, and other kinds of world-view feathers, others say what they mean and mean what they say. Salatin, for one, has always said what he believes — loudly, clearly, passionately, persistently. People listen to him because his actions are consistent with his beliefs and, most of all, because his deceptively simple farm innovations actually *work*.

Salatin's farming innovations skyrocketed to national acclaim with Michael Pollan's *Omnivore's Dilemma* in 2007, nearly 20 years after Salatin had unveiled his newly invented Eggmobile (a portable henhouse for raising chickens on pasture) and described his pastured poultry system at the first conference of the Virginia Association of Biological Farming in Charlottesville, Virginia. I was lucky enough to witness this debut, and I was mesmerized by the young farmer who had invented such an ingenious contraption for his chickens. His presentation had been delayed for various reasons, and the audience shouted down the moderator, who tried to cut him short. When he finished, the entire room erupted in applause, recognizing a leader who was changing farming before their eyes.

A Farm as a Living System

Like most business leaders, Salatin works to reduce labor costs and increase efficiency across his enterprise. But he takes a systems-oriented view of what qualifies as "success." For a new idea to be seen as successful at Polyface, it must be sustainable, which means that it must benefit the whole farm *system*. And to benefit the farm system, it must meet the farm mission of healing land, healing food, healing the economy, and healing culture.

Salatin didn't build his farm system overnight. And his farm is a living system, meaning that it is constantly evolving and changing as Salatin identifies problems and tinkers until he solves them. Over the years, for example, the Eggmobile has been followed by the Feathernet and the Millennium Feathernet, both portable commercial henhouses that improve upon the previous model, giving chickens greater protection from predators and curious cows, keeping eggs cleaner, making better use of space in a smaller footprint, and featuring a self-watering and feeding system to reduce the labor involved — a key feature of all of Salatin's systems. "What just kills you are all the daily chores," Salatin told the *Stockman Grass Farmer* in 2006. His goal is a minimum wage of $20 per hour. "If an enterprise won't return that to our efforts it's not worth having," he says.[210]

The way Salatin operates egg production — which means he and his team know what they're doing and they do it efficiently — only seven hours of labor each week can net $18,750 a season, assuming eggs are sold at the minimum 2010 pastured-egg price of $3.00 a dozen.[211] The chickens are happy doing what is natural to them — pecking at insects, scratching, roosting — while being rotated through pasture, cleaning up after the cows, and contributing to the farm's profitability with their high-quality eggs and meat.

Of course, not every idea works out so well. Salatin had experimented with sheep, but they "competed with the cows, and had to go," reports journalist Mark Neuzil. "And ducks brought in 'to pick flies off of the cow's noses' spent too much time in his ponds."[212]

But Salatin has more successes than failures, and his multipronged and intertwined systems abound on the farm. The natural rooting compulsion of pigs, for example, is harnessed to turn the farm's compost, and the pigs are also rotated through pasture. "The secret to healthy pigs is activity," Salatin told the *Stockman Grass Farmer*. "This is critical for good respiratory function. Also, just as with chickens and cattle, fresh grass is important. With pigs, it is the grass that creates the special flavor."[213] Once the pigs reach 200 pounds, they are taken to a forested area for acorn finishing. Acorn finishing, says Salatin, "completely changed our whole economic model." With it Polyface was able to displace 50 percent of the cost of GMO-free grain with a perennial, unharvested, freely provided tree crop. The acorn finishing system saves the farm time and money, makes the pigs happy, and produces better pork.

Even better, says Salatin, the environmental science behind sporadic forest floor disturbance is very good. Studies have demonstrated that periodically disturbing forest floors is beneficial for silvicultural production. In plain English this means that Polyface pigs are doing good things for the forest, while the forest is creating high-quality, high-nutrient pork.

Building Buying Clubs

Polyface has grown for only one reason, says Joel Salatin. People like its food, plain and simple. It doesn't advertise and never will. It doesn't ship any products and never will. Both of these decisions force the farm to stay local and maintain a quality product. Staying local also means listening and responding to customer ideas and suggestions. And that is what launched the Polyface Buying Club.

Sheri Salatin, the Buying Club manager, says the idea came from two loyal customers, who would come down from Maryland twice a year to visit Polyface and buy hundreds of dollars of food. One day in 2001 they asked what it would take for Joel to make a delivery to them. "Joel just threw out $3,000," says Sheri Salatin, "and three weeks later they called back." Through their friends the two women had gathered together a $3,000 order, and they wanted to know when he could be there.

The Buying Club began with a delivery to one woman's house about every eight weeks. From there it started to build. Someone who lived en route got her church involved, so another delivery was added. That's when Sheri stepped in. Since that grassroots beginning, word spread, and more delivery stops have been added. Eight years later, by 2010, the Buying Club accounted for 45 percent of farm sales, serving 4,100 customers with 29 drop-offs on a run made about once every five weeks.

The Buying Club is a sophisticated system that relies on good communication between the farm and its customers. Sheri Salatin begins by sending a newsletter to her customers, telling them what products will be available and requiring an order deadline of two days before the scheduled delivery. If she tells them, "We have a lot of chuck roast right now, and we need people to buy it," customers respond by finding recipes for chuck roast and placing an order. "These people [are] our cheerleaders," she says. "They [want] us to thrive."

Salatin wishes every farm had a buying club, because it would help bring fresh farm food into metro areas. She even thinks it's possible that a farm could survive entirely on a buying club. But she warns that a buying club is only as good as a farm's integrity, and any farm should always allow customers to come directly to the farm to see how it's operated. She advises that a farm begin with on-farm sales, to establish relationships with its customers; these customers will then become the farm's biggest marketing advantage and tool.

"Our goal is *not* to sell," says Sheri Salatin. "It's just to raise the best food in the world." Polyface does plan and try to predict sales, she explains, but the farm goal is not to sell or meet marketing benchmarks. The goal is just to meet demand. "If we raise the best product in the world," she says, "then people will want it."

Growing New Farmers

As the demand for products rises, a farmer has a choice: keep using just the available resources on his farm — and hope that somehow demand will be met by others — or get creative. Salatin isn't the kind of farmer who easily turns away from a challenge. He has a mission of healing the land, food, economy, and culture. That mission means that, when asked to provide good food to his community, he will try to make it happen.

One of the first ways that Salatin met increasing demand for his products was to boost his summer help for the farm's summer operations. Now, to provide an experience of a lifetime to young people interested in learning about sustainable agriculture, Polyface offers the "Real Farm Experience" for eight summer interns. The interns live 10 minutes away in two apartments paid for by Polyface, receive a small stipend per month, and join the farm family for a Saturday-evening cookout. Summer interns work mainly with the poultry and are expected to put in five days a week as well as every third weekend. "By the time they leave us," says Sheri Salatin, "they know how to raise them from day-old chicks, build the pens, put them on pasture, and market the product." They do learn about other aspects of the farm, but not as intensively, and they also attend weekly workshops where they become versant

in common grazier terms like "cow days." People who seek this internship may be interested in the idea of farming but not yet ready to make a commitment, and a program like this offers a way for them to get a taste of what running a farm would take.

But to grow the farm, Polyface needed more year-round help as well. So Polyface now hires two or three young men as year-round apprentices. While variations on farm apprenticeships are increasingly common as a way for a farm to grow while teaching and supporting new farmers, the Salatins set high expectations for their apprentices and make no bones about being *very* picky about whom they choose. "It's like getting a new brother every year," says Sheri Salatin. "They are immersed in our family and way of life. They do everything with us. When we work 50 hours a week, they do, too." When the apprentices leave, she says, they know how to do *everything* — write a business plan; raise broilers, cows, pigs; and manage the farm.

Today, as demand for his "beyond organic" food continues to rise, Salatin's challenge is to figure out how to continue to scale up business while continuing to heal the land. In the end Salatin didn't have a choice. If he wanted to expand production while continuing to heal the land and produce high-quality food, he needed to expand off-farm. On its 550 acres, Polyface has 450 woodland acres that support 100 open-pasture acres, which in turn support the farm's complex multiculture of rabbits, turkeys, chickens, hogs, and cows.[214] Even working together synergistically, his land and animals have natural ecological limits.

So while Salatin is continually experimenting with ways to improve his operation that will take into account these variables, he is now partnering with other farmers to expand his operation. In 2009 Salatin was experimenting with four different expansion models, all of which offer promise as ways to foster new farmers (see the box on page 224).

The beauty of the Polyface model, Salatin explains, is that all of its infrastructure is portable. This allows Polyface great flexibility in supporting new farmers. And it is happy to be growing what it now calls a "family of farms." With these expansion models, Salatin says, "young farmers — who had no capital, no equity,

Sustainability through Bioregional Food Clusters

Ever creative, Joel Salatin is dreaming up a new model for growing local food systems in a way that would support small family farms, playing to farmers' typical strength in production and providing support for the weak links of distribution, marketing, and customer service. He envisions a system that he calls "bioregional, scalable, sensitive food clusters." To succeed, each cluster would require all of what he considers the six essential components of the food system: production, processing, distribution, marketing, accounting ("Someone has to watch the money"), and a customer. Each cluster could evolve its own regional branding, says Salatin. Immediately I think of Appalachian Sustainable Development (see page 253) in southwest Virginia with its "Appalachian Harvest" regional brand and the Appalachian Sustainable Agriculture Project (see page 243) in North Carolina with its "Appalachian Grown" brand. What Salatin describes is already beginning to happen in some places.

The success of the food clusters will depend on building locally based economies of scale. The cost of distributing small quantities is a real hurdle that all small-scale producers are dealing with, Salatin says. In his own farm business, he sees the benefits of aggregating products from other farms. For example, each year Polyface sells about 3,000 gallons of apple juice from Golden Acres Orchard and is now the orchard's number-one outlet. "We're moving so much apple juice that it pays the gas bill for our Buying Club runs," Salatin says. On its deliveries to restaurants, Polyface also sells vegetables and cheese from other farmers. "When we add other growers' things to our busload, it completely changes the economics of our delivery," Salatin explains.

What Polyface is doing to build economies of scale is a microcosm of what Salatin believes is possible. The ultimate goal, he says, would be for each region to have its own storefront, available to customers as close to 24/7 as possible. Like the Cracker Barrel model, one part of the facility would be a social and dining space, and another part would be the consignment farmers' market showroom.

Salatin's vision for a destination storefront would fulfill five of the six key components of a local food system – all but the production piece. It would enable easy distribution by allowing farmers to drop off their consignment products at any time of the day. It would enable easy processing by hosting a large commercial kitchen – a place to "take beef bones and turn them into stock, or little eggs and turn them into egg noodles and quiche." It would provide the marketing with a large showroom of different local products. It would provide the accounting by having one central cash register keeping track of sales, Salatin says, "so these farmers don't have to sit here with their big, calloused, dirty fingers and punch little calculators." And it would provide the convenience and social space for customers, by offering a local dining space built on locally grown foods.

Salatin explains that it is the overhead that kills little businesses, and a model like this would "keep the cash register humming." The more you can funnel through that cash register, he says, the more efficient the overhead.

This, he hopes, may be a possible future for our local food system: Instead of 100,000 Polyfaces scattered around Virginia, there could be five hundred food clusters. Each food cluster would allow nearby small farmers to put their energies into doing what they do best, farming, by taking care of the other five key food-system components. Size, per se, doesn't matter, says Salatin. If a food cluster meets a need, providing quality local food, it will grow. If it doesn't, it will die.

Growing New Farmers

When Salatin and his family realized that Polyface had reached its natural ecological production limits, they had to come to terms with the concept of growing off-farm to meet growing demand. Discussions ensued. How could they grow off-farm, while also assuring product integrity, helping new farmers become financially sustainable, and still healing the land? Polyface may have a cookie cutter stashed somewhere in its farmhouse kitchen, but it's a sure bet that Joel Salatin doesn't believe in using that approach with anything involving living systems. Farms have individualized needs, and so do the people who run them. Systems fail, people make mistakes. Yet Salatin has a keen desire to succeed – and to help others succeed. The result is an array of approaches that can serve as models for others who are seeking innovative ways to support new farmers.

As you'd expect, each approach is customized to each farmer's needs, goals, and size and quality of land. Each farmer uses Polyface methods to produce products that Polyface sells under the Polyface brand. And these approaches appear to work. By 2010 Salatin reported that Polyface had expanded to feeding more than 3,000 families, 50 restaurants, and 10 retail outlets.[215]

Model A: Debt-Free, by the Piece

Polyface leases another farm with a house. The contract farmer lives in the house, rent free. For eight months of the year, she runs two pairs of Eggmobiles, with 1,800 layer hens, and 15 broiler shelters. Polyface owns the chickens and the infrastructure and buys all the feed. The farmer earns income by the piece, by the salable eggs and salable broilers. She also agrees to move Polyface cows, if and when they are being rotated through the farm. The difference between this relationship and the relationship that industrial-size chicken contract growers have with their industry integrator is simple: Polyface's contract farmer has no capital investment and no debts. Current industry practice requires poultry producers to pay for all buildings, equipment, and infrastructure, while the company owns the birds and provides the feed. This requires a huge up-front investment by the farmer, usually in the form of a mortgage. For the large-scale conventional grower paid by the piece, the industry's risk is minimal, while the farmer's risk is enormous.

With Polyface, however, the contract farmer has virtually no risk, because there is no indebtedness. While she has an incentive to work hard to produce good income, she also can walk away from it, debt free, if she decides to do something else. In this case the farmer figured out that she was making $30 per hour and was so thrilled that she asked to do it again a second year. Polyface wins because it has created an incentivized, autonomous farmer who is paid entirely for her ability to produce results. Though Polyface bears the up-front costs of the hens and infrastructure, the benefit is that if the farmer doesn't work out, Polyface keeps its investment. And the farmer wins because she has no debt and can decide how hard she wants to work.

In her third year, Salatin says, this farmer traded in her broilers for pigs. While she is making a bit less money, she is not working nearly as hard. "The broiler shelters pushed her physically, so the pigs were much easier," says Salatin.

Model B: Lump Sum and Royalty Payments

A new farmer, who worked for Polyface as an apprentice, is renting an apartment in a house on another farm. He lives rent free in exchange for maintaining the lawn. Polyface leases land from this farm for its Polyface beef cow herd, and the new farmer also uses some of this farm's land to establish his own private 10-cow dairy herd. The new farmer moves the Polyface cows every day, in exchange for one lump-sum payment for his service. Polyface wins because it saves more than it spends by not using a thousand gallons of fuel every year just to get to that farm to move the cows every other day. It also wins because the cows are now being moved every day, which, Salatin says, increases the "efficiency of the biomass and carbon sequestration." The new farmer wins because he has a guaranteed lump-sum payment to support his new business and because he is able to sell his milk to Polyface customers. For every gallon of milk he sells, he pays a small royalty to the farm owner and a very tiny royalty to Polyface.

In his second year this farmer adds a pair of Eggmobiles and increases his private dairy herd to 12 cows. The farmer sells half of his product through Polyface and half on his own. Salatin describes this as a synergistic marketing effort, because when the farmer's own clients come to Polyface to pick up milk, they want to buy other Polyface products such as eggs and sausage.

Model C: Own Brand

A Polyface intern decides to become a Polyface subcontract farmer. He doesn't have his own land, but he has some money to invest. He lives on a farm leased by Polyface and here starts his own business. While he purchases his own birds, builds his own brooder, and pays for the feed, he uses Polyface infrastructure – the broiler shelter, chicken pluckers, scalders, and refrigerators, all of which are portable. This means the farmer doesn't have large up-front capital costs, and he hasn't lost his nest egg if he decides one day to do something else. The farmer establishes his own farm brand and grows six thousand broilers under his own label; Polyface markets the birds and pays him by the piece sold. Polyface adds a second label on the packaged birds, indicating that they were grown for Polyface, according to Polyface protocols and standards. And it's another win-win for both parties.

Model D: Lease to Own

Polyface leases a farm where it also rents a house. The new farmer, yet another former apprentice, lives in the house rent free, in exchange for moving the two hundred or so Polyface beef cows on the farm every day. The farmer's family also cultivates its own garden to grow its own food. In this case the farmer also is building his own egg business. Polyface pays for all the infrastructure – the watering system, electric fence, tractor, and four-wheeler. The new farmer purchases the layers (laying hens) and builds his own processing facility; with financing from Polyface he also buys two Eggmobiles and builds a winter hoop house for the hens.

If after one year the contractor and Polyface are still happy working together, then they work out a five-year payback plan on the Polyface $40,000 investment. Within six years the new farmer will fully own his own equipment, tools, infrastructure – everything. And if at that time the farmer decides to buy his own land, he can move with his fully portable infrastructure and start an independent business. Again, it's a win-win. Polyface invests in equipment that it can retain, should the farmer walk away. The farmer gains much needed assistance in building infrastructure and gaining equity and ownership over time. Polyface gains another incentivized, autonomous farmer who can help meet increased demand for its food, and the new farmer gains a guaranteed market and price for his work.

and no experience — are now full-time farming and making earthworms dance."

One thing Salatin has learned is that the solution must not stray from the real goal: healing the land, which requires synergy and a diversity of species. "We don't want to become just a broiler operation or a cow operation," says Salatin. "The whole motive is to have some profitable way to put chickens behind the cows. The whole motivation is land healing."

Localizing the Fast-Food Market

Joel Salatin is pushing to explore these models for developing new farmers because the demand for his products took a quantum leap in 2008, when a nearby Chipotle Mexican Grill — belonging to the fast-food restaurant chain with the "Food with Integrity" motto — began offering Polyface pork.

Chipotle CEO and founder Steve Ells thought that Polyface would be a perfect fit for his restaurant's ethic of serving food "raised with respect for the animals, the environment and the farmers,"[216] but working through all the logistical challenges took "incredible *personal* commitment and integrity," says Phil Petrilli, Chipotle's northeast regional operations director. During the initial phases of working out a system with Salatin, Petrilli traveled to Polyface two to four times a month. Their challenges were numerous, arising from the unique characteristics of pasture-raised pork and the complexities of connecting local sources to a national food-service production system. Consistency of flavor and spice is extremely important for chain restaurants and presents difficulties for chains seeking to localize their products. Jalapeños are sometimes really hot, sometimes not. The pork sometimes has more fat, sometimes less. These issues require conversations and experimentation, says Petrilli. And for the Chipotle-Polyface relationship, they included:

- **RECIPE ADJUSTMENTS.** Chipotle's standard recipe called for almost 100 percent shoulder, but Polyface needed to sell shoulder *and* hams (its Buying Club bought ribs, sausage, and bacon, and restaurants took the loin). If Chipotle hadn't adjusted its mix of shoulder and ham to accommodate Polyface so that the farm could move the whole animal, the partnership couldn't have gone forward. And Polyface pork was different from the pork Chipotle had been using, so Chipotle's standard way of preparing the meat wouldn't work. "Our pork was so much more juicy that our ham was more similar to the shoulder they had been getting," says Salatin. So Chipotle flew Joel Holland, its head chef, to Charlottesville four or five times to adjust Steve Ells's original recipe to get the best results.
- **COOKING METHODS.** For other Chipotle restaurants the pork is aggregated from small family farms, seared and seasoned, then vacuum-sealed and shipped to the different restaurants for finishing with stove-top braising. The cooking method that Holland created for Polyface pork, which tends to have a higher fat content, involved oven-braised seasoning and finishing. This meant that the Charlottesville restaurant had to replace its stovetop with an oven.

- **TRANSPORTATION.** Safe transportation of the meat posed the biggest challenge to the new partnership. Until the prospect of providing meat to Chipotle, Polyface had always used coolers to deliver meat. Salatin couldn't afford to purchase a $30,000 refrigerated van, and Chipotle couldn't afford to have meat delivered without the assurance that it was continually chilled at the proper temperature. The deal looked bleak. Then Heidi Wederquist, in Chipotle's purchasing department, came up with an idea: "temperature log" microchip thermometers, which the food industry uses to audit transport of food, particularly on long transnational runs. The thermometer strips are placed in meat bags to record time and temperature from the meat locker to the restaurant. Not every bag needs a temperature log, but enough need to have one to ensure that the "cold chain" isn't interrupted. The chip is delivered with the meat and can be read by the company at any time.
- **PACKAGING.** A farmer needs to think about packaging that is sturdy and works in large quantities, says Petrilli, such as rigid cardboard boxes that can be stacked, coolers, or durable bags. Originally, the bags used by the slaughterhouse to package the meat were too thin, so Salatin found a heftier bag that wouldn't tear or leak; it couldn't be completely vacuum-sealed, but it could be heat-sealed. "The last thing you want to do is to walk into your restaurant cooler and find blood on the floor. That is a cross-contamination nightmare," says Petrilli.

By the summer of 2009, a little more than a year after it began rolling Polyface pork into its carnitas, the Charlottesville Chipotle had soared to being one of the chain's top sellers of pork. This was an unexpected feat and testimony to the remarkable tastiness of acorn-finished pork. Chipotle and Polyface then began to explore expanding to beef and followed the same protocol — playing with a half dozen cuts to see which would work best and by what cooking method. As of this writing Chipotle and Polyface haven't ironed out the challenges. The cuts that Chipotle wants are the same ones desired by others, so Polyface isn't yet able to piece together selling the "whole animal," as it has with pork. This means the supply isn't as stable, says Petrilli, and Polyface can fetch a higher price for its beef through its Buying Club than what Chipotle can afford to offer. All of this proves just how difficult it is for a restaurant and a farmer to broker a deal, even when the owners want it to happen.

To make a deal like this work, a special kind of visionary leadership is crucial. The farmer and his farm team must be creative, entrepreneurial, risk-taking (within certain reasonable bounds), and fully committed to making it work. The same is true of the company, though it may be arguably an order of magnitude more difficult for an international fast-food company to provide that same commitment, as it answers to a board and thousands of employees at any number of levels who might not agree with the vision. Petrilli emphasizes that a commitment of this magnitude can work only when it comes from the very highest levels of the organization. "If it were not for Ells's relentless commitment to this type of food sourcing, all of the passion that Joel and I had for the relationship would have meant nothing," says Petrilli. "Ultimately, it is Steve who removed obstacles by simply giving myself and a number of others in Chipotle the empowerment to do what we knew was right. That's a pretty special and rare leader in companies today."

"I have nothing but admiration for how that company held our hand and bent over backward to make this happen," says Salatin. "If Chipotle hadn't believed in it, in their hearts," he says, "it would have been too insurmountable, too difficult."

Chipotle's experience with Polyface suggests that there is at least the possibility for change in the food industry. Where there is no will, there is no way. But where this is a will, there is at least the possibility of finding a way. Somehow a national chain restaurant and a farmer found a way to work together to satisfy each other's needs.

Maintaining Integrity

Getting big comes with risks. Cascadian Farm started as a small organic farm in Washington's Cascade Mountains in 1972, and over the next 30 years it grew into a major corporate organic food grower, distributor, and marketer. Owner Gene Kahn was vigorously criticized for selling his farm, going corporate, losing touch with core "organic" values. Michael Pollan wrote in 2001 that Kahn "has become one of the most successful figures in the organic community and also perhaps one of the most polarizing; for to many organic farmers and activists, he has come to symbolize the takeover of the movement by agribusiness."[217] And yet Kahn, now vice president and global sustainability officer for General Mills, argued that Cascadian Farm would have gone completely under if he hadn't sold to Welch's. By joining the mainstream food system and supporting the growth of industrial organic, Cascadian Farm, Kahn suggests, increased access for more people to quality organic food and increased the number of environmentally friendly farms. "Behind every organic TV dinner or chicken or carton of industrial organic milk stands a certain quantity of land that will no longer be doused with chemicals, an undeniable gain for the environment and the public health," he told Pollan.[218]

Certainly, for those who support Walmart organics, strong social justice arguments might be made for the health and environmental benefits of expanding organic food access to the masses. And equally certainly, strong arguments might be made that industrial-scale organics is inherently unsustainable because it is "floating on a sinking sea of petroleum,"[219] and it is socially unjust because of "labor conditions contrary to what any organic consumer would consider equitable,"[220] and it will drive down prices and squeeze out U.S. organic farmers and use its power to weaken the organic standards.[221] The economic pressures for achieving greater efficiencies are on the side of Walmart organics, and the organic industry is aggregating and conglomerating just like other industries.[223]

So entering the fray of working with the mainstream "big boys" is not for the weak-spirited or faint of heart. All the inevitable pressures associated with getting bigger are bound to weigh on Polyface. As the Polyface Farm scales up, working to provide greater quantities of food with integrity to its growing list of customers, including Chipotle, many will be watching.

The Politics of Pricing

Local food is often criticized as being more expensive than grocery-store food trucked in from distant regions. Joel Salatin is clear about one uncontrollable factor that impacts his farm's price structure: regulation. The regulatory climate creates a prejudicial structure that is scaled for industrial-size processing, he says. "A small abattoir doesn't enjoy the economies of scale for overhead and regulatory costs," he explains. "We have a way better product because we have less energy costs, transportation costs, infrastructure costs – because we're not pouring concrete, rebar, and running fans – and we don't have to haul any manure." But the processing costs are what create a pricing structure that favors industrial suppliers: Polyface pays $1.30 per pound just in processing, while the normal supplier pays only about $0.25 per pound. "If we could home process, or whack them up and take them in there ourselves," says Salatin, "we could outcompete the name brands at Walmart."

But this price prejudice may not last forever. A most interesting thing happened when oil prices soared during the summer of 2009. Suddenly, the cost of nonlocal, trucked-in meat increased relative to that of Polyface meat. Even with a higher fixed price for local small-scale slaughtering, rising oil prices drove up the cost of food trucked in over long distances, making local food relatively less expensive.

Today the point at which institutions decide to choose locally produced food is often tipped by customer demand, values, and even health concerns. But when oil prices begin to soar again, as world oil reserves and production continue to decline past their peak, we might expect another tipping point. This tipping point could have the power to transform believers in industrialized, aggregated, globalized agriculture into advocates for food production closer to home.[222]

This future tipping point is simple: lower relative price. When the price of local food approaches the price of long-distance food, then qualities other than price can be expected to become more important in determining purchasing preferences. And the benefits of local, sustainably grown food should become relatively more significant – higher nutritional value because of fewer field-to-table days; higher economic development value because of keeping dollars circulating in the local community; higher community value because of supporting regional working farms and open space; higher environmental value because animal manure won't require transport and application onto fields, leading to less pollution of our waterways; higher educational value because of connections with schools and customers.

Perhaps the motto for local food should be this: if there isn't enough demand for local food now, *just wait a little while*. Joel Salatin's vision of bioregional food clusters may be closer than we think, as changes in our natural resources change the dynamics of decision making. Bioregional food clusters could easily become future food engines, supporting our local and regional economies, environment, and community.

When I met with Salatin, I asked him point blank: How big is too big? Will Polyface risk losing touch with its core values?

As usual Salatin was already one step ahead. When *The Omnivore's Dilemma* became a surprise best seller, Salatin and his family sat down to answer these questions together. They developed specific business practices that would prevent Polyface from going the way of Wall Street. Some of these practices may defy common business wisdom or seem counterintuitive. Yet they reflect Polyface's determination to stay true to its core values of healing land, food, economy, and culture.

"Ethics-Based Anti–Wall Street Contrarian Business Practices"

Guiding Principles Developed by Polyface Farm

No sales targets.

A classic business might set a goal for selling a thousand widgets every month. And then it strives to create markets to achieve that target. Polyface has decided never to set a sales target. A classic business model might suggest that Polyface should set a target to expand by 2015 to supply three Chipotle restaurants. Polyface takes a different attitude. If Chipotle wants Polyface products, then Polyface will try to figure out how to supply it, but Polyface won't be pushing and grasping for that expanded market. "This really changes the way you perceive your product," Salatin explains. "If sales increase, that will be a by-product of good service, good product, and good stewardship – as opposed to us grasping for additional sales."

No trademarks and no patents.

If a model can work for somebody else, Salatin says, "we are not going to circle the wagons and protect our knowledge base, our information base, our customer base." People can come, learn, and go next door and start a competing business. Salatin says their attitude of transparency and openness to competition will protect them from inordinate growth. "If this [attitude] were true in the U.S. marketplace, the big companies would not have gotten so big," Salatin says.

Set a clearly defined market boundary.

Define your market and stick with it. This practice focuses your attention on serving your market, as opposed to building an empire. Salatin suggests that the moment you shift focus from quality to quantity is when you risk losing your business soul. "The day you aspire to become the biggest player on the block is the day you begin disrespecting other players," he says.

Polyface has defined its market by geography and time: a one-way four-hour drive from the farm. This represents a comfortable distance for making a delivery and returning home the same day. No mail orders, no distant bulk deliveries. Salatin recalls that Michael Pollan visited

Polyface Farm only because Salatin refused to ship him a T-bone steak. The rest is history: Pollan featured Salatin as a star of *The Omnivore's Dilemma*, propelling Polyface Farm into the national limelight. "Never underestimate the good things that can happen when you establish a business conviction and then stick with it," Salatin says. People want to patronize businesses that are driven by values, and being values-driven will provide a clarity and emotional freedom that is "palpable."

Use an incentivized autonomous workforce.

Salatin likes to call the farm staff a team. And as Polyface grows, instead of growing employees, it wants to grow a team of autonomous collaborators. Salatin has chosen the word "autonomous" carefully. Autonomy means not complete independence but collaboration, with each collaborator able to decide how much or little he or she wants to do. Incentivized means no salaries, and each collaborator's earnings are directly proportional to production.

"I especially dislike hourly wages because it's so hard to measure either success or performance," says Salatin. "It's very subjective. It becomes my word against your word." He tells me the story of a blue-collar worker he met at a local copper fittings manufacturing plant, who was being paid by the piece and "going like a one-armed paper hanger." This man was earning $60 per hour, and that's what Salatin calls "incentivized." The man could choose to work hard and make an extraordinary income – or not.

No initial public offerings (IPOs).

Growing a business through huge cash infusions can compromise quality, innovation, and the valued relationship with your intended market. Salatin believes a business should seek creative ways of financing its growth through local relationships – investors who know and care. This kind of slower, organic growth is more sustainable and helps businesses avoid growing too fast and losing sight of their core values. "Meteoric rises usually result in meteoric falls," Salatin says, "so beware fast cash and the imbalance it usually creates."

No advertising.

Again, the idea that no money should be spent on advertising flies in the face of standard business wisdom. But Salatin argues that advertising runs counter to the Polyface mission. If customers come only because of word of mouth, then it forces Polyface to make its customers happy. Instead of spending money to tell people how great Polyface products are, "we completely walk by faith in our customers that they will evangelize for us," says Salatin. "We cast ourselves on their recommendations for where this business goes." Polyface would rather depend entirely on word of mouth, as a way of ensuring that the farm never sacrifices quality.

Stay within the ecological carrying capacity.

Healing the land is a mission that requires sustainability. Polyface believes a healthy farm producing healing food is able to manage its own waste in productive ways, on-farm, and this clearly defines a farm's carrying capacity. Salatin argues that concentrated animal feeding operations (CAFOs) violate this principle by generating excess waste that the farmland cannot metabolize and must be transported off-farm. The simple decision of refusing to haul manure or waste anywhere motivates Polyface to devise creative solutions for managing animal manure in ways that support healthy land, animals, and community.

A second aspect of ecological carrying capacity encompasses the broader community. Polyface believes that a healthy farm producing healing foods will also support the ecology of its local community. Salatin believes that industrial processing plants "destroy the community ecology" when poor working conditions and low pay mean the industry must rely on low-paid foreign workers who often require additional support from the community.

Polyface believes that a healthy farm doesn't externalize costs to society; it doesn't create environmental damage, increase risks of food safety, or increase a social burden on a community.

People answer the phone.

Polyface is committed to never having a robot answer the phone. Salatin believes the human touch is not only more personable and satisfying for a customer-oriented business, but it also can be more efficient than the fancy answering systems that annoy and alienate people. He argues that farmers are in the relationship business and that Westerners are starved for human contact. "Shame on me if I shortcut this human touch and force my patrons to talk to a robot," he says.

Stay seasonal.

The Polyface team rejects the idea that life should be a treadmill. It believes in following the rhythms of nature, allowing the farm work cycle to reflect this natural ebb and flow. Polyface seeks to balance three seasons of intense work with one season of rest and recharging. "The assumption that scaling up the corporate ladder requires us to sacrifice our families and marriages is an unrighteous, evil axiom in America," Salatin says. The joy in farming is lost when it becomes a 24/7 operation with no seasonal respite. Salatin believes this is why so many industrial-scale farmers discourage their children from staying on the farm – it's no way to live. Polyface faces great pressure to scale up, become the new "Tyson" of pastured poultry, producing chickens year-round. But this would shift the farm into a year-round treadmill. Salatin advises businesses to build in breathing time and create time for rest and recreation.

Quality must always go up.

"How many businesses have grown up to be very lackadaisical big business? A lot!" says Salatin. He explains that his family doesn't ever want customers to come buy their food just because of the brand name. Polyface is proud that the quality of its chicken and beef is better today than it was years ago, even as production increased from 300 to 20,000 broilers and from 20 to 500 cows. Striving for continual improvement in healing the land will drive continual improvements in producing healing food. This is their bottom line: Always deliver quality food. Period.

Lessons Learned
Scaling Up to Meet Large-Volume Demand

**Join forces with other small-scale suppliers
to provide the quantities needed.**

Local growers often can't supply the quantities desired
by restaurants and other large-volume food-service
companies. Ever since the Chipotle Mexican Grill developed a relationship with Polyface Farm, all sorts of small
farmers have been calling the restaurant chain to try to
sell their products. But Phil Petrilli, Northeast regional
operations director for Chipotle, says that working
with lots of different farmers takes a great deal of time,
and it's not a worthwhile investment. He suggests that
small farmers consider aggregating into co-ops like the
Niman Ranch and Coleman Natural models, in which
small-scale, independent farmers and ranchers raise
livestock using sustainable (often organic) methods and
market their products under the co-op brand. Country
Natural Beef (page 261) is another co-op aggregating products from small-scale suppliers. Having that
recognized brand name benefits small farmers, Petrilli
believes. Also, co-op members share advertising and
marketing costs and are able to provide a host of benefits to members, such as group health insurance for
employees and national accounts for supplies. And, of
course, by aggregating products, a co-op can sell to
markets that require a volume greater than any one of
the co-op members could supply.

The amount needed might be less than you think.

Depending on the product, Petrilli suggests that an
individual farmer doesn't have to be able to supply *all*
that a restaurant needs to make it worth discussing. An
individual farmer being able to provide 30 to 40 percent
of the restaurant's lettuce for one month of the year, for
example, could be a significant contribution to the restaurant's business, particularly if the restaurant is pitch-
ing itself as a purveyor of local foods.

The nonprofit Appalachian Sustainable Development (ASD) in southwestern Virginia (page 253)
agrees. Contrary to common opinion, ASD says that
the supermarkets it works with have required neither
year-round supplies nor huge quantities of product, and

they have been very patient with ASD's ability to provide limited amounts. "If we can only provide 90 pounds
of peppers, they're willing to work with us," says founder
and former director Anthony Flaccovento. What supermarkets desire more than anything else is reliability and
predictability, he says. "If we promise them a quantity
and can't deliver that quantity, then they're frustrated."

The bottom line is that farmers cannot begin the
conversation if they can provide only a one-week supply of a product. If they can find a way to increase
supply through aggregation or by scaling up their own
production to meet 30 to 40 percent of a monthly or
annual supply, then they have the basis for beginning
a conversation.

Set a timetable for scaling up.

The challenge for small farmers is to scale up without
compromising quality, says Joel Salatin of Polyface
Farm. To provide the supply needed by a large institution means more infrastructure — more electric fence,
more paddocks, more waterlines — and more expense
and manpower. And with livestock, production follows
breeding timelines. As Salatin says, working with the
biological time clock is not the same as manufacturing
paper clips; you can't simply speed up the machinery.

Chipotle's Petrilli suggests that starting small is the
better approach, allowing you to work out the kinks in
the system before scaling up. "If we had demanded an
overnight supply, it could have been a train wreck for
Joel," says Petrilli. "And it could have been a train wreck
for Chipotle too — we might have ended up with the
wrong cuts or tough meat."

Salatin and Petrilli sat down together to develop
a timetable for introducing Polyface meat into the
Charlottesville restaurant. Salatin needed to triple his
pork supply and figured it would take a full 12 months
before he could supply 100 percent of the restaurant's
pork. Petrilli says they started very small, taking about
only 30 to 50 pounds of Polyface pork a week. "It was
less about serving Joel's pork than seeing if his locker
plant could butcher and package the meat the way it

was needed," Petrilli explains. "Both parties needed to know it could be done."

After this very small trial, Salatin gave Chipotle all the pork he could until he could fully ramp up production to the desired 300 pounds a week.

The experience of Applachian Sustainable Development underscores this point. ASD was able to get its feet on the ground as a distributor for small farmers throughout the region by first working with small to medium-sized grocery stores. Smaller stores have standards that must be met in terms of produce quality, but they tend to be more manageable for a young distribution system than the regulations and standards enforced by larger chains. "They allowed us to cut our teeth on this – everything from cooling, packaging, and grading," says founder Anthony Flaccovento. Over time, as ASD learned the ropes and streamlined its distribution network, it was able to approach larger grocery chains and successfully negotiate a partnership.

Create strategies to sell the whole animal.

"Nobody can do this unless you have a symbiotic patron chain – a diversified patron portfolio," says Salatin. When a large-scale customer wants only specific cuts of meat, the farmer needs to line up buyers for the remaining cuts, or he'll rapidly go out of business. Chipotle had wanted to buy Polyface chicken, but Salatin couldn't do it, because a symbiotic system wasn't yet in place. Chipotle wants only dark meat, Salatin explains, meaning that if Polyface wants to sell chicken to Chipotle, it also needs a large-volume buyer for chicken breasts, and building a bank of customers for that quantity of chicken breasts will take time.

In contrast, Chipotle nicely rounded out Polyface's portfolio for hogs. The Polyface buying club wanted ribs, bacon, and sausage, while white-tablecloth restaurants would take as much loin as he had to give, and he could sell the hams and shoulders to Chipotle. And the future looks good for beef as well. Chipotle wants to use the middle portion – round roast, top round steak, sirloin tip – which is Polyface's slowest mover, meaning Chipotle would be a helpful addition to Polyface's patron portfolio. Salatin recalls the fate of Buffalo Ridge, an artisanal beef producer in Virginia that went bankrupt

because, Salatin says, "they just kept selling the top end and had a million pounds of ground beef in cold storage in Richmond." If livestock producers don't move the whole animal, they can't be successful.

Tell the story to your customers.

Chipotle knew it was on to something when its Charlottesville customers began to ask, "What are you doing differently? This is delicious!" Customers had no idea they were eating the new Polyface carnitas burrito. Chipotle had wanted to fly under the radar with Polyface meat until they knew it was going to work, Petrilli explains. But management hadn't anticipated the backlash they would face when customers brought others back to the restaurant only to find that they weren't getting the same unusually good carnitas.

"This is probably the single biggest lesson we've learned," says Salatin. "If you're going to get serious and play this food game with an institutional outlet – whether it's a university dining service, Chipotle, or whatever it is – if you're going to step up, past someone's home freezer, you've got to be ready to put yourself out for continuity. Or you're in a bait and switch, which can jeopardize the entire operation because the institution doesn't want that kind of negative P.R."

"We didn't do a good job of telling our customers what we were doing," agrees Petrilli. Chipotle had not told its Charlottesville customers about its plan to slowly introduce Polyface meat, so customers didn't understand that they were part of a cutting-edge experiment. "Our customers are more disappointed and puzzled when they're not told. Because they want to know. So we've gotten better at telling the story." The Charlottesville Chipotle began posting a sign announcing when Polyface pork was featured and providing information on the story of their relationship. Customers became more understanding.

Petrilli says he's learned that it's important to champion the cause and take credit for supporting local farmers. And it's also important to educate your employees. When they did, Chipotle employees began asking, "When can we go out and visit Polyface Farm?" Petrilli shakes his head in wonder and asks, "How often are restaurant employees eager and excited about visiting the farmer?"

Lessons Learned
Running a Buying Club

Define your delivery parameters.

Polyface Farm in Virginia (page 220) has been running its successful Buying Club for years, bringing deliveries of Polyface products to surrounding local communities. The farm defines "local" as within one day's drive, there and back. Translated into practical terms, it has set a four-hour limit for making deliveries, which means its farthest Buying Club drop-off is Annapolis to the north and Williamsburg to the east. As part of this calculation — being able to return to the farm in the same day — there is a natural limit to the number of deliveries. Polyface's Buying Club manager, Sheri Salatin, says they have settled on a limit of three drops in one day. In 2009, with 29 drop sites, the Buying Club operated on a five-week rotation schedule.

Engage and reward your customers.

Showing your appreciation to customers is an incredibly important part of creating a relationship with people, says Sheri Salatin. She rewards customers who bring in a friend by giving them a free item; Buying Club customers who sign up a friend for the delivery service, she suggests, can be given $10 off their next order. A tool Salatin uses to stay connected to her customers is a customer-only blog, accessible through a private link. There, customers can post recipes and engage in lively, no-holds-barred exchanges. Just because they all like Polyface food doesn't mean they agree on everything, she says, and Polyface happily supports a vigorous discussion of important issues.

Use thermal bags, not plastic bags, for packing individual orders.

Polyface packs individual orders in bags and transports the bags in coolers. One lesson learned "the hard way," says Sheri Salatin, was to avoid using plastic bags, which tear easily. Now, Polyface packages its products in reusable custom thermal bags, which feature a zipper closure. Polyface tags the bags with sheep tags, an easy and effective way to label them.

Label coolers with letters, not numbers.

Another hard-learned lesson, says Sheri Salatin, is that a lettering system for the prepacked coolers is far more efficient than numbers. The letter call signs — "Delta Bravo" for DB, or "Zulu Charlie" for ZC — are simply easier to hear than numbers. With this system, Buying Club staff can quickly and reliably provide the right coolers to the right customers.

Use software that can integrate online orders to prepare invoices.

Until 2007, the farm was still doing invoices by hand, which was tedious and could result in mistakes. Salatin says that transitioning to an electronic accounting system was a big step forward, especially since it works seamlessly with their online order system. Today the Polyface Buying Club accepts online orders but not online payment. Farm products do not all weigh exactly the same amount and can only be invoiced when the individual product is weighed, so the amount due is not known until the delivery is prepared with individual products. At that time, when the weight is entered into the program, the software automatically generates an invoice.

Support other local farmers.

Sheri Salatin says that Polyface encourages its Buying Club customers to buy more locally — meaning that they should support any farmers that are more local than Polyface! And the Buying Club won't compete with more local farmers; it doesn't deliver to places where there are already adequate local food sources.

Building a Sustainable Farm Business
Hickory Nut Gap Farm

In **western North Carolina**, Hickory Nut Gap Farm is a working farm consisting of 350 acres, protected by conservation easement, that supports production of pastured beef, lamb, pork, and poultry, as well as eggs, fruits, and vegetables. Under the stewardship of Jamie and Amy Ager, and with the support of other family members who have begun their own enterprises, the fifth-generation family farm is being transformed into an environmentally and economically sustainable enterprise that also helps sustain the community through a broad range of activities and events that are both educational and fun. The farm's mission is to "connect sustainable agriculture practices, our family history, and our customers by sharing the family farm experience and serving as an example of healthy land stewardship."

Hickory Nut Gap Farm is all about family. In the rolling mountains of western North Carolina, the lush and verdant southern gateway to the Blue Ridge Mountains, Hickory Nut Gap Farm is proving that young, fourth-generation family farmers can succeed by following a new formula: selling high-quality, ethically raised meats to local markets. Hickory Nut Gap Farm lies off the beaten path, at a bend in the country road where it's hard to see the full reach of the farm. But across the road from the farm store, steers are munching away, their lowered heads nearly invisible in tall grass that reaches almost to their bellies. In a higher field pigs have gravitated to the shade of trees, where they are splayed out in cool dirt. Down the road a piece and around another bend, chickens are pecking at bugs in a field, kept near their mobile home by a lightweight mobile fence.

Jamie Ager is among the 23 descendants of Jim and Elizabeth McClure, a Presbyterian minister and his visual-artist wife who bought the property in 1916 and began restoring the old inn and surrounding farm. Today Jamie's mother Annie Ager and her five siblings own the farm, and Jamie and his wife Amy lease the farm from them. The family has placed a permanent conservation easement on the farm's 350 acres to preserve it as a working farm for all generations to come.[224]

The family is deep into farming. While Jamie and Amy run Hickory Nut Gap Meats, his mother Annie and aunt Susie run Hickory Nut Gap Farm Camp; a cousin and her husband run a nearby organic farm; and another cousin runs his own sawmill while his wife operates the North Carolina Organic Grain Project. A third cousin is cultivating a biointensive garden and permaculture orchard, while her husband is writing as a Kellogg Food & Society Fellow with the Jefferson Institute for Agriculture and Trade Policy.[225] Though other farmers may retire and sell their farms to development, this family is not about to sell its prized asset.

Pigs rooting in the woods, cows chewing grass and clover, and chickens pecking at insects, grass, and clover — this is how animals at Hickory Nut Gap Farm live their lives. Jamie and Amy Ager are among the new generation of "grass farmers" who are bucking the late-twentieth-century trend toward monocultures and intensified animal operations. Instead the Agers raise their animals entirely in pastures, on diets the animals would naturally eat. By increasing animal diversity and managing multiple species in high-density, short-duration pasture-grazing rotations, the Agers are practicing a new land stewardship and sustainable agriculture ethic.

Efficiency: The Green Revolution vs. Whole Systems

After World War II, the Green Revolution aimed to eliminate hunger and feed the world by increasing yields and making food production more efficient. This revolution was made possible by a convergence of federal policies that subsidized commodity crops and scientific brainpower. Led by plant breeder Norman Borlaug, the Green Revolution pushed crops into higher and higher yields. The new hybrid dwarf wheat put more of its energy into producing full, fat grains. New hybrids matured faster and enabled farmers to grow two crops instead of one. These new traits, coupled with cheap synthetic fertilizers to push plant growth, and herbicides and pesticides to wipe out all competition, produced nothing less than a miracle. Yields skyrocketed, doubling and tripling in a matter of a few decades. We measured farm efficiency by increasing yields, decreasing labor costs, and increasing economies of scale for processing and distributing food.

"Our agricultural system is an amazing feat," says Jamie Ager. "The bigger industrial model has been incredibly efficient. It's good and bad — now we're realizing some of the negative ramifications of that industrialization. On a community level, there's a disconnect between people and food. There's a primal need for knowing where your food comes from that wasn't taken into consideration."

Even as we were transitioning to this intensified agriculture, some people had doubts. Evidence began to accumulate that intensified crop and animal production had unanticipated costs for our environment and our health — polluted groundwater and streams, damaged soils, loss of valuable topsoil, and higher rates of cancers among people exposed to pesticides.

Some pioneering thinkers began to turn their attention to developing new kinds of efficiencies. On Cape Cod John and Nancy Todd founded the New Alchemy Institute in 1969 to develop a self-sustaining system for growing food in a small space, without harming the environment and without dependency on external energy and materials.[226] In 1971 Robert and Ardath Rodale bought a Pennsylvania farm to develop sustainable organic agricultural practices for maintaining soil and environmental health.[227] In Colorado Amory and Hunter Lovins founded the Rocky Mountain Institute in 1982 to develop practical whole-system energy-efficient solutions for building design, transportation, and renewable energy generation.

Today concerns for efficiency are giving way to concerns for the whole system — ecosystem management, sustainable agriculture, and full life-cycle analysis and cost accounting.[228] Today in places where these holistic approaches are taking hold — in agriculture, building design, city planning, engineering, corporate sustainability — efficiency is gaining an entirely new meaning.

As one example of this new ethic, Amy and Jamie Ager are practicing a kind of farming that restores and builds efficiencies in the form of healthy animals that don't need antibiotics or growth hormones; fields that don't need synthetic fertilizers; and food that doesn't clog our arteries but provides healthy nourishment. These are what I call *whole-system natural efficiencies*[229] — because they build on our world's natural systems, and because they benefit from whole-system stewardship.

In agriculture, whole-system natural efficiencies are realized through farming stewardship that supports the health and well-being of the entire system. A simple example of this stewardship is to mimic what occurs in nature. By following a cattle rotation with chickens, a farmer can mimic one of nature's systems in which birds follow the path of grazing animals to pluck insects from the piles of manure left behind. The farmer lets chickens do what comes naturally, eating a diet of insects, grass, and clover. In turn, the chickens contribute to soil fertility and tilth by breaking up and distributing piles of animal manure, and they interrupt the cycle of parasite infection by eating the larvae before they have a chance to reinfect the grazing animals. The chickens are healthy, the grazing animals are healthy, and the farm's soil also benefits. Surely this is efficient. This efficiency values all parts of a naturally occurring system and seeks to enhance outcomes for the whole without compromising or creating negative impacts on individual elements. There is an inherent core value of reciprocity: while one part of a system supports the whole, the whole supports the one part — all for one, and one for all.

Building the Farm Business

A visit to Joel Salatin's Polyface Farm (see page 220) in 1998 inspired Jamie Ager to begin considering the possibility of continuing his family legacy of farming. Both Jamie and Amy were attending Warren Wilson College at the time, and Amy was actually working on the college's farm.

"I grew up on the farm and liked doing it," says Jamie Ager. "But I was never encouraged to do it, because nobody in the family saw it as a reasonable livelihood." With his college visit to Salatin's Polyface Farm, Jamie saw that farming could be profitable. Salatin was making a living by operating a different kind of farm than what Jamie's parents and grandparents had run. And all the other pieces started to fall into place, too. Amy was learning how to market local food at the college farm, and Jamie had the family land and experience working with livestock.

So in their senior year, the two wrote a business plan. They combed through three years of his mother's farming business books to figure out what had or hadn't been profitable. The farm hadn't been run as a business to support a family, and the Agers were disappointed to discover there wasn't any part of his mother's existing business that could support them. So they came up with a plan: They would work in the family apple orchard for immediate income, and inspired by Salatin, they would begin a pastured poultry operation.

After their first year the Agers were able to spin off their own little beef cattle business. They were determined to make farming their main livelihood, so they found a way to begin acquiring the cattle owned by Jamie's mother. Instead of being paid by his mother for his labor, Jamie Ager would acquire value in the animals. By the end of the summer, Ager owned and managed the entire cow herd.

Jamie's next decision to grow the new family farm was counterintuitive. He decided to sell off his newly acquired herd. He wanted a different breed that would do well on grass. "Most feedlots in the U.S. want their animals to gain weight fast, which requires a larger-frame animal," he says. "Our challenge is to create quality beef with a lower-calorie diet, so we focus on a smaller, moderate-frame animal." Hickory Nut Gap Farm is in the "fescue belt," so Ager began by buying South Polls from Alabama, a breed that does well on fescue.

Because of her experience working at the Warren Wilson College farm, and because Jamie's family farm already had an established name in the community, Amy Ager says it was fairly easy to start building their new business. They started small in 2000, by selling at the farmers' market in Fairview, a small community outside Asheville. Amy says that starting young, right out of college, made it easier because they didn't yet need a lot of cash flow and because they had a strong social support network through Warren Wilson. The Agers worked their tails off during three seasons, she says, and used the winter season to learn and plan the next season.

Finding Creative Ways to Expand

Soon the Agers realized that the market was so strong for their grass-fed meat that they couldn't keep up with the demand. The next business decision was whether to keep doing what they were doing or to expand. "Jamie is our big-picture thinker," says Amy, "which allows us to stay ahead of the curve."

The farm's location set natural constraints on how much beef it could produce: only 90 of the farm's 200 acres are pasture, and the Asheville area offers little pastureland for lease. At the same time, if the Agers couldn't expand, they would never be able to produce the quantities of meat needed to break into the wholesale markets. So the Agers expanded in a different way: by partnering with other farmers.

Today the Agers' business model is to raise their own animals, increase their supply by buying animals from other growers who will raise them according to Hickory Nut Gap Farm standards, and market directly to the consumer. Through this business model the Agers are helping other farmers transition to more local markets and sustainable farming practices. The Agers tell me about one farmer, Sam Dobson, a sixth-generation dairy farmer, who was ready to branch out and do something different from his family dairy business. Now Dobson provides sustainably raised grass-fed beef to

Hickory Nut Gap Farm, which enables him to successfully supplement and diversify his dairy operation, which he is moving toward organic milk. Located two hours away from Hickory Nut Gap Farm, in a more rural part of North Carolina, Dobson is able to easily expand his operation by leasing pasture from farmers who are aging out of farming. He buys the calves, raises and finishes them on grass, and delivers them to the Hickory Nut Gap processor. From there Hickory Nut Gap markets his meat through Asheville outlets.

The Agers sell only their own beef at the local farmers' market, and they sell their partners' beef to wholesale outlets such as restaurants and groceries. All the poultry and pork they sell is still raised entirely at Hickory Nut Gap Farm. The quality of the beef is consistent and "awesome," says Amy Ager, and people are responding with increased demand.

Systems for Post-Industrial Agriculture

Jamie believes that the success of Hickory Nut Gap Farm reflects on the failure of the industrial agricultural system. "We've created a lot of cheap food in the world," he says. "Everywhere you look, it's efficient, and it's based on oil. But the industrial model lacks the human touch, the artisanal touch, and it doesn't help our spirit to connect to the food we're eating."

"Knowing where your food comes from makes you feel safer," Amy adds. "I feel comfortable with everything we eat — knowing how it was raised, how

it was handled, how it was slaughtered. It's the same kind of happy feelings your son has when he plants a sunflower seed and the joy he has at seeing it grow."

The march toward industrialization and greater efficiency is not going to end anytime soon, the Agers predict. They talk about how Brazil and India are going into ultralarge production on a scale that's difficult to conceive. Jamie Ager saw 70,000- to 80,000-acre farms in Brazil, very efficient, producing lots of food. He thinks industrial agriculture in this vein has a built-in mortality, because of its dependency on nonrenewable resources. He predicts this system will still be in place in 50 years but not in 300.

The Agers see themselves as part of a movement that is preparing for the era of post-industrial agriculture. "Our goal," says Ager, "is to start thinking through and developing a new system." The Agers are young leaders, inspiring and assisting others to follow suit in farming for local markets. They are developing unique systems of stewarding their land and animals. And they are also developing community networks and systems to reconnect people with food, farmers, and land.

Jamie doesn't romanticize his work. But he does say, "Good, clean food is worth it — and it will allow farmers to stay on the land." He dismisses the claim that "we can't feed the world on organic," saying nobody knows if it's true or not. "Our goal should be to get people to grow locally." Smaller, local plants process smaller quantities and are infinitely more safe, he says,

Grass-Fed Beef: The Healthy Meat Choice

Compared with grain-fed beef, studies show that pastured, grass-fed beef has a healthier nutritional content.[230]

- Less total fat, including less of the saturated fats that have been linked to heart disease
- 78 percent more beta-carotene
- 300 percent more vitamin E (alpha-tocopherol)
- 400 percent more vitamin A
- More thiamine and riboflavin

- More minerals calcium, magnesium, and potassium
- 75 percent more total omega-3 essential fatty acids
- A healthier ratio of omega-6 to omega-3 fatty acids (1.65 vs. 4.84)
- 500 percent more conjugated linoleic acid (CLA), which has been shown to combat cancer, body fat, heart disease, and diabetes
- More vaccenic acid (which can be transformed into CLA)

equating to smaller risks for our food supply. And, he continues, not only does the centralized food system present greater risks for food safety, but it also presents an easier target for intentional contamination. In our new world of living with the daily threat of terrorism, less reliance on a centralized food system is just common sense.

Lessons Learned
Making a Living on a Farm

Create a diversified business plan for long-term stability.

A farmer who treats the farm as a business, creating and following a business plan, is likely to enjoy a greater chance of success. Diversification is an essential component of a farm business plan. It isn't smart to have just one outlet for your produce, says Amy Ager of Hickory Nut Gap Farm in western North Carolina (page 234). A diverse marketing strategy will pay off in the long run, creating the stability a business needs to weather the ups and downs in each market. Hickory Nut Gap Farm has created a "three-legged stool" plan, wherein they aim to sell one-third of their products wholesale, one-third to restaurants, and one-third through direct sales at the farmers' market and their own farm store.

A diversified approach like this can help grow a loyal clientele. "If people come out to our farm and like what they buy, they want to be able to buy it in town," says Amy. "And if they buy our meat in town or enjoy it at a restaurant, they should be able to come out to our farm to see how it's done, to meet us personally."

To reach larger markets, small-scale farmers can pool their products.

Many farmers and organizations across the country have found that the best way for small-scale farmers to access larger markets is to pool their products. Some farmers have found success coming together as a cooperative, marketing their products under a single brand name. The Country Natural Beef cooperative (page 261) is one example. Another is the Rainbow Farmers' Cooperative of Milwaukee (box, page 76). In most cases, cooperatives handle the logistical details for farmers, from marketing to transportation and storage, letting the farmers focus on farming.

But a cooperative isn't the only aggregation model that works. Gladheart Farms in Asheville, North Carolina (page 87), shows that a farm can double as a wholesale distributor, marketing and selling products for other farms. And Appalachian Sustainable Development, based in Virginia (page 253), shows that a nonprofit working as a distributor for a network of small-scale farmers can be tremendously effective in building a local food system to support farmers.

The experience of Warren Wilson College (page 165), also in North Carolina, suggests that a wholesale distributor dedicated to serving food-service programs in large institutions is a promising business opportunity. A campus food-service program needs a consistent, dependable supply of food, whether it is serving a few thousand meals per day, as at Warren Wilson College, or tens of thousands of meals per day, as at larger state universities. Regardless of its dedication to serving local food, no food service that uses such massive quantities of food can afford the time to coordinate two cases of peppers from one grower, five from another, and nine from another. In this situation, a large food-service operation would highly value a central distribution center that aggregates products from local growers and can provide reliable quantities of local food. Ian Robertson, dean of work at Warren Wilson College, suggests that a wholesale service providing locally grown food to institutions would play a dual role: the classic wholesaler role of asking the institution what it wants and needs, and the additional role of educating the institution about what is seasonal and locally available at different times of the year so that the institution can adjust its menus.

Develop markets for "seconds."

One key for farmers is having multiple and diverse markets for produce that doesn't make the grade for commercial contracts. Supermarkets won't usually take, for example, cucumbers that are slightly curved, or zucchini that is slightly too fat, as their customers buy first on looks and second on quality. Generally, in any given year, as much as 30 to 40 percent of the produce a farmer grows won't be suitable for selling at the supermarket, says Anthony Flaccovento, former director of Appalachian Sustainable Development (ASD; page 253). So, until supermarkets and their customers change their expectations, farmers who sell local organic produce to supermarkets will have a huge amount of food that doesn't quite make the grade.

The logical solution to increasing farmer profitability is to find markets for these "seconds." ASD's first step was to develop the Healthy Families program, in which churches and civic groups buy these seconds from farmers at reduced rates and then donate the food to local services that feed the hungry, such as food banks. In its six years, Healthy Families has been able to provide an average of 60,000 to 90,000 pounds of produce each year to needy families. And, in the process, the program helps sustain local farmers; instead of writing off nearly half their produce as a loss, farmers are able to obtain a greater return for their effort and investment.

ASD has also expanded the market for seconds by selling to institutions – university and hospital food-service programs – that don't care if a cucumber is curved, the zucchini too fat, or the tomato unusually shaped. And more recently ASD added a third market, a partner who buys seconds to make salsa. Each of these markets pays a different rate, says Flaccovento. And while the lowest rate is paid by those who buy to donate food to the local food bank, that rate is still enough to allow farmers to at least break even.

Finding multiple markets for seconds is an important step to improving the economics of farming. Flaccovento makes a compelling case *against* asking farmers to donate their seconds to a good cause: donating seconds is one more way that farmers, the people who already bear the greatest risk in the system while also performing the hardest physical work, are asked to be the ones to sacrifice. And his argument resonates strongly with the philosophy of sustainability: efforts to grow a sustainable local food system will succeed only when farming becomes an attractive and profitable enterprise, worthy of our time, respect, and dollars.

For financial sustainability, build a CSA.

Many farms, whether for-profit or nonprofit, have found community-supported agriculture (CSA) to be a financial asset. Edwin Marty, executive director of Jones Valley Urban Farm in Birmingham, Alabama (page 158), suggests that a CSA may offer the "single biggest potential" for bringing stability to a farm's budget. While a CSA requires marketing and management, Marty says, it provides multiple benefits, such as building connections and long-term relationships.

A farm located near large businesses might consider offering them a CSA. Gladheart Farms' operator, Michael Porterfield, offered a deal to a large multinational industrial manufacturer: if the manufacturer could sign up more than 20 people in the CSA, Gladheart Farms would provide a weekly delivery to its lobby and give employees a 10 percent discount. The manufacturer – and its employees – enthusiastically agreed. The local hospital is interested in offering a similar deal to its employees. If a university offered a deal like that to all its employees, Porterfield suggests, a farmer might be able to sell the entire CSA to one place of business. Porterfield believes this is an avenue for CSAs that offers much promise.

Extend year-round income with a small nursery business.

At Greensgrow Farm in Philadelphia (page 96), Mary Seton Corboy reports, the greenhouses were initially used to grow hydroponic greens. Then, to gain the attention and business of the influential Philadelphia horticulture clubs, the farm transformed its greenhouses into a full-fledged nursery and began cultivating flowers. Today Greensgrow offers a broad range of nursery plants that sell from early spring through late fall, including fruit trees, shrubs, vines, grasses, perennial flowers and hanging baskets, and vegetable starts. The nursery brings in as much as 30 percent of the farm's revenues over the first two to three months of the year. It also brings in customers who become interested in other aspects of the farm, such as the CSA, bolstering overall sales.

Identify and build on your farm's "unfair advantage."

Every farm has its own "unfair advantage," says Amy Ager of Hickory Nut Gap Farm. Every farm has some unique characteristic. In the case of Hickory Nut Gap Farm, the Agers capitalized on the farm's heritage and name recognition in the community as a platform for marketing new grass-fed meats. When demand for their products began to outstrip their ability to supply them, the land-locked Agers – unable to expand on their own land – partnered with other farmers, who agreed to work under the guidelines established by the farm. In this way they were able to expand while retaining the trusted Hickory Nut Gap Farm name.

"You can make a living on 2 acres, if you're really creative," says Ager. "But you can also not make a living on 2,000 acres, if you're not creative." Contrary to the adage of Earl Butz, secretary of agriculture under President Nixon, who advised farmers to either "get big or get out," Ager says that farm size does not predict or determine success. She believes success is determined entirely by the creativity of the people who are running the farm. Decide what your farm's unfair advantage is, she advocates, and build on it.

Agritourism builds community and profit.

Many farms, including Hickory Nut Gap Farm in North Carolina, Polyface Farm in Virginia (page 220), and Greensgrow Farm in Philadelphia, have found that opening their gates to agritourism is an effective way to diversify their income stream. Not only can agritourism serve as a financially stabilizing factor but, depending on the kind of agritourism offered, it has the potential to become a major source of profit. Agritourism might mean offering an old-fashioned hay ride, planting a corn maze, or hosting special events such as chef-cooked dinners or food festivals.

The disconnect today between people and the sources of their food has important consequences for farms that open their gates to the public. For many people a visit to a farm will represent a completely unique set of experiences, say Amy and Jamie Ager of Hickory Nut Gap Farm. It may be the first time a child has seen a cow chomp on grass, touched a pig, held a baby chick, or picked a pumpkin. And this means a family may view the farm as a "place to roam free and do what we want." The Agers have seen parents open a gate to allow a child to get closer to a bull or ram, not realizing the dangers, or let their dog loose to roam free.

Faced with this, some farmers might decide to permanently lock their gates. The Agers believe it is well worth it to allow visitors, as it not only helps educate people but builds relationships with their customers. So the Agers have learned that everything about the farm must be approached with a beginner's mind. They use every opportunity to educate visitors about the farm to ensure a safe and fun time for everyone. In advance of an open-farm event, the Agers suggest creating a guided event by preparing educational signs about everything relevant to your farm:

- The role of gates
- Life cycles of each animal
- Information on the different breeds
- Information on how each animal is raised
- Information on each crop and the crop cultivars
- Information on the larger ecosystem, such as bees and pollination
- What to avoid, such as the location of electric fences

This makes the farm a great learning experience for the entire family. "And it's better than a movie," says Amy, "because you can hang out a lot longer." In addition, so that farmers can get the farm work done while also allowing for public interaction, the Agers have learned several important practices:

- Make sure you have insurance policies to cover public visitors to your farm.
- During public visit times, rearrange the location of animals so that they are kept within nonelectric woven fences. The public should not be allowed to get close to electric fences, and should be told their location, to ensure safety.

- Clearly define the boundaries of your farm and the boundaries of what is open to the public.
- Hire extra helpers for parking.
- Do the little things needed to make the public more comfortable — weed-whack, put up directional signs, or mow a walking trail through the field.
- Bring in additional activities to increase the fun for children, such as face painting.
- Because a farm is such a new experience, people do not always feel comfortable on a farm, say the Agers. So it's important to do everything possible to help people feel comfortable and have a positive experience.

This kind of interaction with the public may not be for all farmers, the Agers admit. But the Agers' decisions show they care about more than growing grass-fed meats for their community; they also care about being a resource and growing mental, physical, and spiritual connections with their community.

At Polyface Farm, the Salatin family has been forced to learn how to manage an explosion of public interest. They've developed a set of guidelines that allow the farm to continue operating while valuing their customers. To begin, they open the farm store just one day a week. Customers can make purchases on other days, but they need to call in advance. The Salatins also have established clear parameters for their open-door policy. From December 15 through February 28, the farm is closed to visitors, except on Saturdays. From March through December 15, the doors are always open. People can come for a free self-guided tour and look around; Polyface feels this is important to maintain transparency. The farm also offers guided group and individual tours of the farm from March through October, for a fee.

8: Infrastructure
Building Local Food Networks

Sometimes the most important work is not the most visible. The face of local food is most commonly portrayed by bucolic images: the farm landscape with barns and sheds; the pumpkins, apples, chickens, and cows; the beneficent chef putting finishing touches on a gazpacho made of local tomatoes or a tart of local potatoes and shiitakes. And let us not forget the children in gardens, beaming at their first carrot pulled from the earth, or the impossibly small chick cupped in their hands.

Rarely do people see or think about the go-betweens, the persons or organizations that connect the dots of local food infrastructure. These go-betweens may be the most unsung heroes of community food movements. And like invisible glue they can be the force that actually creates and holds a local food system together — and enables it to thrive.

Throughout our nation, as the food industry consolidated and streamlined, local distribution networks slowly dissolved. Local slaughterhouses and butchers, local granaries and mills, local canneries and packing houses, local dairies and creameries, and local fruit and farm stands — all became faded memories of an earlier time. Now, as the demand for local, fresh food grows, communities are realizing that they need to re-create many of these key pieces of infrastructure.

Where there is a will, there often is a way. And when there are new markets, new businesses can emerge. A sustainable community food system requires weaving a web of connecting threads. These connectors can be all sizes and shapes, serving communities as small as city neighborhoods and as large as the entire United States. Regardless of their size, these innovative enterprises are stimulating their local economies, growing new "green" jobs, returning profits to local producers, circulating food dollars within local economies, and creating markets that support greater numbers of small-scale producers and farmers.

"Infrastructure" may seem a cold, impersonal word implying technical and mechanical systems that require the attention of engineers. Yet nothing could be further from the truth. It is here where just one person with a good idea can almost single-handedly transform the face and future of a community food system. In my own community Kate Collier, an entrepreneurial woman with guts and business gusto, decided it was high time for farmers and institutions to have an easier way to connect. Rather than 50 farmers each making 20 deliveries to different stores and restaurants — a thousand stops overall, demanding gas and precious time away from the farm, as well as a thousand chunks of receiving time from store and restaurant personnel — she envisioned a community service that would streamline this task. Collier founded the Local Food Hub, a community-supported nonprofit that collects and delivers farmer produce. By aggregating this produce in sufficient quantities to broker deals with the local university, hospital, and other institutions, the Hub has immediately and profoundly altered our community's access to fresh, locally grown food.

This is community food infrastructure with a face and a heart. Each story in this chapter begins in a similar place — with people who have passion and vision for increasing their community's access to fresh, healthy food, and who have found creative ways to connect the dots.

Facilitating Farm-to-Market Connections
Appalachian Sustainable Agriculture Project

with research by Robin Proebsting

The face of agriculture in western **North Carolina** is being transformed by the work of the Appalachian Sustainable Agriculture Project, as this former tobacco-growing region evolves into a thriving center for small sustainable food farms. Centered in **Asheville**, and supported in large part by grants arising from the 1998 tobacco settlement, ASAP is a nonprofit founded in 1995 whose mission is to "help local farms thrive, link farmers to markets and supporters, and build healthy communities through connections to local food."

A business can be successful (and sustainable) only when it has a market to sell to. Without a market a farmer can't survive. Marketing food might seem like a simple proposition: Take a sample of your produce to the local markets, and see how much they'd be willing to buy and at what price. Then figure out if it's worth your while, and if it is, deliver what you promised.

But farm business has probably never been quite that simple, and today's markets are not easy to understand or penetrate. Neighborhood groceries are often owned by national or international food conglomerates that aggregate food grown by farmers all over the world. And connected to these companies is an intricate network of distributors who aggregate food shipments to supply hospitals, schools, retirement homes, restaurants, and grocery stores.

Today, if farmers are not serving the industrialized system of aggregation and distribution — if they want to stay small and serve their local community — they face many challenges. While there is the obvious challenge of long hours of hard physical work, many challenges have little to do with knowing how to grow crops or raise animals. They concern the business of farming: identifying the right crops or products that in smaller quantities will bring in a profit; finding markets that will consistently buy smaller quantities; branding, packaging, transporting the products safely to avoid spoilage; and meeting food safety, liability, and

insurance requirements. Many of those challenges can seem so insurmountable that even the children of farming families, those who might have the most farming savvy and the most to lose by leaving their family heritage, are nevertheless leaving the farm behind.

But farmers love to farm, community organizer Charlie Jackson insists. "If you can help them at a few critical points, they'll make it work." With that premise, Jackson founded what would become the Appalachian Sustainable Agriculture Project (ASAP) in 1995.

Jackson, a North Carolina native and longtime fan of Helen and Scott Nearing (whose early book, *Living the Good Life*, played a role in inspiring the 1970s back-to-the-land movement), began his work with farmers in western North Carolina, whose long heritage in tobacco farming was coming to an end as the industry shifted tobacco production overseas and to larger farms. These farmers were facing fearsome losses that would have dramatic impacts on their families, as well as on the vitality of their communities. And because of the mountainous terrain, the farms of the southern Appalachians were smaller, averaging about 80 acres, or less than one-quarter the national average. They could not rely on crops that require high-volume production.

At first Jackson thought the challenge was to figure out which crop farmers could grow with their existing equipment to bring in the same high value as tobacco. With this goal ASAP began the "broccoli project." It didn't take long for Jackson to see why the plan wasn't going to work. The farmers were innovative and could easily adapt to growing broccoli, Jackson recalls, but they hit a brick wall in finding markets. Because grocery stores had become part of a vast multinational network, with inventory and ordering processes dictated by national headquarters, they no longer needed or wanted locally raised broccoli. There was little mechanism or incentive for the local stores to buy local food.

Jackson's proverbial lightbulb went off. ASAP decided to let farmers do what they do best, instead concentrating its own efforts on challenges farmers were least equipped to handle: creating markets. In the 10 years since that turning point, ASAP has learned how to push demand and create markets where there once were few to none. Jackson leads an expert and passionate staff who operate on a budget of more than $1 million. Together, as change-agents pushing the frontier of creating local markets for local farmers and reconnecting people with good food, they are focused on learning — and sharing with others — how to effect long-term change in the infrastructure of local food supply.

Building Demand and Markets

Jackson emphasizes the necessity of doing some basic research on your local food system to identify the options for marketing local food. This sort of assessment doesn't have to be academic or exhaustive, but it does need to provide enough information for you to understand what your region has — its strengths and assets, weaknesses and opportunities — so that you can begin to make strategic decisions.

When ASAP began talking about "local food," the first gap they discovered in the local food system was simply the concept: it was so foreign that ASAP needed to focus its efforts on education and creating demand. One of the easiest and most effective starting points for creating demand for local food, Jackson says, is to produce a local food guide. "People get it when you talk about the benefits of local food," says Jackson. "Their next question is, 'Where do I find it?'" As one of its first initiatives, ASAP produced the region's first

local food guide and distributed it for free. Advertising in the guide helps pay for its production. "The power of the guide cannot be underestimated," Jackson maintains, noting that many of the farmers he works with claim that the guide is their number-one marketing tool.

Jackson maintains that the key to building markets for local food is all about building relationships. Meet with people, he says. Have conversations. Listen. Learn about their particular needs and interests. Ask community organizers, members of local organizations, heads of local businesses, administrators of local schools — anyone and everyone — whether they would have an interest in and a market for local food, if a supply could be arranged. The goal is to explore, educate, and see where possibilities for markets open.

Above all, Jackson warns, don't tell people that their current food choices are "bad" or preach about what they should be doing. Focus on the benefits of local food, for the people who consume it, the people who provide it, and the community as a whole. In this kind of marketing, a positive approach is far more powerful and helps build trust and relationships.

After producing the local food guide, ASAP began to help develop and expand local farmers' markets, and it continues to assist them today with developing marketing plans and designing their messages and weekly ad campaign; it even offers them matching funds for development and expansion. ASAP also believes that local food should be for everyone and works with farmers' markets to develop ways for people on food assistance to use their electronic benefit transfer (EBT) cards or coupons at the markets. The group holds workshops and trainings to professionalize the markets, and it even created the Asheville City Market, a demonstration market that it manages in downtown Asheville.

Building Farm Business

As demand for local food grows, so will farmer interest in growing food for local markets. And this is where ASAP fills a second important gap — helping new or existing farmers build their capacity to serve the local market. ASAP program director Peter Marks, with many years' experience in small business development

Food Assistance Programs at Farmers' Markets

Federal and state food assistance programs these days most often work through the use of electronic benefit transfer (EBT). Users are given an account into which their cash benefits are deposited each month, and they access their account through a debit card. This system can be a hurdle for farmers' markets wishing to serve low-income populations, as EBT requires secure Internet access, which can be technically difficult as well as costly to establish. And farmers usually accept checks or cash only.

There are various ways to overcome this challenge at farmers' markets. Most markets offering EBT operate on some kind of scrip system, wherein EBT users exchange their cash benefit for tokens or certificates, which vendors treat as cash. ASAP advises that at any farmers' market that uses such a scrip system, the booth or table where users redeem their benefits for scrip should be clearly marked and should have signs explaining the process. ASAP also publishes "Best Practices for Accepting EBT at the Farmers Market" (see the resources section), a pamphlet on how to reach out to EBT populations and how to prepare the market to process EBT transactions.

---

in the private and nonprofit sectors, heads up this initiative, helping new farmers get off the ground and experienced farmers expand into new markets.

Marks helped Michael Porterfield, of Gladheart Farms in Asheville (see page 87), prepare financial and farm production projections and write a small grant application for his farm. Porterfield says he has farming friends all over the country, and "nobody has a resource like ASAP!" Continuing, he explains, "I can call [ASAP] with questions about marketing, or buyers for stores. They're in the business of connecting people, and they'll even make the initial contact for you, to introduce you." ASAP also connects Porterfield with other farmers, he says, so that he can troubleshoot about farming issues, such as dealing with a disease in tomatoes. "We had a mission before we even came into contact with them," Porterfield says, "and they've helped us realize that mission."

ASAP works with all local farmers equally, whether organic or conventional growers. "We don't think change should come just from the farmer," Jackson says. "It should come from the market to be sustainable." If people start demanding more organic food, Jackson says, farmers will step in to meet that demand. And, in fact, many farms in the region are now shifting to using sustainable and organic methods.

Growing Confidence in Local Food

In the process of growing markets, ASAP discovered that farmers weren't the only ones needing a facilitating hand. Retailers also needed assistance in working with farmers. Now a large part of ASAP's work is filling the third major gap in the local food system: serving as a relationship broker for the retail market — groceries, restaurants, hospitals, and other sellers (or potential sellers) of local farm products.

ASAP found that retailers were concerned about working with local farmers on several fronts, most importantly in terms of consistency, quality, food safety, and being inundated by farmers new to the retail business who wouldn't be familiar with their requirements for adequate high-quality product. Almost like a dating service, ASAP began to arrange meetings between retailers and farmers and helped make sure the matches were appropriate. Unlike a dating service, however, ASAP does *not* charge for its services, nor does it take a cut of the deal. ASAP brokers only the relationships; it doesn't broker the deals. (ASAP gets funding for this work from a variety of other sources, such as tobacco settlement funds to help former tobacco farmers become more economically viable and partner-retailers who may demonstrate their support of the service by becoming a sponsor or paying for marketing materials.) Every year now ASAP holds a marketing conference where retailers and farmers can meet each other.

People from all corners of the food system have learned that they can call on ASAP for help, and through its work ASAP has learned that a rubber-stamp approach doesn't work. Sometimes, after sitting down with a buyer to learn what he or she wants, ASAP may take the buyer out to visit a farm or may just share the buyer's interest with different farmers and let them contact the buyer directly. Sometimes, knowing there is a new buyer and market, ASAP will work with farmers to help them understand what would be required and crunch the numbers to see if the retailer is the right market for them. Sometimes the retailer and farmer meet at ASAP's annual marketing conference and work everything out on their own; in these instances, ASAP might provide marketing assistance in the form of brand-name "posters" about the farmer for signage at the store. Each situation is different.

In concert with brokering these relationships, ASAP began to explore the idea of creating a local "brand" that would assure quality through annual certification. This would be a win-win-win: retailers would be assured of consistently high quality, farmers would gain access to new markets, and the public would have a way of knowing it was buying something truly local.

Creating a Certified Local Brand

When ASAP's local food guide and educational efforts took off, people began to ask for local food.

But unforeseen problems began to emerge as stores jumped on the "local" bandwagon: There was no unified definition for local. Stores were deciding for themselves what they would call local, creating confusion and disagreement among consumers and producers. Does local mean anywhere in North Carolina? What about food from just a few miles away over the border in Tennessee? Can a store advertise "local fish" trucked in from the coast hundreds of miles away? What definition works best, and for whom?

ASAP soon realized that "local" needed a commonly accepted definition to build value and trust among consumers, retailers, and growers, thereby benefiting everyone, as well as ensuring that the money spent on "local" food stayed with local producers. Otherwise the entire notion of "local" was at risk. After much research and discussion, they agreed that the general consensus in the nation defines local as "within 100 miles." Looking at their own cultural, agricultural, and geographic identity, ASAP was pleased to find the 100-mile criterion would work well in their region. Running along the spine of the southern Appalachian mountains of Virginia, Tennessee, North Carolina, Georgia, and South Carolina, 100 miles forms an oblong circle, with Asheville in the center.

ASAP then developed a certification program for farms within its local region, along with a trademarked brand name, "Appalachian Grown," and an attractive logo to foster recognition. ASAP now certifies Appalachian Grown products from an increasing number of farms every year; in 2010 it was able to boast

Bright Idea: Foster Economic Development with a "Foodtopia"

Economic development is not always connected to farm and farm tourism development. In North Carolina ASAP is actively facilitating connections between economic development, tourism, and farms to make the Asheville region a farm and food destination. Some regions, especially those that have lost their manufacturing base, may find food a profitable pathway for economic development, particularly if it connects local farms and small businesses in a comprehensive food-oriented development strategy, argues Peter Marks. As an example of what can be done, the city of Asheville now promotes itself as the "World's First Foodtopian Society" and boasts that it has 17 farmers' markets, 250 independent restaurants, seven microbreweries, and a long history of "field to table." Its website (www.explore asheville.com) offers a range of ways to interact with the Foodtopia — local food adventures, Foodtopia events, ask a local farmer a question, profiles of local Foodtopians, food-oriented getaways and packages, recipes, and more.

432 certified farms. It develops displays and signage to tell the story of each farm and farmer, to help create relationships of value and convey the authenticity of Appalachian Grown. The group does have to work with the stores — individual businesses as well as larger chain groceries — to make sure that noncertified products aren't mixed with and sold as Appalachian Grown products. Sales of Appalachian Grown products nearly tripled from $10 million in 2007 to $28 million in 2009. Over the course of its project, ASAP estimates the program has branded a total of $53 million in food sales.

Facilitating versus Competing

While ASAP uses a strategy similar to that of co-ops such as Country Natural Beef (see page 261) in creating its own brand to build recognition, value, and trust, three things set ASAP apart: it is not a producer, it does not make money on the sale of Appalachian Grown products, and it doesn't operate its own distribution network. Nor does the organization ever intend to do any of those three things. It offers a community-based, not a commodity-based, model for marketing local foods.

ASAP is fortunate to have sources of funding that enable it to play this unusual role of a purely facilitative go-between. What would happen if its sources of funding were to dry up? Jackson invokes change theory to explain why, even if it were to face economic hardship, ASAP won't ever get into the business of distributing food.

For change in the food system to be sustainable, Jackson argues, it must be supported in the marketplace. This means the impetus for change must come from consumers, not farmers or producers. So ASAP works from the bottom up, focusing primarily on raising consumers' awareness about the value of local food. As consumers begin to ask for local food, their demand transforms the chain of food supply — a change that will be more sustainable.

Because of its core belief about how sustainable change happens, ASAP purposely works within the existing system. It doesn't want to create a new distribution system that would compete with existing distribution businesses. Though Jackson acknowledges that some communities may not have a good distribution network in place and may be forced to develop or encourage their own distribution system for local food, western North Carolina is endowed with a robust and growing infrastructure, he says, and ASAP doesn't want to re-create the wheel. The sustainable way to drive change here is to support the existing chain of commerce, as well as new entrepreneurs, in getting local farm products to market.

Fresh Food for All

Most people, ironically, do not think of the places that care for our sick, elderly, and children — hospitals, senior centers, and schools — when they think of good, healthy food. But with ASAP's help, many such venues in western North Carolina and neighboring regions are beginning to join the local food movement. By featuring fresh local foods and offering food- and farm-related educational programming, these centers can improve community health while also sustaining farms.

Through its farm-to-hospital program, ASAP has helped some regional hospitals organize CSA pickups for their employees as well as farmers' markets right in hospital corridors or parking lots. Some hospitals are even bringing fresh, local food into their cafeteria food service, though incorporating local food into hospital patients' diets remains a challenge, because dietary restrictions govern the content and preparation of their food. But given the momentum and interest, it is beginning to make its way onto patient trays.

As with all its programs, ASAP wants to do more than just provide food. Its farm-to-hospital program also offers cooking classes, wellness programs, and field trips. "Hospitals have the potential to act not only as a resource to treat the sick, but as a community role model for proactive healthy choices," ASAP says on its website.[231]

ASAP has also joined with the Buncombe County Council on Aging to launch Project EMMA — its motto, "*Eat Better, Move More, Age Well*" — an effort to bring local, nutritious food as well as exercise to seniors. With a grant from Blue Cross Blue Shield,

ASAP has created a vegetable and herb rooftop garden on a downtown apartment building, so its seniors can enjoy gardening and fresh food. The group hosts walking trips for seniors to downtown tailgate markets and takes seniors on field trips to farms. And to increase affordability of local food, qualifying seniors can receive vouchers through the USDA Senior Farmers Market Nutrition Program to buy fresh fruits and vegetables at local tailgate markets.

Its most ambitious program in this vein, however, may be Growing Minds, a Farm to School program that works throughout the region and is also the Southeastern regional lead agency for the National Farm to School Network. The program focuses on helping teachers create and use school gardens, take kids on farm field trips, and cook in the classroom to achieve lesson goals and competencies they are already required to teach. "The goal is not to add anything to [the responsibilities of] already overburdened teachers," says Emily Jackson, director of the program. (See page 150 for more on Growing Minds.)

Growing New Farmers

"There's little affordable land in the mountains anymore," says Charlie Jackson, the director of the Appalachian Sustainable Agriculture Project. He is reflecting on what may become ASAP's next phase of work. He thinks ASAP's success in creating more demand, and therefore more markets, means that its next step will need to be creating more supply — i.e., more farmers. This is perhaps the hardest nut in the local food system to crack. With 40 percent of farmers at 55 years or older, farmers are an aging, and dying, breed.[232] "Most farmers don't have heirs that want to come back to the farm, Jackson says. "Or they have multiple heirs who split up the farm."

Some organizations, like Nuestras Raíces in Holyoke, Massachusetts (see page 123), are working on growing new farmers through incubators. Others, such as Southside Community Land Trust (see page 65), offer term-limited four-year internships that incorporate both farm and business training. But incubators are not necessarily the answer, because graduating farmers still need to find land. And unless you're looking for abandoned land in places like downtown Detroit, land is still expensive most everywhere.

Jackson isn't sure how to crack this nut. But he's got an idea. Why not try to attract farm heirs *back* to the farm? The vibrant local food scene in the region already attracts young farmers, young farmers who don't have access to land. The children of farmers are leaving, going to college, finding careers off the farm. But some children are coming back to the farm, Jackson says, because they want to farm the family land. He points out the obvious: Children of farmers already have access to the land. They already understand the value of farms. They already understand the quality of that kind of life. And when they return to the farm, they bring fresh ideas and perspectives. Jackson believes this is where organizations like ASAP can make the biggest difference in creating the next generation of farmers. Because these young people already have access to the land, luring them back to the farm with a new business model and local markets is probably the easiest, first strategy — the lowest hanging fruit.

Peter Marks says that ASAP's strategy is building on the work of long-standing farm-support groups — Farm Credit, Farm Bureau, and Future Farmers of America — who work hard to encourage and recognize the younger generation who choose to come back to the farm. ASAP is now working with a number of young people who have returned to their family farms and become successful farmers, and is developing a strategy to help others make the same decision. Jackson wants to get their stories out into the schools in the surrounding rural communities so that others will understand success is possible and realize they should consider farming as a career.

In other words, Jackson is beginning to think, once again, about messaging — how to communicate the idea that farming today doesn't have to be a back-breaking dead end. Most young people leaving their family farms are discouraged firsthand from staying on the farm by their own parents, Jackson says. So his idea is to expose young people to different voices — voices of creativity, energy, success. He will do this by bringing young farmers into the schools, where they can tell their own stories, firsthand. That will be more effective than having anyone else talk to discouraged farm heirs, Jackson says.

Another idea, he says, is to create longer-term pathways to land ownership. There are a lot of owners of large farms in the region, he says, who no longer want to farm but do want to keep the land in farming. Jackson believes these farms offer pathways for new farmers to gain ownership. Long-term leases, living directly on the part of the land being farmed, and building equity in the property are all options worth exploring.

ASAP's track record is impressive, so when ASAP's multitalented staff decides to tackle a new problem, it's a good bet that we will all benefit from its work. If ASAP decides to crack this nut of creating our next generation of farmers, I'll place my bet now that in another 5 to 10 years communities throughout the nation will have creative and proven models to apply in their own regions. And that, we can all agree, will be a good thing, because having new farmers means America will still be growing its own food. And more, new farmers growing for local markets will help grow local economies, will introduce fresh foods into community schools and institutions, and will create relationships that sustain community land and spirit.

Finding the Courage to Dive In

The food system is incredibly complex, says Charlie Jackson. Though ASAP knows more and has been working in it longer than many others, Jackson emphasizes how much he and his staff *don't* know. "We're still really beginners at understanding the distribution of food," agrees Peter Marks. I ask them how this is possible. If *they* don't comprehend their local food system, after working in it for over a decade, how can anyone hope to?

Jackson says they are constantly discovering new players in the system. They might suddenly learn of a distributor, whom they've never even heard of, who has been operating in the community for many years. "Pull on one thing, and you'll find it's connected to everything else," Jackson says, paraphrasing John Muir.[233] There are no isolated pieces in a food system. If a hospital wants to start bringing in local food, ASAP is often asked to work with its food distributor. This leads to conversations — finding out whether the distributor is already carrying products from local farms and what else it might be willing to carry. This, in turn, leads to learning about the distributor's other institutional clients. And this leads to conversations with those institutional clients about whether they, too, might want to begin offering local food. One new relationship leads to many more new relationships. All are connected, and new connections are constantly being discovered.

"People are getting excited about these ideas, and trying to dive in, and many are really clueless," Marks says, then laughs. "We're only slightly less clueless." It's clear to me, however, that ASAP is successful, respected, and growing. It is anything *but* clueless. ASAP has one of the longest and strongest success records of growing local food demand, markets, and

Assistance for Community Food System Assessments

Because of its long and successful track record, the Appalachian Sustainable Agriculture Project is now being asked to help other regions expand their local food economies. There are many ways to do this work, and the method selected should reflect the community's goal. Most efforts begin with a community food-system assessment. ASAP conducts assessments that are designed to be practical and implemented. ASAP director Charlie Jackson says that the goal is to develop markets for farmers. ASAP will begin by asking core questions about community demographics, current knowledge of agricultural and food trends in the community, and community needs and assets for supporting local farms and building local markets. To explore the possibility of working with ASAP to expand your local food economy, contact ASAP directly.

farmers. And perhaps one of the reasons they are successful is precisely because they *don't* think they know it all. They are constantly learning and open to learning more. I get their point: success depends on a positive, learning attitude mixed with humility. This is how change happens: with good people diving in, willing to learn, willing to grow new markets and relationships, willing to help their community grow.

Lessons Learned
Building a Market for Local Food

Market with redundancy and repetition.

"Local Food: Thousands of Miles Fresher" is the mantra of the Applachian Sustainable Agriculture Project (ASAP) in Asheville, North Carolina (page 243), which works to build markets for local food. ASAP director Charlie Jackson calls this the "gateway message" to a deeper understanding and commitment to local food, and he says people eventually get a simple message, if it is put out over and over. One reason ASAP has been so effective is that it repeats its message in multiple media and social networking paths — an online calendar, press releases, Twitter, Facebook, a blog, a listserve, and an electronic newsletter. Other cool marketing tools ASAP has developed include:

- **BUMPER STICKERS.** ASAP has given out hundreds of thousands of bumper stickers carrying the message "Local Food: Thousands of Miles Fresher." Given a limited messaging budgets, a bumper sticker is an affordable method of branding for small organizations.
- **LOCAL FOOD GUIDE.** ASAP says this is among the most powerful tools to build awareness of and interest in local food. The group expects to give out its millionth copy in 2011.
- **REGIONAL FOOD MAP AND TRIP PLANNER.** This map of southern Appalachia identifies tailgate markets, farm stores, roadside stands, farms to visit, you-pick farms, wineries, restaurants serving local food, grocery stores selling local food, and B&Bs and farm lodgings. ASAP distributes the map for free.
- **FARM TOUR.** ASAP annually organizes tours of more than 40 farms in six counties over a weekend. It provides the farm tour map and an admission button

and tells people to "gather friends, fill a car, choose the farms you want to visit, and plan a route. Come with a cooler and your appetite."[234] The tour features farms with vegetables, meats, trout, cheeses, honey, fresh bread, artisan ice cream, yarn, soap, crafts, and a garden labyrinth, as well as demonstrations of border collie shepherding, apple butter making, maple syrup tapping, wool spinning, milking, and hydroponics.

- **HARVEST FESTIVALS.** ASAP helps regional farmers put together and promote harvest festivals, celebrating everything from apples to pumpkins to the entirety of the fall crop.
- **"GET LOCAL" CAMPAIGN.** ASAP's "Get Local" campaign focuses on one seasonal food to feature each month of the year, such as squash in July, apples in September, and trout in February. ASAP identifies farms willing to supply that food, distributors willing to carry it, and restaurants willing to feature it in chef-prepared dishes, then develops marketing materials to promote that food at local eateries and farmers' markets.
- **RESOURCES FOR TEACHERS AND PARENTS.** ASAP produces *Hayride*, a guide to farm field trips for teachers, and recently published the *Local Food Guide for Kids*, with accompanying lesson plans and a "kid-size" bumper sticker.

Become a reliable source of information about local food.

Newspapers and television and radio stations are all strapped for money and staff, so one way to market your work successfully is to become a reliable source of information for the local media. Beginning marketers, says Charlie Jackson, should work with local media on a regular basis to build credibility and legitimacy. ASAP began providing a weekly news release to the largest local newspaper with information on the week's fresh produce at local markets. If you provide the article, newspapers, particularly the smaller ones, will often run it in entirety and will also provide a link to your website. Radio stations, too, will often make announcements for you. ASAP makes a big effort to provide good, reliable, up-to-date information, and Jackson believes it has paid off.

Be patient.

Change doesn't happen overnight, ASAP staff say. So if you're trying to build a local food system by facilitating relationships between farmers and buyers, it's important that you're willing and prepared to work slowly and be in it for the long haul. Relationships and trust take time to build.

Develop a certified brand name for your region's local food.

Certification of a local-food brand name provides buyers with the assurance of local provenance and high quality. In building buyers' confidence about local food, it permits transactions between farmers and distributors that otherwise wouldn't be possible. ASAP has had success with its "Appalachian Grown" brand, and Appalachian Sustainable Development in southwest Virginia (page 253) has seen similar success with its "Appalachian Harvest" brand.

ASAP's Charlie Jackson explains that the first step in developing a certification process is to decide your criteria for "locally grown." Then develop a process for verifying that farmers meet these criteria that makes sense for all parties, considering such issues as the paperwork required and the frequency of on-farm visits. Next, develop agreements for buyers or distributors, and meet with them to discuss the terms and obtain signed agreements. Finally, as part of the certification and marketing process, develop signature signs and marketing materials that can be used at the point of sale, to create buyer recognition and build a customer base. These materials often include "farmer profiles" with photos of the farm, farm family, and fields or animals, as well as a brief story about the farm and what makes it special. ASAP provides the design, and the outlet or store pays for the materials and printing. Also, ASAP has found that its brand is marketed more effectively when labels are placed on individual items, in addition to the overhead display label. The bottom line is that the certified brand must provide value to all involved.

Every local-brand certification will necessarily reflect the distinct qualities of its region. Still, you can take advantage of the long hours that ASAP spent working out the complex issues of what it means to be "Appalachian Grown" by looking at ASAP's certification agreements (available on the organization's website) for farmers and other sellers who are marketing fresh and/or processed products.

The Appalachian Center for Economic Networks in Ohio (page 256) has a similar program. Its "Food We Love" brand doesn't take the place of specific product branding but serves as an umbrella brand under which a variety of local products are sold. ACEnet distributes and maintains retail displays of "Food We Love" products. Retailers are guaranteed that the products are of local provenance, and the program gives small-scale producers access to those retail markets.

Tell your story and that of your producers.

Sharing the story of farmers and other producers personalizes their products. ASAP does a good job of implementing this strategy by providing to sales outlets colorful posters that highlight each farm, farm family, and products. Iowa State University recommends this step in its 2004 brochure "What Producers Should Know About Selling to Local Foodservice Markets."[235] The brochure suggests that the story include:

- Farm mission statement
- Length of time as a community member
- Level of community involvement
- Production practices
- Farm and family history
- Pictures of the facility and product(s)
- Special recipes with the products

Develop a process for monitoring displays in supermarkets.

Though Appalachian Sustainable Development (ASD) provides marketing materials with farmer profiles to supermarkets for their Appalachian Harvest displays, a lot of information can get lost in the translation. Some farmers may not care how their produce is presented at the store level, as long as they receive a good price when they bring it to the packing barn. But handling the produce well in the marketplace is vital, warns former director Anthony Flaccovento, because poor displays can create misconceptions about local, organic food and injure the long-term market. Buyers might begin to think that organic produce is supposed to look terrible.

The reality of selling to supermarkets is that your product is controlled by the store's produce staff. Even if your branding is solid, a store that doesn't handle your produce properly can create a bad image for your brand. One possible solution, suggests Flaccovento, would be to find one or two partners who could work at the store level – perhaps a store buyer or your own representative who can work with the store produce managers to ensure that your produce is presented in the highest-quality manner, someone who can build relationships with store managers so they feel it's worthwhile to handle your produce properly.

Serve a diverse array of customers.

The economics of distribution suggest that the fewer delivery stops made, the more economical and therefore the more profitable the system. But when the core goal is to develop a market for local food, other factors may be equally important. Appalachian Sustainable Development (ASD) sells its locally grown organic produce to larger supermarket chains but at the same time continues to sell to smaller grocers in the region. "This is an important *service* to the community," says former director Flaccovento. "It's good for building relationships and community support."

Similarly, ASD includes in its customer base several CSAs that are located along its supermarket delivery route. For example, a partnership with Lynchburg Grows (page 116) is mutually beneficial. The Lynchburg Grows CSA provides an outlet for ASD's high-quality seconds, and ASD's reliable supply relieves the pressure on Lynchburg Grows to keep CSA customers happy with full weekly baskets.

Consider a nonprofit distribution model.

The Appalachian Sustainable Development experience offers a nonprofit model for local food distribution, distinct from traditional for-profit cooperative models. "We are serving a business function," says Anthony Flaccovento. "But we are also filling the nonprofit functions of educating farmers, educating the community, walking farmers through certification program." These are all functions that a traditional sustainable agriculture advocacy group or a Cooperative Extension might usually fill.

Yes, a for-profit cooperative model could work, especially if it was a for-profit with a social conscience like Organic Valley or Country Natural Beef (see page 261). A for-profit could probably have pulled together faster and more efficiently, and capitalized it better, explains Flaccovento. "It took us many years to get to the point where we have the cooling systems, the branding, packaging systems, the software to handle it. It would've been better if we'd had all this in Year 3 instead of Year 8."

But the nonprofit model offers another kind of benefit. Nonprofits don't give up when they don't make money, says Flaccovento. A for-profit that didn't make the money, had problems assuring the quality, or had problems getting enough growers to meet the demand would likely have lost sight of the mission. A for-profit might have quit or said, "Let's just get one farmer with 100 acres to grow everything for us."

By contrast, a nonprofit does not lose sight of the social and environmental mission, says Flaccovento. "A nonprofit sticks to the mission and vision until it is able to learn, grow, and make ends meet." Of course, Flaccovento's view of the nonprofit supposes broad community support is available to keep the nonprofit afloat during its years of learning and growth. But until it becomes a new normal for all businesses to be socially conscious and responsible to their communities, Flaccovento's point is compelling: a nonprofit distribution system offers one effective model for growing the local food system.

From Tobacco to Tomatoes: Reestablishing a Local Food System
Appalachian Sustainable Development

In the rural Appalachian region of **Virginia** and **Tennessee**, the nonprofit Appalachian Sustainable Development (ASD) works to facilitate the revitalization of rural economies through ecologically sustainable food and forestry, with, as its mission states, "development that is sustainable and beneficial for nature and people, for culture and community." As part of its mission to facilitate the development of a robust local food system, ASD promotes the growth of organic and sustainable farmers through education and training, market development for local food, and distribution services.

Forget about the organic arugula and lavender. If you want to build a local food system, you need to grow ordinary food for ordinary people, says Anthony Flaccovento, founder and former executive director of Appalachian Sustainable Development (ASD). Squash, melons, beans, meats, eggs — these are not just food staples but food *system* staples.

ASD's work is evidence that the local and organic food movement is not necessarily elitist, as some insist, but in fact can specifically address issues of social justice. Indeed, it was Flaccovento's roots in social justice that led him to found ASD in 1995. Running the Catholic Diocese's office of social justice in eastern Kentucky after he graduated from college, Flaccovento found himself in the thick of local fights over "jobs or the environment," a chronic lose-lose proposition epitomized by the national debate over the Northwest spotted owl. Flaccovento didn't believe the issue required a draconian choice. He thought it was possible to do both: create jobs *and* protect the environment. With the founding of Appalachian Sustainable Development he and his cohorts — a "mishmash of strong environmentalists and economic developers" — aimed to revitalize the struggling regional economy while also protecting the environment. "We took stock and saw that we were all losing in this game," he told PBS in 2007. "What we've been trying to do in the ensuing 12

years is to redefine what a healthy economy is and what economic development is."[236]

In rural Appalachia of Virginia and Tennessee, ASD had its work cut out for it. A region of both intense beauty and poverty, Appalachia needed to reinvent itself for a post-coal and post-tobacco era. As a matter of principle, ASD believed that any effort for economic revitalization should build on the region's existing strengths, so farming and forestry were natural arenas in which to focus its efforts. Over the years, ASD's approach to fostering sustainable farming and forestry has evolved, as it has learned what works and what doesn't, and it now offers a model that others in the nation are seeking to emulate.

When ASD began working with a few local farmers, it started very small, selling their organic, mostly boutique produce only through a CSA and to local restaurants. "It was a small market," recalls Flaccovento, "and we weren't changing the complexion of farmers." ASD trainings and workshops about organic vegetable production were not attracting traditional farmers. And tobacco farmers didn't have sufficient incentive to make the transition to organic vegetables. "You couldn't make a profit or grow a new food system on arugula alone," Flaccovento says.

Local longtime tobacco farmers were skeptical about growing organic vegetables for good reason.

During the 1990s, as they saw the industry shift its support to tobacco farmers overseas, they had begun exploring ways to keep their families and communities from entering economic freefall. However, pilot projects to help tobacco farmers transition to "niche" products, such as strawberries and bell peppers, were largely unsuccessful. Taking note of this cautionary history, coupled with its own failure to make serious inroads in building a local farm economy with boutique crops like arugula, ASD decided to reject the idea that you could only make money by growing specialized food for specialized markets. Displaying a rare combination of guts and business acumen, ASD decided that it would help tobacco farmers transition to growing ordinary food — squash, beans, tomatoes — while at the same time developing reliable local markets for these crops.

Taking on Distribution

Any community trying to grow a local food system faces a similar dilemma: customers won't surface unless there are farmers providing the products, and farmers won't provide the products unless there are customers. Flaccovento pushed through this classic wall by working as a broker, negotiating contracts to provide organic produce to local markets and even to supermarket chains. It was a high-risk proposition, particularly since Flaccovento hadn't yet lined up the farmers to grow the produce that he was promising to the stores. But his stubborn faith paid off. Once he had contracts in hand, some farmers were willing to make an investment in growing the vegetables.

"If we'd done a careful analysis, we probably never would've done it," says Flaccovento. But ASD believed in creating business opportunities that would meet the triple bottom line; that is, that are economically, socially, and environmentally sustainable. And given that food and farming are huge components of our ecosystem, ASD was motivated to find ways to improve agricultural practices while also improving the economic outcomes for farmers.

Starting small, ASD learned the ropes of the distribution business for a couple of years with locally owned retailers. Helping the new distribution business succeed became a community effort. One tobacco farmer offered up his former drying barn as a place for ASD's grading operation, and community volunteers pitched in to pour a concrete floor, buy a cooler, and help with the sorting and grading process. Thus ASD — and its trade-name brand of Appalachian Harvest — was born.

Like the Appalachian Grown brand initiated by the Appalachian Sustainable Agriculture Project (page 243), the Appalachian Harvest brand serves as an umbrella name for all the small-scale producers who distribute through ASD. "Our brand hasn't taken the world by storm," says Flaccovento. "But it's a solid brand that has achieved what we wanted — hometown, not too slick but credible, and [with] an emphasis on local and organic. And that's a good fit."

After gaining experience, ASD made another bold move and approached larger regional grocery chains — Food City, Ukrops, and Whole Foods — with its Appalachian Harvest produce. The chains were interested but wanted produce that was certified organic. This, in turn, motivated ASD to begin the process of working with local farmers to help them transition to certified organic methods.

Recruiting Farmers to Organic

As ASD soon found, building a local food system takes many years of sustained effort. Even when ASD could offer large and highly secured markets for organic produce that would bring in more revenue per acre than tobacco, Flaccovento explains, the process of recruiting conventional farmers to organic production was slow. Farmers might attend ASD workshops and seminars to learn about the opportunities that organic production offers, but it would often take two to three years of interaction before they would become willing to take the plunge to transition to organic.

"Tobacco has been so much the centerpiece of our economy for so long that, even with a relatively profitable crop, it's hard to move away from tobacco," says Flaccovento. He explains that tobacco farmers have "heard this tune before," and southwest Virginia has served as the testing ground for many failed ideas and projects that were supposed to help farmers lift themselves out of poverty. "So there's a real reticence to accept new ideas, because they've been there, done that," explains Flaccovento. "Each bad experience adds to a farmer's hesitation. Farmers have to be risk-averse, as they have cash-flow problems. Younger farmers may be less risk-averse as a group."

In fact, Flaccovento says, one way to grow a local food system is to increase the number of new local farmers. While encouraging existing farmers to expand their acreage under cultivation is certainly an option, Flaccovento believes a local food system will be more secure as the number of new local farmers increases, to ensure that local farming has a future after the aging population of current farmers retires. Some attrition should be expected, of course. As with people entering any new profession, a certain percentage should be expected to fail and/or decide that farming is not for them. No doubt about it, farming is an extreme profession, both physically hard and psychologically stressful, fraught with daily unknowns and hazards. Finding a way to help new people enter and stick with the farming profession is where a community organization like ASD can make the difference between failure and success — by offering the new farmers valuable mentoring and assured markets to provide a reliable income.

Some ten years after its first efforts to grow a local food system, ASD has exceeded all expectations. In 2010 it boasted a network of about 70 organic farmers, about three-fourths of whom are former tobacco growers, and another 30-plus beef producers. And this network is providing seasonal local organic produce to about 600 stores in the region, from small mom-and-pop chains to larger regional supermarket chains.[237] Farmer income is highly variable, says Flaccovento, but a well-managed operation might gross $8,000 to $10,000 per acre, compared with grossing $4,000 to $5,000 for growing tobacco. And because the costs incurred in growing vegetables are lower than those for tobacco, the farmer's net income is proportionally higher.

Working in a rural mountainous region, ASD is not by any stretch of the imagination "close" to the markets of a large metropolitan region. In fact, growing a successful local food system is arguably far more challenging in a remote region than in a more populous one. Nevertheless, as ASD has slowly expanded its work in farmer education, networking, and brokering distribution, it has pioneered a model that isn't bound by a particular geography. Its recipe for success includes patience and responding to the evolving desires and needs of the region. It also includes a willingness to take risks, along with an abiding faith in people. ASD has shown that if you can secure markets for fresh food, and if you can build networks of trust, farmers — and their communities — will thrive.

Fostering Food Entrepreneurship
Appalachian Center for Economic Networks

coauthored by Megan Bucknum

Founded in 1985, the nonprofit Appalachian Center for Economic Networks (ACEnet) in **southeastern Ohio** is essentially a community development organization that uses a variety of tools such as technical assistance, loans, and joint marketing to foster new small businesses in the art, retail, technology, and manufacturing sectors, as well as in food and agriculture. Its mission is to "build networks, support innovation, and facilitate collaboration with Appalachian Ohio's businesses to create a strong, sustainable regional economy."

ACEnet provides a host of services in business training, business consulting, and product development support. For food ventures ACEnet staff help entrepreneurs with issues ranging from procurement of ingredients and preparing recipes for mass production to nutritional analysis and food labeling and packaging. A sister organization called ACEnet Ventures provides qualified applicants with access to capital, filling in the financial gap between a creative idea and a thriving business.

ACEnet clients also have access to the Food Manufacturing & Commercial Kitchen Facility, a shared-use facility boasting a commercial kitchen, a packing room, a warehouse for dry, cold, and frozen storage, and a thermal processing room. This facility operates 24 hours a day, seven days a week and can be rented at an hourly rate. ACEnet gives highest priority to clients in its primary service area, which includes the

Appalachian Ohio counties, by giving them a 50 percent discount on rent to ensure that the center is accessible to that area's large low-income population. The rent and client fees cover the operations of the business incubation center and also contribute to the kitchen facility's annual equipment and maintenance expenses.

One challenge for ACEnet's clients has been distribution into larger retail markets. To give its clients a leg up in marketing, ACEnet has created an umbrella brand, "Food We Love." ACEnet clients who qualify have their product distributed under the "Food We Love" label; ACEnet negotiates to install a display unit of the products in stores that small business owners might otherwise have difficulty getting into, and ACEnet staff restock the unit as necessary.

ACEnet's economic impacts have been significant, demonstrating that food entrepreneurship is a worthy avenue for economic development. In 2009

Bright Idea: Build an Umbrella Brand

The ACEnet "Food We Love" label is an umbrella brand for a range of products manufactured by its clients. The label assures retailers of a local provenance for the products, and the umbrella branding gives a wide range of small business owners access to a larger distribution and marketing network. In ACEnet's program each product retains its own label; the products as a group are sold from retail display units that display the "Food We Love" brand.

The concept of an umbrella brand is widely applicable. Whether sponsored by an agency devoted to small business development or by a cooperative of small business owners, an umbrella brand can act as a creative marketing tool for a wide variety of products.

the Athens facility alone hosted 25 artisan businesses, with an additional 150 food-manufacturing businesses using its Food Manufacturing & Commercial Kitchen Facility. Leslie Schaller, director of programming, estimates that its current tenants contribute as much as $8 million to the regional economy in annual product sales, as well as 129 full-time and 87 part-time jobs. This is no small feat, especially in hard economic times. Since 1996, when the Athens center first opened, more than 260 kitchen-facility tenants have made products ranging from breads and noodles to salsa and sauces.

There is no strict graduation policy, and entrepreneurs can stay at the Athens facility indefinitely.

While some business incubators languish for lack of interest, ACEnet's combination of technical support services and physical infrastructure has proven successful, creating new businesses, new jobs, and value-added opportunities for some of the region's agricultural products. In fact, in 2006 ACEnet opened another business incubation center, this one in Nelsonville, Ohio, to provide mixed-use space for manufacturing and distribution companies.

Lessons Learned
Building a Business Incubator

Include all stakeholders in the planning stage.

Before getting started, says Leslie Schaller, director of programming for ACEnet, it's critical to consult and involve the people the program is intended to benefit. Their skill sets, talents, ideas for untapped economic markets, and creative impulses will drive the demand for the project's services. An inclusive planning process that helps them develop their interests unveils and builds the community's social capital.

Similarly, it's helpful to engage investors on an ongoing basis; many ACEnet leaders and contributors are connected to national social and business networks, creating an even larger pool of resources for ACEnet clients. It's equally important to engage local officials, with policies that could affect a project, and also to educate consumers about food policies, inspiring them into action to facilitate positive policy change.

Most important for any food-related business incubation center is compliance with health codes. Creating a commercial kitchen facility that will be shared by many people with varying skill levels is a complex task. ACEnet took a collaborative approach from the onset, involving state health inspectors in the facility's design process to ensure approval in the project's final stage.

Tap into existing infrastructure.

Schaller advises new community food projects to coordinate with resources that already exist in the community.

Warehouses and agricultural resources are examples of important local infrastructure that can help root the food project in the community and encourage the community to feel connected to it.

Create tiered goals for varying time frames.

New projects can risk failure by adopting goals so large that they overwhelm and stifle the original creative motivations. When ACEnet sought to create a business incubator, there were few examples and no how-to manuals explaining how to achieve this goal. Schaller believes that much of the group's success can be attributed to their decision to create a tiered goal structure, divided into one-year, three-year, and ten-year goals. Tiered goals allow people to stay engaged over the long term, she says. Balancing the desired end results with smaller goals along the way allows the momentum of small-goal successes to help carry the team forward toward the end goal.

Structure fees to attract the desired clientele.

ACEnet business was founded to assist the people in the rural region of Appalachian Ohio. As part of its strategy to empower this underserved population, ACEnet identifies the 32 counties in this region as its primary service region and offers a hefty 50 percent discount on kitchen-facility rent to its inhabitants.

Growers, Millers, Bakers: Rebuilding a Local Grain Economy
Local and Heritage Grains

coauthored by Regine Kennedy

Across the nation, the grassroots food movement is expanding into the production, milling, and distribution of whole grains. Though they may be one of our oldest sources of nourishment, appealing to a sense of deep heritage, grains are a relative newcomer to the local food scene. The leaders and groups described here are pioneers in a broad-based effort to reestablish locally grown and milled grains as a prized staple in our community food pantry.

Advocates of local food decry the centralized, mass production of food, based on the availability of cheap fuel, in favor of regional, smaller-scale food production that contributes to local sustainability and food security. Grains are still a common missing link in the local food equation, however. Over the past hundred years, as wheat and other grain production and processing consolidated in the Midwest, in other parts of the country, grain fields have become a thing of the past. Local mills have shuttered their doors and, if still standing, are often considered relics worthy only of historic preservation. Yet without grains — the foundation of our diet in both calories and nutrition — a local food system can never be fully resilient.

Grains are still a relatively rare component of local food initiatives. Whether to foster land stewardship, promote healthy eating, support local economies, or revive heritage foods, people across the country are rethinking the impact of their daily bread. But growing a local grain economy is not simple and requires simultaneously growing a new infrastructure. Just as local meat initiatives may require the creation of small or mobile slaughterhouses, and local vegetable and fruit initiatives may require local refrigeration and commercial kitchens and canneries, a local grain economy requires growers, millers, and customers.

In Oregon the Southern Willamette Valley Bean and Grain Project is working to build a year-round local food system through a variety of strategies that include educational workshops, connecting buyers and farmers, and fostering local food resources. Central to its effort is the desire to restore cereal grains as a significant regional crop and a viable edible alternative to grass seed, the dominant crop in the Willamette Valley.[238]

Founded by Harry MacCormack, the Bean and Grain Project began growing bean and grain cultivars in small hand-harvested test plots and since has moved to larger combine-harvested plots, sharing lessons along the way with other growers through research papers and workshops. Dan Armstrong, one of the project leaders, argues that "with the increasing price of petroleum products and thus food freight costs, it is becoming increasingly apparent that we would be wise to grow more food crops locally, and grains are a very suitable substitute for grass seed. Though wheat will not grow everywhere grass seed does, wheat is arguably the best substitute, and as recently as 30 years ago, it was the dominant crop by acreage in the valley."[239] Cereal grains — including wheat, maize, oats, barley, rice, rye, triticale, and millet — compose "more than half of what humans eat worldwide," explains Armstrong.

Just as modern tomatoes are bred for high yields and the ability to survive the rigors of transportation, modern wheat is bred for high yields and uniformity.

And just as locally grown tomatoes are considered to taste better and have a richer nutrient profile, and heirloom tomatoes to be suited to specific soil and climate conditions, so too are the local grains considered more nutritious, while heritage, or landrace, cultivars are better suited to specific soils and climate conditions.[240]

Regional organizations — such as the Northern Grain Growers Association, a farmer-grown organization in Vermont that works with the University of Vermont, and Northeast Organic Wheat, a consortium of heritage wheat teams across four states — are helping to raise awareness about local grain heritage by identifying which varieties grow best in specific parts of the country. The North Carolina Organic Bread Flour Project is working to close the distance between the baker and the farmer, to create a viable grain market that will "provide a tangible level of security and sustainability" for the local grower, miller, and baker.[241]

That North Carolina initiative inspired similar work in Virginia, where the Central Virginia Grain Project is working with local farmers to build a seed bank for heritage wheat varieties. To move the Virginia project along more quickly, experienced farmers are planting proven wheat varieties on multiple acres and then sending the wheat to a small local mill. By these efforts farmers are strengthening the local economy while also building the seed bank.[242]

But rebuilding a local grain economy takes time. To retain more of the natural nutrients in grains, smaller mills are preferable, as their grinding process is better able than heavier, industrial mills to retain more of the wheat germ, a rich source of nutrients. So while the flavor and nutritional quality of heritage wheat may be greatly desired, and while the demand for grains grown by local farmers is increasing, a constraining factor is still that few local mills are available.

One critical step in rebuilding this infrastructure is the creation of grain CSAs, which can help reduce the risk for small grain mills while enabling consumers to access local grains and support local farmers. In British Columbia a group of farmers and community activists founded the Kootenay Grain CSA, with the intent of growing grain sustainably in the Kootenay region for local consumption.[243] In western Massachusetts

Wheatberry Bakery was motivated by the 2008 "catastrophic" rise in flour prices to create a more economical local source of grains. Offering shares of beans and grains, including heritage wheat, corn, beans, spelt, and rye, the bakery created the region's first grain CSA, which grew to 115 shares by 2009 and in 2010 distributed 10,000 pounds of organic local grains. The bakery also provides another part of the missing infrastructure by offering daily fresh milling of grains at its Amherst shop and also selling affordable small home mills to CSA members.[244]

In California the Huasna Valley Farm makes its sustainably grown heirloom wheat flour available through a local CSA.[245] And in Oklahoma the Oklahoma Food Cooperative links family farms to consumers through its website, offering a host of locally grown and made products, including wheat berries and wheat and barley flours from three different Oklahoma farms, as well as a variety of breads made with locally grown and milled flours.[246]

Bakers are natural leaders in the effort to revive regional grains, as they can create a steady demand for locally grown and milled grains as well as raise awareness among consumers. In the Northeast and the San Francisco Bay Area alike, a few notable bakers are leading the way in creating demand for local grain. The percentage of local grains used in the loaves varies, and the bakeries are finding it necessary to educate their customers about how the bread may change during the year depending on which varieties are available.[247] At Berkeley's Saturday market a "Local Loaf," made entirely from heirloom whole wheat flour grown and milled by local Full Belly Farm, is in high demand and commands a higher price than sister loaves.[248] Its creator, Eduardo Morell of Morell's Breads, writes that some of his breads "are totally unique" and "offer an alternative to those who are sensitive to modern wheat, and are purchased enthusiastically by many customers who thought they had to give up bread."[249]

In the Northwest the Cascade Baking Company uses only Food Alliance–certified unbleached flour from a Washington farmer cooperative. It also sells grain flour from Shepherd's Grain, a co-op of regional grain farmers.[250]

Similarly, in western Massachusetts the Hungry Ghost Bakery is providing leadership for restoration of local grains through its Wheat Patch Project, in which it is distributing seed to bread lovers who are willing to grow test plots of wheat, with data collection by a local college, to help determine varieties that grow well in the Pioneer Valley. Hungry Ghost argues that its project is "more than a gimmick but a radical approach to food production, economic participation and agricultural re-integration." Rather than using "road weary" and "carbon heavy" flour, Hungry Ghost believes it makes good economic sense to reinvigorate the production of local grains and that it also is agronomically possible, noting that some records show Massachusetts as the "site of the first wheat harvest in North America in 1602."[251]

Grains are gaining strength as an important consideration in building local food security. Interests in land stewardship, healthy eating, vibrant local economies, and food heritage all converge around grain production and processing, just as they do around other parts of the food system. CSAs, grain co-ops, and local bakers are beginning to offer consumers a way to complement their local meats, eggs, fruits, and vegetables with local grains and loaves of local bread. Through these leading-edge initiatives, one day local grains, while earning a place at our table, may also become valued contributors to local economies.

Giving Small-Scale Ranches Efficiencies of Scale

Country Natural Beef

Encompassing ranches across the **western United States**, Country Natural Beef is a cooperative of small-scale beef producers focused on producing meat sustainably, humanely, and profitably.

In the 1980s, ranchers Doc and Connie Hatfield were looking for a way to raise beef cattle naturally and humanely, while maintaining an ethic of land stewardship and also supporting their family. In an age of conglomeration and massive Midwest feedlots, they wanted to hang on to their family ranch and make a living running it. Some of their neighbors felt the same. In 1986 the Hatfields and thirteen other ranching families joined together to form Country Natural Beef, a beef marketing cooperative. Since then the co-op has grown to include nearly 100 ranches across the western United States, selling beef to retailers nationwide.

Each Country Natural Beef ranch pledges to follow the group's "Raise Well" and "Graze Well" initiatives. "Raise Well" deals with animal welfare, from how the cattle are handled to what they are fed and how they are slaughtered, while "Graze Well" advocates environmental responsibility, promoting land and water stewardship through sound grazing principles and restoration and preservation projects. All of the co-op's production practices are certified and audited by reputable outside agencies to ensure that the beef is produced humanely and sustainably.

The co-op's ranchers are passionate about their role as stewards of the land, sustainable communities, and family ranches. While they may not always agree on everything, they share a core commitment to independence and family. Being both independent and part of a group may seem a contradictory state, but the Country Natural Beef ranchers know that on their own they cannot provide a year-round supply to a group of customers, so they use the co-op as a tool to collect, finish, and market the boxed beef from the cattle that they raise and own until the day they are processed. And the business model is hugely successful, allowing families to make a living from a small-scale family-run ranch.

When we think of natural, humanely raised meat, for most of us a small, diversified family farm springs to mind, with a few head of beef cattle grazing serenely on green pasture amid fields of grain, barns of hay, and the gentle cacophany of chickens, tractors, and other barnyard life. That picture, while wonderful, cannot feed the nation. It's operations like Country Natural Beef, with more than a hundred thousand head of cattle, spread out on ranches managing from several hundred to several thousand cattle a year, that have a scale that can impact the marketplace. The extent of that impact is evolving; with support from consumers, it may grow to define meat in our future food systems — a commodity and consumable, yes, but also a production system that supports local communities, enables ranching families to make a living, and promotes sustainability.

Lowell Forman was one of the fourteen founding members of the Country Natural Beef cooperative. CNB now includes nearly a hundred ranches across the West, all focused on producing naturally raised, humanely treated cattle. CNB provides ranchers a way to thrive independently, while supplying restaurants and retailers with top-quality beef.

Above: Doc (left) and Travis Hatfield walk an early-season group of cows and calves into the corral for measuring and tagging. All cattle run through Country Natural Beef are documented from birth through slaughter and packaging to assure the best quality control.

Following page: A hands-on connection to the cattle defines life on the Hatfields' ranch.

Top: Working with the Audubon Society, the Hatfields set aside wetlands on their ranch in order to allow migrating waterfowl seasonal access to critical nesting habitat. *Bottom left:* Travis Hatfield works on a solar panel that's replacing diesel-powered pumps on the wells on the ranch. *Bottom right:* Connie Hatfield repairs a post-and-rail fence on the ranch.

The Yamsi Ranch holds five of the six springs that form the headwaters of the Williamson River. In the 1950s Dayton Hyde thought to fence off the creek for over 20 miles through the ranch, preserving what is today one of Oregon's best trout fisheries.

Joe Jayne grew up working summers on Yamsi Ranch in Oregon, one of the Country Natural Beef member producers. He has now come back to the ranch full-time, and feeding several hundred cattle is one of his many winter duties.

Joe's younger cousins, Elizabeth Gerda Hyde and Henry Walton Hyde, ride along with him learning the ropes. Providing an economically viable option for future generations to take over the ranching operations is a conscious goal of Country Natural Beef.

Some details at the Yamsi Ranch are what you might expect on a working ranch, while others bust stereotypes. Here a pair of Vise-Grips allow passengers to roll up a window in an old ranch truck — thrift and utility exemplified — while energy-efficient, modern-day compact fluorescent bulbs provide illumination in the ranch house's chandelier.

A Farm-to-Restaurant Network
Farm to Table D.C.

Founded in 2008, Farm to Table D.C. is an innovative pioneer in bringing local food to restaurant menus in **Washington, D.C.** This restaurant cooperative works with regional farmers to provide them with a reliable and profitable market for their products, while also providing D.C. chefs with reliable sources of quality food to serve their clientele.

In Mie N Yu, one of Georgetown's most ethnically distinct restaurants, the chef is creating a twenty-first-century cuisine based on old principles — he's going whole hog. Executive Chef Tim Miller uses his study of anthropology to research and create authentic cuisines of the ancient Silk Road cultures, and he uses his culinary arts training to reinvigorate ancient culinary traditions of using the whole animal. With the support of Chef Miller, restaurant manager Oren Molovinsky is pioneering a radical approach to bringing local food into restaurants. Since 2008 he has been buying, transporting, and serving millions of dollars of Virginia meats in DC restaurants — whole animals at a time.

Restaurants generally order specific choice cuts of meat for their preset menus — rib eye, New York strip, tenderloin — and leave it up to the farmer to market all the other cuts. This practice presents one of the biggest challenges for local farmers who want to sell to the lucrative restaurant market: either move *all* the cuts of meat, or go bankrupt. And it's so difficult to move the whole animal, Molovinsky says, that most farmers give up on selling directly to restaurants.

Molovinsky has been in the restaurant business for decades and has always believed in buying food fresh from the source. Back in the 1980s, however, sourcing direct meant buying directly from the manufacturer, not the farm. As the local food movement took hold in the 2000s, Molovinsky became more interested in supporting local farmers. Eventually, a representative of the Virginia Department of Agriculture and Consumer Services arranged for him to visit several Virginia farms, where Molovinsky heard the same story over and over again: Virginia farmers were shipping their live animals on trains out to the Midwest for processing.

This made absolutely no sense to Molovinksy. "Why are we buying our meat from the Midwest or Northeast, with a label [that says] it's naturally grown? Maybe it is natural and maybe it isn't, but we have no control and no way of verifying the product source. Meanwhile, right here in Virginia we have great farmers with great stories, and we're not even supporting them."

Later, in a conversation with an Australian international meat distributor, Molovinsky learned the distributor was selling its middle cuts to the United States — rib, loin, short ribs, rib eyes, New York strips, tenderloins — and shipping all the other parts to other nations around the world: top round to Korea, for instance, and chuck roast to the Middle East. Similarly, Molovinsky says, the U.S. beef market keeps middle cuts for U.S. consumption and exports the rest. This world market system makes no sense to him, either. "This is very costly," says Molovinsky. "If we can learn to use *all* the animal cuts, then we could keep all the cuts of U.S.-grown beef here. Then we can stabilize the prices, and we don't need to buy meat from Brazil or Argentina or the UK."

Molovinsky wanted to turn around this upside-down system so that it was right side up. He says he was fortunate — his restaurant has a large meat budget, a large corporation doesn't govern its decision making, the owner gave full carte blanche, and the chef was willing and creative. Together Molovinsky and Miller decided to "just do it": they would find a way to buy meat from local Virginia farmers.

The Restaurant Cooperative

Over the next four months, the manager and chef did the math and worked through all the logistics of how to bring in local meat. Step by step, they experimented with different cuts to reengineer the menu so that it could move all parts of the animal. Together they were initiating a major culture change in the restaurant business.

Traditionally, chefs place an order to the distributor at midnight, at the end of the shift, Molovinsky explains. The distributor places the product on a truck at six in the morning, and it arrives at the restaurant at seven. Restaurants have become accustomed to this "just-in-time" inventory. Molovinsky needed to figure out a system that would provide similar flexibility to the chef while taking care of the farmer's need to sell the whole animal and also allowing the time needed for local processing and dry-aging in the old-fashioned way, as opposed to the industrial processing method of vacuum-sealing meat in plastic and aging it in a cooler.

Molovinsky quickly determined that Mie N Yu couldn't do it — at least not alone. Transportation costs were going to break him. Through his affiliation with the D.C. restaurant association, Molovinsky approached other restaurants, which turned out to be just as eager to buy local meat. The other restaurants didn't have staff to manage the logistics of picking up meat at the Virginia farms, and farmers weren't willing to bring their meat to D.C. They were all eager for a solution, and they let Molovinsky know that, if he would do the legwork, he could count them in. The Farm to Table D.C. restaurant cooperative was born.

Numerous meetings with farmers, truckers, and state agency staff led Molovinsky to decide that the easiest solution to get off the ground was to use existing distributors. Half a dozen restaurants pooled their orders, and Molovinsky contracted with the distributors to pick up and deliver the meats to D.C. Each restaurant could make arrangements with its own farm, and Molovinsky would coordinate the orders with the distributors. And if restaurants wanted to add a new product, Molovinsky would work with the Virginia agriculture agency to find a farm that could supply it.

The cooperative began operating in October 2008, and in its first 18 months, it spent more than $1 million, all of it supporting small Virginia farmers, slaughterhouses, and distributors. Molovinsky schedules a steady supply of meat from different farmers, whose animals are all on tightly interlocked delivery schedules. The cooperative has been a real blessing for farmers, says Catherine Tatman, who raises sheep, cattle, chickens, and eggs at Hilldale Farm in central Virginia. Her risk has dropped significantly; instead of raising animals to be sold at auction for an unpredictable price or selling parts of the animal while not knowing how she would sell the rest of it, now Tatman knows she'll sell the whole animal at a guaranteed price, better than she could get at auction, thereby increasing her profit margin. She also knows that if she wants to increase the size of her flock she likely has a ready market to support that growth. Molovinsky has created a new supply chain that is a win-win for all parties.

Local farmers aren't the only ones to benefit from the restaurant cooperative, Molovinsky notes. By March 2010 the cooperative had transitioned to using just one distributor—D.C. Central Kitchen (DCCK), a nonprofit that, among other things, assists people recovering from homelessness, addiction, and incarceration by giving them culinary training and placing them in jobs. (For more about D.C. Central Kitchen, see page 278.) DCCK had been eager to break into the local food arena, and the restaurant cooperative contract finally gave it the needed break. DCCK could now send its refrigerated trucks out to the country to pick up the restaurant meats, and with its transportation costs covered, it could afford to pick up local bulk produce.

Thanks to this synergistic relationship, DCCK became the largest single buyer of produce in 2009 at the Shenandoah Valley Produce Auction, a weekly wholesale market outlet for Mennonite and other valley farmers. To manage the sudden influx of huge quantities of fresh produce, DCCK added work shifts to its all-volunteer canning operation, which provides locally grown food for the hungry. DCCK expects to double its purchases at the produce auction in 2010.

"It's unbelievable," says Molovinsky with a happy chuckle. The restaurant cooperative has benefitted the D.C. community in so many unanticipated ways.

Cooperative Criteria

As Farm to Table D.C.'s coordinator, Molovinsky is not paid for the hundreds of hours he puts into logistical coordination, but he saves his restaurant the ordering and shipping costs that would be incurred if he were ordering through a regular distributor. For a new restaurant cooperative to get off the ground, Molovinksy says, someone needs to step forward to do the work — and that person must be completely transparent about the operation, as well as trusted by all the hoteliers and restaurateurs.

"A single restaurant cannot do this alone," says Molovinsky. "A cooperative is the only way to do it. The [concept] of buying local meats does not work with the model of any distribution company, no matter how small or large. So you can't deal with a typical meat or seafood distributor and rely on them in a sustainable way. You have to create your own logistics network." At the same time, however, Molovinsky suggests that it is helpful to engage existing distributors in some way, when possible, for special deliveries or other purposes, to avoid creating adversarial relationships. Nearly two years after making the fateful decision to "go local," and after a very challenging first year, Molovinsky says the cooperative operation is now stable and can continue to grow.

The cooperative also stabilizes costs for restaurants and makes local grass-fed meat affordable relative to conventionally grown all-natural meat. "This makes *financial* sense," he argues. "Right now, because of the economy, this is a tough time for the farmer to sell whole animal. But at some time the prices of other meats will shoot back up, and our farms won't raise their prices on us. They're giving us true costs — grass doesn't cost more now than it did three years ago!"

For Molovinsky a critical benefit of the cooperative is that restaurants can control and verify sources. "We don't have a big budget to travel to Australia or Argentina or the UK. It's a lot easier to jump in the car and drive to the Shenandoah Valley." Molovinsky

worked with Catherine Cash, staff of the Virginia Department of Agriculture and Consumer Services, to develop very specific animal welfare rules and criteria for labeling meat "all natural" for farms selling to the cooperative. These criteria are important for complete transparency, says Molovinsky. He wants his restaurant customers to be able to visit the farmers and see how the animals are raised.

The criteria are only as good as what's actually practiced on the farm. For example, one cooperative's rule is that only meat from animals who were never treated with antibiotics can be labeled "all natural." But some farmers feed their animals antibiotics simply by default, because the antibiotics are contained in the feed bought at standard farm-supply stores and labeled with obscure language that makes it difficult to identify. Molovinsky says the cooperative is trying to implement an annual "auditing" process that will be simple to conduct and free to the farmer but will also have some teeth in it. The restaurant cooperative plans to work with farmers to assist them in meeting the animal welfare criteria.

"This is all about relationships, about knowing where we're buying a very important product," Molovinsky says. "Before, I just wanted to know my distributor, and we were always loyal to our purveyors. Now we're building the same kind of relationship with the farmers." Having a relationship with the farmers does create occasional difficulties, because Molovinsky says they become like family. So if the restaurant feels it has been overcharged, or if a farmer has an issue with the restaurant's payment, it's not always easy to talk about. Molovinsky shrugs, indicating there has to be some give and take in these relationships.

The Farm to Table D.C. cooperative will likely continue to grow. Without an additional investment Molovinsky says he can imagine continuing to serve as the coordinator until it doubles in size to about 30 restaurants. He estimates an expanded 30-restaurant cooperative would spend about $2 million per year, with about 75 percent going to local farmers, 20 percent to local processors, and 5 percent to local trucking. If the cooperative were to invest in a staff for logistics and administration, then Molovinsky could focus on his strong suit, which is enlisting more farmers and helping more chefs learn how to cook "whole animal."

In the meantime, as Farm to Table D.C. solidifies its business model, Molovinsky is also expanding his efforts to help other businesses buy and distribute local food. He is assisting Harry's Tap Room, a multilocation restaurant in the Washington metropolitan area, in learning to source organic and naturally raised ingredients from local farms and to create a menu around the whole animal. He is also working the other side of the equation by helping local meat distribution expand into the local markets. The Capital Meat Company now buys whole animals from Maryland and Virginia farms and distributes cuts individually to their customers. Another distributor, Saval Foods, a large food distributor for the Baltimore region, has ventured into local beef by offering naturally grown beef from Grayson Natural, a farm cooperative in southwest Virginia. Saval is hoping eventually to market other beef cuts, to transition to the whole-animal approach.

Restaurants as Change Agents

Molovinsky has been a remarkable change agent, almost single-handedly creating innovative pathways for farmers, restaurants, and distributors to build new, mutually beneficial relationships. "It's *not* difficult, it's just different," says Molovinsky. "You have to change your habits and behaviors and procurement behavior. But that's all it is: a change. And once the change happens, it's not a big deal."

The way to appeal to chefs, says Molovinsky, is through the unique qualities of local meats. Buying directly from the farm is very appealing to a chef because it's unique and is appealing to the *guest*. Molovinsky laughs, then says, "Everyone knows that chefs have big egos and would like to be able to say, 'We're the only ones in the world who sell this.'" So buying locally — patronizing a specific farm that nobody else does or creating a unique dish using locally raised meat — creates a competitive advantage.

Similarly, the motivation for upscale diners to buy dishes with locally raised meats is not necessarily that the meat is *local*, according to Molovinsky. In his experience diners perceive that a dish associated with a specific farm and farmer has a higher *quality*. The interest in higher quality as well as a unique dish is what leads diners to support their local farmer.

"I get frustrated with restaurants who *say* they do a 'local' program, when all they do is go down to the local farmers' market." That's dipping into the consumer market of choice cuts, says Molovinsky. He believes that restaurants should be doing it the right way — giving the farmer real support by buying the whole animal and leading the way in developing whole-animal cuisine.

"We can change this industry!" Molovinsky argues. "Chefs can change this whole movement *faster* than consumers, because their buying power is immense. Purveyors will do whatever we want them to. The product is there. We just need to let the purveyors know we want it."

Lessons Learned
Building a Restaurant Purchasing Cooperative

Work with restaurants that commit to using whole animals.

Agreeing to buy whole animals ensures a viable market for farmers, making them more willing to sell to and work with a restaurant cooperative. Most important, says Oren Molovinsky, manager of the Farm to Table D.C. restaurant purchasing cooperative in Washington, D.C. (page 271), the cooperative works best when *all* the participating restaurants are using the whole animal. If one restaurant wants only specific cuts of the animal, the cooperative then faces the same kind of logistical nightmare of moving different cuts of meat as the farmer would have had. "It's just too difficult," says Molovinsky.

Another important piece in the network is creating a "slot" – or scheduled appointment – with a slaughterhouse, so the farmers can be assured that their animals will be accepted by the processor at the appropriate time. Having some variation in the number of animals to be processed at any one time is not a problem when restaurants are taking whole animals; when they're taking just certain parts, Molovinsky says, a sudden increase in an order for, say, loin leaves the slaughterhouse or farmer scrambling to find buyers for the rest of the animal parts – not always an easy task. Molovinsky says that this whole-animal system is an "incredibly valuable commodity."

It also is important that the buying group be likeminded about the definition of "natural" – whether that means antibiotic free from birth, grass fed, grain or grass finished, cage free, or other such things. The Farm to Table D.C. cooperative has gravitated toward farmers that raise exclusively grass-fed animals. There are a few farms that don't fit this model, and the cooperative is willing to allow member restaurants to buy from them as long as the restaurants limit their advertisement of the meat to "local" and not "natural."

Help farmers develop terms of payment appropriate for the food-service industry.

One of the challenges for a restaurant cooperative is teaching farmers to deal with the food-service industry. Farmers are used to taking their animals to auction and receiving a price, which they may or may not be happy with, but they're paid on the spot, says Molovinsky. When farmers work with restaurants, they will receive a higher price for their product, but they need to work on payment terms of 30 to 60 days. The other side of the coin, says Molovinsky, is that it's important to screen the restaurants to ensure that they are solid financially. While a large food service can afford to write off losses, a farmer cannot. So the farmer needs to be able to trust that the cooperative will pay for the animals. Both sides need to learn patience and trust, says Molovinsky, as invoices do get lost and there are inevitable challenges as a cooperative gets off the ground.

Engineer the menu to move the whole animal.

Whole-animal cuisine impacts the fundamental nature and pricing of a menu. Buying the whole animal means, in effect, that every part of the animal, chuck roast or steak, costs the same. And menu pricing also affects the quantities sold. So recipes and pricing need to work together.

For every steak sold, Molovinsky says a restaurant needs to sell five to six times that amount in chuck roast, bottom round, and eye of round. So the chef needs to become clever at engineering dishes on the menu that enable customers to buy the desired quantities of those cuts.

Certain pieces of the animal are not delivered to the restaurant. The processor keeps the neck, tail, hooves, cartilage, outer fat, and skin and sells them to other industries. The restaurant receives about 50 to 60 percent of the animal. Molovinsky says they are still challenged to find creative uses for the kidneys, heart, and liver – which are great items, he says, but hard to put into an appealing dish. If the restaurant isn't able to work these items into the menu, it writes them off as a loss.

Mie N Yu needed 8 to 12 months to work out a solid formula for using and moving all cuts of the animal, in appropriate quantities. It doesn't always have enough steak, and at particular times in the week it is unable to offer dishes with the steak cuts. The restaurant has learned that when the reason is made clear to them, customers are supportive. Molovinsky says that his restaurant's program has reached an "optimal" point, because it has figured out how to make local meat profitable and can increase its purchase of animals as needed.

Highlight the source of meat on the menu, rather than the cut.

The fact that food is local can draw people to a restaurant and also to a particular dish, says Molovinsky. And when a restaurant shifts to whole-animal cuisine, it also needs to shift its strategy in how it sells the dishes. Instead of highlighting the individual cut, the traditional restaurant marketing approach, it becomes important to call out the *source*. The menu at Mie N Yu highlights that its meats are all natural, provides information about the individual farms, and for each dish lists the farm sources. The restaurant's top-selling dish is made from chuck roast, a testimony to a chef's ability to create crowd-pleasing cuisine from different cuts.

Mentor chefs to teach them to accommodate whole-animal buying.

Molovinsky works with chefs to help them take steps toward being able to handle the whole animal. For example, if a restaurant thinks it might be interested in joining the cooperative, Molovinsky encourages the restaurant to take one whole lamb and run it as a special. Then Molovinsky helps the restaurant create several dishes for the different cuts — a dish for the shoulder that can't be prepared in the same way as leg or rack of lamb and a way to use the trimmings as well. Until the restaurant is able to use the whole lamb, it is not able to advertise that it is buying the "whole lamb," only that it is supporting the specific farm.

For the well-known choice cuts, such as filet mignon, Molovinsky says the restaurant has an opportunity to create a unique dish on its menu, to name both the cut *and* the farm of origin, which allows the restaurant to charge a premium price for it.

Molovinsky encourages chefs to rethink all aspects of their kitchen. Take stocks, for example. Over the past 20 years, Molovinsky says, restaurants have sought a variety of ways to reduce labor costs, including buying demi-glace to make stocks, rather than making them from scratch. At Mie N Yu, before the cooperative was born, demi-glace costs averaged about $120 per week and were thought to save the restaurant about 20 hours of kitchen labor each week. When they began to consider using the whole animal, Molovinsky and executive chef Tim Miller rethought this standard wisdom and realized that they weren't actually seeing those savings in labor; instead, they were simply reallocating those hours to other tasks. Now Molovinsky saves $120 a week by *not* buying demi-glace, instead using bones to make a fantastic high-quality in-house stock. "If we throw out 10 pounds of bones," says Molovinsky, "it's like throwing out $700 in cash!"

Molovinsky is excited about what this program does for chefs. Once chefs learn how to work with difficult cuts — that they can braise the shoulder, slice it thin, or make a nice sandwich out of it — their creativity is unleashed. "It helps the chef be a chef again!" says Molovinsky.

Teach traditional slow-cooking methods at culinary schools.

Through the 1980s and '90s, culinary schools moved away from slower methods of cooking to fast cooking methods, according to Molovinsky. So chefs may not be familiar with slower methods of cooking, such as braising and roasting, which are important for tougher cuts of meat. Molovinsky says a major step for the local food movement to enable restaurants to buy whole animals is for culinary schools to become a source of innovation and information — to teach younger chefs how to work with all parts of the animal, including slow-cooking methods.

Budget sufficient time to establish the program.

Getting the D.C. restaurant cooperative off the ground was a significant effort for six to eight months. Molovinsky spent long days visiting the Virginia farms and processing plants, covering two or three farms in a day. Occasionally, he needed to plan overnight trips as well. "Farmers want to talk and feed you," says Molovinsky. "It's very time-consuming, and you need to visit a lot of farms." Not all the farms you visit are going to work, Molovinsky warns. If a farm already sells a lot of produce through a farmers' market, then its price point is likely going to be too high for restaurants. But you will learn from every farm you visit, he says.

After the first six months of identifying producers and processors, Molovinsky needed another 10 to 12 weeks to create a secure pipeline, which involved a lot of planning and phone calls to piece together the puzzle of different farmers and their animals. Once the system was put into place, Molovinsky said he needed about half a day each week to manage the program and work out the kinks. After the first six to eight months of full operation, Molovinsky says, the program smoothed out, and his coordination time shrunk to about two hours per week.

Obtain assistance from your state department of agriculture.

The Virginia Department of Agriculture and Consumer Services helped Molovinsky at every stage while he worked to establish the cooperative. "They helped me find good farmers who were honest, who had good reputations, who had good products," says Molovinsky. Each state has its own agency for agriculture, which should be able to connect you with small farms that want to sell to restaurants, small processing plants for the meat, and local trucking or distribution companies, as well as helping you establish criteria for participation in the program.

Put your cooperative's needs and requirements in writing.

Molovinsky suggests that you put together a written description of your program. Initially, he would tell farmers verbally what they were looking for, but it became easier to provide it to them in writing. Eventually, with the help of the Virginia department of agriculture, Molovinsky created a set of written criteria for participation in the program (available on the program's website; see the resources). He suggests that all programs create a similar set of principles that will guide their decision making.

When possible, connect with farmer collaboratives.

The D.C. restaurant cooperative has had great success working with farmer collaboratives, according to Molovinsky. The benefit of working with a collaborative is that you only have to talk with one person to access a number of farms. Farmers are traditionally hard to contact, as they are outside working during the day, and some farmers don't even have phones. For farmers who don't belong to a formal collaborative but would like to be able to sell whole animals to restaurants, Molovinsky suggests that they elect someone in their community to represent the local farms who can be available to market and manage the distribution.

Food as a Tool against Poverty, Hunger, and Homelessness
D.C. Central Kitchen

D.C. Central Kitchen (DCCK) is a sprawling organization of interlocking programs, all aimed at combating poverty, hunger, and homelessness. Through its outreach, education, and jobs training programs, DCCK serves a wide-ranging population in the **D.C.** area, and it has spun off an independent food-recovery program that is being replicated at colleges across the country.

The DCCK guiding mantra is the two-pronged "combating hunger — creating opportunity." The fight against hunger manifests in programs such as First Helping, in which outreach workers head out onto the streets to distribute breakfasts to the homeless, at the same time working to introduce the homeless to what DCCK calls its "continuum of care," or ever-evolving network of support and empowerment programs. Another program involves repurposing leftover or surplus food from food-service businesses and "seconds" (unsalable produce) from farmers; DCCK collects this food and uses it to produce meals that it delivers to various social service agencies.

In all its efforts to relieve hunger, DCCK's focus rests on empowerment. Though feeding people may be a worthwhile goal in and of itself, DCCK was founded on the premise that hunger is a by-product of a larger issue: lack of opportunity, whether for jobs, good wages, affordable housing, education, appropriate health care, or any of the host of other benefits that larger society may take for granted. Lack of opportunity can lead not just to hunger but to poverty, unemployment, homelessness, drug abuse, and incarceration.

So the second prong of DCCK's mission, creating opportunity, focuses on empowerment. The group offers a 16-week culinary training program tailored to at-risk populations, such as the homeless or recently incarcerated. Graduates are eligible to be hired into one of DCCK's catering programs, or DCCK may help them find placement in a local food-service business.

Working out of its own kitchen and on-site dining services facilities, DCCK's Fresh Start program caters meals at Washington Jesuit Academy (WJA), a private middle school for at-risk boys, as well as Next Step Public Charter School, the University of the District of Columbia, and seven D.C. public schools. The kitchen staff for Fresh Start are graduates of DCCK's food service training program, and, says executive chef Allison Sosna, they often serve as unofficial mentors for the students, sharing with them stories of their own struggles.

The staff have disparate backgrounds. Caterer's assistant Derek Nelson had been in a drug rehabilitation program when he learned about DCCK from a friend. He applied as a way to get back on his feet and fell in love with being in the kitchen. He was helped, Nelson says, and now wants to pass along that service to others, through the common denominator of food. Cook Derwin Gaines, on the other hand, had been an electrician before he was hurt and retired; he saw DCCK as an opportunity to retrain. Sous chef Howard Thomas had been a drug dealer, but after two friends were murdered while in a car that he also was supposed to be in, he decided to change his life. He heard about DCCK through a friend, and though at first he laughed at the idea, he eventually decided to give it a try. The training, Thomas says, was the best thing he ever did. Perhaps prep cook James Weeks, who is also a minister, says it best: DCCK, he says, is a way to reach out to people in need. And whether it's the people preparing, serving, or receiving the food, DCCK is doing just that.

Executive chef Allison Sosna fell in love with food while studying in Europe. Returning to America, she wanted to combine this new passion with socially responsible work. At D.C. Central Kitchen, running the Fresh Start Catering program, she helps others get started as chefs, while cooking fresh, farm-to-table food for local schools in the Washington, D.C., area.

Sous chef Howard Thomas preps vegetables for Fresh Start Catering. He says that the training and experience he's received from D.C. Central Kitchen has been invaluable. He even had one of his own original recipes featured on a seasonal postcard produced by DCCK to promote the organization's work.

Former electrician Derwin Gaines had always enjoyed cooking, and when he was hurt on the job and forced to retire, he took the opportunity to enroll in D.C. Central Kitchen's educational program. He is now a cook at Fresh Start Catering.

Students at Washington Jesuit Academy in Washington, D.C., get farm-fresh meals made on the spot by Allison Sosna's crew at Fresh Start Catering.

Latasha Washington first heard about DCCK when she saw a flyer for it. She'd always loved cooking, so she thought she'd give it a try. She graduated in DCCK's seventy-ninth class and was hired to work in the Fresh Start Catering program. She'd someday like to start her own catering company, calling it "Art Is Catering."

James Weeks has had several jobs with D.C. Central Kitchen. He currently works as a prep cook and delivers meals from the Fresh Start Kitchen at the Washington Jesuit Academy to several other schools in the D.C. area.

DCCK's Campus Kitchens Project takes advantage of surplus food and resources to feed at-risk communities from 30 (and counting) school and university kitchens across the country. At Gonzaga College High School, students coordinate food donations and deliver meals to residents of the Sibley Plaza housing project in downtown Washington, D.C.

Joelle Johnson (at right), the buyer for DCCK, works with Dennis Showalter, the auction manager for the Shenandoah Valley Produce Auction, to buy food directly from local farmers, most of whom are Mennonite. Buying bulk produce and seconds (produce with cosmetic imperfections) at auction makes it possible for DCCK to afford fresh, local food for all of its programs.

Lessons Learned
Building a Nonprofit Community Food Project

Foster independence for each program.

In Portland, Oregon, Janus Youth (page 105) — a huge nonprofit organization by any standards, with an operating budget of $9 million — finds success by operating on a modular system. In this system each of its programs can operate nearly completely independently, giving them freedom to do things creatively. This approach reflects a trend among businesses and other organizations that need to be able to respond quickly to new pressures and developments. Those bound by a more top-down control structure often find themselves falling behind their competitors, mired by their own bureaucratic constraints. Janus Youth believes it has retained its creative "edge" by adopting this modular way of doing business.

Staff fully and creatively.

Nonprofits are famous for stretching the limits of their staff. But excessive multitasking, says Amber Baker, program director for Janus Youth's Village Gardens program, is ineffective. On the Janus Youth food farm for teens, one staffperson is dedicated entirely to farm management and another to youth support. Having the youth supervisor out on a tractor all week, says Baker, would render him ineffectual. Ideally, all staff, regardless of their specific job, understand the core program goals — for Janus Youth, that's positive youth development.

Baker also believes it's important to staff a project from within the community as much as possible. Janus Youth hires staff for its garden program from the community the garden serves. Over the years, the group has also developed a strong partnership with AmeriCorps, has attracted former Peace Corps volunteers, and has hired students right out of school — who, according to Baker, are "blown away by the pace and nature of the work."

Pay staff a living wage and benefits.

Edwin Marty, cofounder of Jones Valley Urban Farm (page 158) in Birmingham, Alabama, says they used to have trouble keeping a farm manager because they weren't paying a sufficient wage. When a farm manager left, they had to find and train a new farm manager, which always meant more time, which means more money. When they raised the pay by $10,000 and added health insurance, suddenly they were able to keep their staff, which was better for everyone.

Stay focused.

Nonprofits often try to do a million and one things, because one good idea inevitably leads to two or three more good ideas. But a nonprofit that pursues every good idea that crosses its path will soon run itself ragged. To prevent this, creating a core mission and staying on a double, short-term strategic plan is vital.

Village Gardens' experience shows the benefit of reflection on the mission and strategic plan. Amber Baker, program director, recalls the heady excitement in the early days of the project, when they wanted to re-create its glowing success in every neighborhood in every community in Portland. Now, she says with a slightly rueful laugh, "we've realized how much work and energy all this takes. And there's so much more we can do to make things better, *right here*." For Village Gardens, staying true to the mission meant evolving to meet the needs and energies of the community residents, not expanding to take on ever greater numbers of communities. Always expanding, Baker says, is not always the right or best path.

Translate achievements into dollars that are meaningful for policy makers.

In Milwaukee, Growing Power (page 74) has seen increasing visibility and success over the years. One reason is its ability to create partnerships and to translate its achievements into dollars that are meaningful for policy makers, according to Jerry Kaufman, president of the board of directors. A community food project will gain better standing in its community and be able to attract more partners if it is able to translate its impact on the community into specific numbers, such as the number of jobs it has created or the income it generates. As more communities become aware of the once-hidden costs of the industrial agricultural system and its negative impacts on public health, Kaufman suggests it is empowering for a community food project to tell the story of how it delivers quality food to people who are prone to obesity or diabetes, and how it contributes to achieving public health goals.

Growing Power offers a tremendous example of how to translate success for local policy makers. Kaufman rattles off numbers in quick succession: Growing Power is generating close to $200,000 gross income on less than 2 acres, with only 1 acre in actual food production. This translates to roughly $5 per square foot. Growing Power is proving that an urban farm can be a profitable, entrepreneurial enterprise that generates significant income and creates green jobs — it's grown to 50 staff. "And we haven't laid off anyone in the tough economic times," says Kaufman. What's more, Growing Power attracts hundreds of visitors to Milwaukee — some to attend Growing Power workshops and others to one of the several national conferences it has sponsored. Kaufman advises that, while your city or town may not be receptive to establishing a partnership at first, as it sees the beneficial economic spin-offs of an urban farm, it may become a more willing partner. After seeing what Growing Power has been able to achieve, Milwaukee is now interested in positioning itself as the "urban agriculture capital of the U.S.," says Kaufman.

Diversify your income stream.

Virtually every nonprofit I've ever talked or worked with agrees: grants can create a very difficult cycle of dependence. Nonprofits talk about how "chasing the money" can lead to "mission drift." Instead of dependence on grants, the ideal nonprofit creates fiscal stability and flexibility by building diverse sources of income — a diversified portfolio. This is not easy. Boston's Food Project (page 136) has a budget of nearly $3 million. While a significant 28 percent of this budget comes from individual donations, 40 percent from private foundations, 11 percent from revenues (CSA shares, farmers'

markets, and honorariums), 9 percent from corporations, and 8 percent from miscellaneous sources, with less than 3 percent coming from government grants, it requires five full-time development staff to keep this money wheel turning.

"Too many nonprofits are too dependent on grant writing, which can detour you from your mission," declares Growing Power's Kaufman. Over the years, Growing Power has diversified its income through the development of training programs, consulting services, and the sale of a diversity of products such as greens to restaurants, weekly market baskets, compost, worms, and honey. In addition to these income streams, Growing Power makes a conscious effort to raise individual and corporate contributions through an annual fund-raising appeal, which can raise as much as $80,000 to $90,000, or 5 percent of its $1.8 million (and growing) annual operating budget.

At Troy Gardens in Madison, Wisconsin (page 41), an early multi-year W.K. Kellogg Foundation grant of close to $500,000 proved a boon and a detriment. The grant provided much needed stable funding for staff, programs, and site development to enable the Community GroundWorks program to come into its own and thrive. However, with this money in hand, the organization didn't focus on expanding its capital base and fund-raising efforts. This is a common nonprofit lesson. Nonprofit core services generally require funding rather than generate revenue, and to support those services the nonprofit must raise capital, either by fund-raising from sponsors and donors or by implementing services that *do* generate revenue. While focusing on one to the exclusion of the other can put the nonprofit at risk, finding the right balance between the two is often the path to sustainability. After the Kellogg grant ended, CommunityWorks faced challenging financial times, and the group worked hard to find additional funding sources and to redefine and expand its mission to include revenue-generating activities. "Community GroundWorks has come through with flying colors," says Marcia Caton Campbell. "The organization is flourishing, with strong programs that reach more widely into the community as well as drawing people to the site."

Connecting Food to a Sense of Place
Diablo Burger

Diablo Burger calls itself a "local foods based burger joint." In northern **Arizona's** hip community of **Flagstaff**, Diablo Burger offers only locally raised open-range beef and uses other local ingredients as much as possible. The story of why and how this burger joint got off the ground underscores its desire to "connect the well-being of our community to the sustainability of our landscape through gastronomy . . . which is just a fancy word for cheeseburger."

A tiny burger joint in downtown Flagstaff, Arizona, is making a very big point. "At Diablo Burger, we say that it's our intention to connect community and ecology through gastronomy — which we think is just a fancy word for cheeseburger," says the restaurant's owner, Derrick Widmark. "We want to connect our community to the landscape it's built on through the food we eat." Widmark is doing just that. Through the simple medium of a burger joint, the Flagstaff community is becoming connected to ranches that have been raising cattle in the Diablo Canyon area east of Flagstaff since the late 1800s.

Eating Beef to Support Ranchers

In 2006 Widmark had just begun working for the Diablo Trust, a grassroots collaborative nonprofit founded in 1993 by two longtime ranching families, the Prossers and Metzgers, to manage and conserve healthy ranch lands in a collaborative and environmentally sound manner. The name itself, Diablo Trust, set the direction for the group's work. Not meant in the fiduciary sense, the word *trust* was chosen to reflect the unusual vision that a wide range of stakeholders would come together, in trust, to help preserve and manage the land area.

As the trust's communications and program coordinator, Widmark was learning about the trust's latest innovation, the Diablo Canyon Rural Planning Area, an unusual collaborative effort to preserve historic ranching operations in the face of changing economic conditions. Adopted in 2005 by Coconino County after a two-year planning period, the Diablo Canyon Rural Planning Area is nothing less than a plan to protect traditional ranching and open space — both of which are at risk of being permanently lost — while also preserving healthy ecosystems.

Widmark's restaurant is an example of one strategy suggested by the plan — creating value-added beef by direct-marketing to local restaurants. "Beef is the largest cash crop in northern Arizona," Widmark tells me. "But until now, it's all largely gone elsewhere." This made no sense to him: there was plenty of beef being raised all around Flagstaff, but not an ounce of it to eat. He thought Flagstaff folks should be able to eat the beef being raised around them, and he decided to do something about it.

He credits Gary Nabhan, whom he names the "patron saint of the local food movement, not just in the Southwest but around the world," with creating his "light bulb" moment. (On the day that I visit, I realize how much he esteems Nabhan's influence when I see that the menu features a special "Nabhan Burger," with grilled nopales (cactus), pepper jack cheese, and ancho chile mayo.) It was Nabhan who had brought in a speaker from an Idaho conservation group, Lava Lake Land and Livestock, to meet with Diablo Trust; he told them that as sales of their lamb to high-end restaurants in Sun Valley increased, so did awareness and appreciation of their conservation work. Widmark says, "I realized that if we could feed people local beef, there would be a lot fewer people asking, *What the [heck] is Diablo Trust?*"

A Plan to Preserve Working Lands

For regions facing the possibility of losing working farmlands, the model of the Diablo Canyon Rural Planning Area demonstrates how to explore and think through a range of options for preserving these lands and regional character. The challenges faced by the Diablo Trust ranching families are not unique: farmland and farming heritage are being lost to development throughout the country. The Diablo Canyon plan explains, "When economic conditions become unfavorable to ranching . . . working landscapes tend to be converted to other uses such as housing developments. . . . When that occurs, landscapes are fragmented, there is a loss of open space and habitat values, and our cultural and historical values are diminished. In other words, Coconino County becomes less unique and special, and more like everywhere else."[252] To develop an effective regional plan, the Diablo Canyon RPA model suggests it is essential to create a partnership dialogue that brings together agricultural property owners, county government, and the local economic development commission. Together, these groups can develop strategies for keeping traditional farm and ranch lands economically viable.

He opened Diablo Burger in 2009, and it quickly became a hip burger joint. Over the course of a week, depending on the season, the restaurant uses the equivalent of one to two whole cows from the nearby Diablo Trust ranches. While this represents a very small percentage of the ranches' overall cattle sales, Widmark says that everyone involved believes that it is but a first step into the local market, with Diablo Burger serving as a "tasting room" for this quality product. Through serving local food Widmark is building awareness and appreciation for the region's rich ranching heritage and helping a region become more connected to its food and landscape.

Serving Local to Support Local

Diablo Burger's motto is "All About Local": the burger toppings come from nearby farms, as do the greens, and the bread and cookies from a bakery in Phoenix. "The logistics of using local suppliers are much more challenging," says Widmark. "But the rewards are much greater. When people hand over the product to us, they're giving us something with a lot of pride, hours of work, sweat and toil. I tell my staff that we have a personal responsibility to these producers, that it is our obligation now to take their product and represent it as best we can."

Diablo Burger is also all about promoting local environmental and economic sustainability. To help keep the money spent at Diablo Burger circulating in the local economy, the restaurant accepts only cash. "Studies have shown that cash stays in the community," he explains. Credit card fees are a form of economic "leakage," and cash is more likely spent among local businesses and individuals, thereby creating a multiplier effect that feeds the local economy. Widmark works hard to spend his money locally, and he is proud to point out that as much of the restaurant as possible was built by local craftsmen.

Another unusual feature of the restaurant is that, for environmental sustainability, it doesn't have a dishwasher. "Water is the most precious resource on the Colorado plateau," says Widmark, "and dishwashers use tremendous amounts of water." Instead, even though it comes at a higher cost, Diablo Burger uses compostable serving baskets and utensils made of bagasse, a sugarcane by-product, to generate as little trash and use as little water as possible.

It's too soon to determine whether the restaurant will be successful, says Widmark, but he believes the restaurant has struck an important chord — helping people feel more connected to place. "People live in Flagstaff because they feel a connection to the landscape. When people feel like the landscape is feeding their families, and that eating this food helps to preserve open space, healthy watershed, and wildlife habitat, that's hard to walk away from." Though the restaurant had been in operation for only five months at the time of my visit, Widmark was able to say he hadn't run a single ad, yet somehow the word had spread. Tourists were even beginning to come in, pointed there by locals. For Widmark there is no higher compliment or testimony that Diablo Burger is filling an important need than for visitors to be sent to Diablo Burger for an experience that is specifically and locally authentic to Flagstaff.

Food: The Ace in the Hole

Diablo Burger is not the sole answer for keeping the large ranches of northern Arizona economically viable, says Widmark, but it is pointing the way to one part of the solution. "The economies of scale that industrial food has to offer just can't be beat. So if we were able to do what Diablo Burger does at a much bigger regional scale, it would make a bigger difference for these ranches. That's the issue."

The biggest challenge for scaling up in the Flagstaff area is the same I've heard in other parts of the country: the lack of certified local meat-processing and storage facilities. These are not issues that can be solved quickly, as the coordination of state and federal inspection is complex, time-consuming, and costly. Increasing awareness and pressure for changes at the federal level to create a streamlined process for smaller facilities that serve local food markets may eventually help local food systems overcome this persisting challenge.

Challenges aside, Widmark believes that we have passed the tipping point and that local food is not a passing fad. "Without cheap gas and oil, the industrial food system doesn't work. And I don't think the pendulum on energy costs is going to swing back in the other direction in my lifetime." His goal with Diablo Burger is to help solve the puzzle of how to ensure that local food is affordable. "If the food is right here, it shouldn't cost more. We can cement the long-term sustainability of the local food system if it is affordable." Widmark has gone to great lengths to ensure that his burger is truly affordable. He tells me that people come in all the time thinking it is going to cost more and are pleasantly surprised. A couple from Phoenix wrote in his comment book, "We pay more for fast food, and this is so much better!"

"This movement isn't about nostalgia," he says, then points to the walls of his restaurant. "Do you see one single piece of artwork of cowboys, cows, or ranch lands on our walls?" Rejecting the idea that local food is about romancing the past, Widmark is clear that, if anything, it is about romancing the future — in a practical, pragmatic way. "Once urban communities begin to see rural areas as *part of our greater community*, not just for another subdivision or golf course, but as a place that feeds our families, then we will value those lands differently. So it is my belief that food is the ace in the hole that will help make the difference for working landscapes to survive as open space, providing us with food and water and contributing in such a wide range of ways to our greater quality of life."

Bright Idea: Pooling Deliveries

One method for increasing the efficiency of obtaining local products is to pool resources with other businesses that use the same products. For example, Diablo Burger has teamed with several other restaurants in Flagstaff to share the weekly four-hour round trip to McClendon's Select organic farm in Peoria, near Phoenix. Previously, the restaurants had either driven down to the farm independently or picked up produce from McClendon's when it came to the local farmers' market. Now when McClendon's comes to the weekly farmers' market, it adds a preordered pallet for the restaurants. Each week a different restaurant pays the $50 delivery charge for the pallet, picks up the produce off the back of McClendon's truck, and delivers to the partner restaurants.

It may take time and effort to set up an efficient partnership, but it can pay off quickly. "It took us nearly two years before we set up this current deal," says Paul Moir, owner of Brix, a high-end Flagstaff restaurant. "One of our big struggles in Flag is getting the food up here. So it made sense to partner up. Now each [of] us picks up the pallet from McClendon's every third week instead of every week."

Gleaning: Keeping Food on the Table
The Society of St. Andrew and Marin Organic

The Society of St. Andrew, founded in 1979, with headquarters outside **Lynchburg** in **Virginia's** Blue Ridge mountains, runs the largest nation-wide gleaning operation, repurposing large quantities of unused food to feed those in need. North of San Francisco, another organization, Marin Organic, operates an efficient farm gleaning operation to bring locally produced organic food into Marin County schools to improve youth nutrition. Together, these two organizations offer insights into how we can reduce massive waste and inefficiency in our food system, to help ensure that everyone has healthy food to eat.

During World War II Parisians hoarded and "gleaned" every tiny thing for reuse because of extreme scarcity of wartime resources, a Frenchman once told me. He said they recycled used staples, paper clips, nails, broken zippers, even eyelets in worn shoes — anything with metal. While I don't know if this was an exaggeration or a real ethic of Parisians, I'll never forget my sudden awakening to the enormous wastefulness in modern life. When something breaks, most people in developed nations throw it away and buy a new one. When people don't finish the food on their plates, they usually scrape the leftovers into the garbage, from where it is sent to a landfill or incinerated, or it is ground in a disposal and sent through a sewer system into our waterways. Few people in our nation of abundance live by the adage that my mother and father drummed into my head — *Take only what you will eat!* — which was always followed by, *Don't leave the table until you clean your plate!* America may be a land of plenty, but it is also the land of plenty of waste.

The Green Revolution has been accompanied by another revolution of profligate wastefulness. Michael Waldmann, executive director of the Society of St. Andrew (SoSA), summarizes it succinctly: if all the food that is wasted annually in the United States were gleaned, *we could feed all 40 million of America's hungry*, including more than 13 million children. What's more, a University of Arizona study concluded that

reducing the rate of food waste could save U.S. consumers and manufacturers tens of billions of dollars each year."[253] The story does not end there. According to the Environmental Protection Agency, food waste — uneaten food and food preparation scraps — accounts for over 14 percent of the municipal solid waste stream and totals more than 34 million tons per year.[254] While our industrial food system seeks to improve production efficiencies, a 2008 Stockholm International Water Institute (SIWI) study concludes that losses between field and fork may reach as high as a whopping 50 percent![255]

Imagine: would any other business losing 50 percent of its product between processing and consumption consider such losses acceptable? SIWI suggests one reason for these staggering losses is the increasing distance that food travels from farm to fork. A second contributor to this wastage, it suggests, is the increasing number of people who handle food along the way — transporters, processors, wholesalers, retailers, chefs, and consumers themselves — each contributing another piece to the waste stream.

I'm struck by the odd paradox of an agricultural system that seeks to engineer still greater production yields to feed the world yet pays little attention to rampant wastage of those yields. It brings to mind someone trying to fill a bathtub, crying for more water, more water, without noticing that the drain has no plug.

Those who have seriously studied the issue of hunger seem to agree: we do produce more than enough food to feed all the people in the world. Since Frances Moore Lappé first discussed this in her 1971 best seller, *Diet for a Small Planet*, our food production yields have continued to climb. But hunger has been neither eradicated nor, some argue, even abated. Whether you consult the UN World Food Program or a specialized nonprofit such as StopTheHunger.org, the numbers on hunger only deepen the paradox: somewhere from 900 million to over 1 billion people — or 15 percent of the world — are undernourished, while even more people (17 percent) are overweight, and 5 percent are fully obese. The common diagnosis among many experts is that the paradox — a world that produces enough food to feed everyone but fails to do so — results from poor distribution systems and politics.

In the grassroots food movement, a new kind of recycling is taking hold, encompassing a variety of gleaning strategies to make sure good food doesn't end up in the landfill. Ken Horne, cofounder of SoSA, says that when considering how to feed more than 40 million Americans who face hunger each day, gleaning is a "simple, common sense solution."[256] In fact, with the power of the grassroots food movement behind it, I believe that gleaning will one day become as common and natural as recycling, a part of everyday life from grade school to corporate America.

The Society of St. Andrew

The Society of St. Andrew is now the nation's largest gleaner of fresh produce, with offices in 20 states, donors in 32 states, and deliveries of gleaned food in all "lower 48" states plus the District of Columbia. This nationally recognized program was founded in a tiny place in central Virginia's Blue Ridge Mountains called Big Island, with a population hovering around 1,000, by Ken Horne and Ray Buchanan, with the blessing of the United Methodist Virginia Conference.[257] Since then, more than 30 years later, the society has grown into a lean nonprofit, putting almost 96 cents of every donated dollar toward food for the hungry and demonstrating that gleaning is an effective way to reduce waste and feed our hungry.

Over the years SoSA has innovated a variety of programs, all aimed at saving and channeling good food toward the hungry. Through its Gleaning Network, founded in 1988, the society coordinates as many as 30,000 to 40,000 volunteers each year who gather in fields to glean 15 to 20 million pounds of fresh fruits and vegetables that were overlooked during the harvest or considered unsalable, running an average of ten events with nine volunteers every day of the year.[258] SoSA is also plugged into the food distribution network of shippers, brokers, and food processors, who donate tractor-trailer loads of food that may not be commercially marketable but are still good, fresh food. SoSA receives, packages, and transports this food to nearby hunger relief agencies. In 2010 these tractor-trailer loads of donated food accounted for nearly half of the total 28.1 million pounds gleaned by SoSA.

Doing a little calculation on the back of a napkin, using SoSA numbers, I figure that each pound of food provides an average of three servings. So to feed America's 40 million hungry with three meals per day is 42 billion servings, or 14 *billion* pounds of food. But with SoSA and others we may be well on the way toward achieving this number, simply by increasing the efficiency of our food system through gleaning. The nonprofit organization Feeding America alone distributes more than 2.5 billion pounds, or nearly 19 percent of what's needed.

Other programs innovated by SoSA are the well-known Hunters for the Hungry program, piloted in 1991 and spun off into an independent program two years later; the Washington Area Gleaning Network, spun off into an independent program in 1993; the international nonprofit Stop Hunger Now, spun off into an independent program in 1998; and the first National United Methodist Hunger Summit in 2002, bringing together church leaders with hunger workers to move the effort forward. And through its Seed Potato Program, launched in 1983 when a Virginia Eastern Shore farmer offered them a dump-truck load of potatoes, SoSA now donates seed potatoes every year to needy families in rural Appalachia and on Native American reservations, to enable them to grow their own potatoes.

Grassroots Foraging

Saving good food from being wasted inspires passionate responses, especially when that food is hanging in plain sight. All it takes is someone to notice the red cherries, golden peaches, or black plums dangling from the limbs overhead, or the wild leeks, stinging nettles, or fiddlehead ferns near your feet. In California, artist Asiya Wadud had a vision of building a fruit-bartering system that would also build neighborhood relationships. In 2008 Wadud began riding her bicycle through Oakland neighborhoods and asking fruit tree owners if they might like to barter their Santa Rosa plums for other fruits grown by neighbors, such as sour cherries or apricots. Forage Oakland was born.

Early in her explorations, Wadud learned to articulate her philosophy. If a homeowner's fruit is likely going to waste, is it fair game to take? Some people might feel comfortable making off with the fallen fruit, but Wadud believes it is important to request permission – to build a neighborhood's awareness about its own food resources, while also building trust and a sense of community.

In Seattle, Washington, Solid Ground, an antipoverty nonprofit, runs a sophisticated, finely tuned fruit-gleaning program called Community Fruit Tree Harvest. In 2009, for example, volunteers gathered 19,600 pounds of apples, plums, and pears from Seattle's yards. Only prime-condition fruits are given to the food bank; bruised or rotten fruits go to compost. Of course, volunteers may taste and take fruit, as well. When I joined a plum tree gleaning expedition, the old plum tree on a quiet residential street in Seattle's University District was laden with fruit but needed only a few volunteers to pick it clean in under an hour. The plum pickers shared different motives – reducing waste while helping Seattle's hungry but also less obvious motives of feeling more connected to nature, having quiet meditative time high in the tree, remembering and reminiscing about childhood activities, and meeting others and building relationships.

In San Francisco, Iso Rabins founded ForageSF (page 37) in 2008 to turn people on to the "forgotten food system" of wild foods. Foraging local wild foods not only reduces carbon miles, he claims, but builds the local food economy. In fact, Rabins hopes to create a sufficiently high demand for wild foods that foraging can become a full-time sustainable occupation for him and others. Rabins is introducing people to wild foraged foods through wild-food walks in the region, wild mushroom "adventures" and workshops, acorn classes, local fishing tours, and even a CSF (community-supported foraging), in which he delivers a monthly box of all-wild foraged foods. As an example of the kinds of foods that can be foraged, one forager writes on Rabin's website that he's able to provide 20 pounds of curly dock each week, along with 10 pounds each of miner's lettuce, watercress, and wild nettle, plus California Bay nuts, wild turkey, and English plantain. Only an experienced forager could offer up such steady quantities.

The idea of a CSF is radical enough. But when Rabins discovered that he couldn't sell at a farmers' market because a vendor must be the primary *producer* of the food sold (and foragers collect but do not produce wild foods), he created his own market. In December 2009 he held the first "Underground Market" at a private home in San Francisco's Mission District. The Underground Market is a private, members-only club, whose only requirement is that members sign a liability waiver that specifies they understand they will have access to foods produced in home kitchens that haven't been inspected. Membership in the club is free, and the entrance fee to each market event is minimal. At his first event Rabins featured eight vendors and attracted nearly two hundred buyers. Within six months, The Underground Market exploded to 70 vendors and attracted more than two thousand people. The market features food vendors who are not quite ready to jump into selling at a farmers' market or a store but want to test out their products – baked goods, jams, and vegetables, as well as hot prepared foods such as soups, burgers, pulled pork sandwiches, casseroles, Mexican cuisine, Jamaican cuisine, and much more. Rabins told the *Washington Post*, "We think of ourselves as an incubator."[259]

Marin Organic

Just north of San Francisco, Marin Organic, a nonprofit founded in 1999 to preserve organic farming as a way of life in Marin County, is pioneering an extraordinary gleaning program to bring organic food to school lunchrooms. Through its volunteer Marin Organic Glean Team, which meets weekly at local organic farms to glean fruits and vegetables, Marin Organic in 2010 provided organic food for as many as 10,000 students each week, benefitting 36 public and private schools (more than half of the county's schools), three outdoor education programs, and 15 community centers. All told, since it began in 2005, Marin Organic's gleaning program has distributed more than 150,000 pounds of organic food within Marin County.

From a fiscal perspective the gleaning is a win-win for both local farmers and schools. "Gleaning makes it viable for schools to offer healthy, local organic food while staying within their budgets," Helge Hellberg, former director of Marin Organic, told the *Marin Independent Journal* in 2009. "This way, they can order what they can and subsidize that quantity with gleaned and free food. Even though schools have tiny budgets, they provide an additional revenue stream for the farmer and it connects schools with farmers."[260]

As much as 20 percent of what is grown on local farms may be available for gleaning because it isn't pretty enough to sell to restaurants or markets, claims Marin Organic. The organization routinely gleans a wide variety of organic vegetables and even meats and dairy products, such as eggs, yogurt, and even ice cream. In one case mislabeled organic yogurt was gleaned from a local dairy, and another farm donated leftover organic olive oil. After farmers have completed their own harvest, the Marin Organic Glean Team combs through the fields for the "seconds," or produce that isn't exactly the right size or shape but tastes just as good.[261]

"Marin Organic's gleaning program has been used as a blueprint for similar projects in farming communities from the Midwest to Hawaii," Hellberg concluded. And it deliberately led the way by founding National Gleaning Day on September 20, 2010. With National Gleaning Day and ongoing community gleanings throughout the year, Marin Organic continues to encourage communities, farms, and schools across the nation to glean healthy food for children.

A Simple Recipe for Gleaning

Faith Feeds, a cooperative of individuals and faith communities founded in 2010 to alleviate hunger in Kentucky, offers the following easy recipe for forming your own gleaning operation.[262]

1. **FIND A SOURCE OF FOOD.** Identify and contact local growers of fresh produce:

- Vendors at local farmers' markets
- Farm, orchard, or you-pick operations
- CSAs
- Community gardens
- Individual gardeners

Explain your purpose, and ask if they would like to participate. Find out what they can donate, when they can donate it, and what's the best way to pick it up.

2. **FIND A RECIPIENT.** Identify and contact local agencies that distribute food to the needy:

- Local food banks
- Places of worship that feed the hungry
- Shelters
- Halfway houses and group homes

Find out what they can use, when they can use it, and what's the best way to get it to them.

3. **RECRUIT VOLUNTEERS TO DELIVER FOOD.** Invite people to be a part of this simple activity:

- Friends and neighbors
- Worship groups
- Work colleagues
- Families

Tell them what needs to be done, and ask what they would like to do. People usually find it easier to say "yes" to specific requests.

Some basic rules:

- Be respectful of property.
- Don't give away food in such poor condition that you would not eat it, make it into soup, or preserve it yourself.
- All food should go to those in need.
- Write thank-you letters to donors.
- Start simple, and keep things as simple as possible.

Notes

Unless otherwise noted, all comments from the staff of the enterprises profiled in this book, and information about those enterprises, derive from personal interviews, conversations, and correspondence with the author, contributing writers, and research assistants in 2009–2011.

1   National Gardening Association, "The Impact of Home and Community Gardening in America" (2009 survey), http://www.gardenresearch.com/home?q=show&id=3126.

2   These arguments and comments are drawn from a series of e-mails written to the "Foodplanning" Listserv in September 2009 by a professor of planning who is also a self-described proponent of local food.

3   Dan Barker, *Queen Jane* (n.p.: CreateSpace, 2009), 14–15.

4   David Schwartz, "A Gift of a Garden: Green Activist Dan Barker Is Seeding Many Lives with Hope," *Smithsonian* 28, no. 6 (September 1997).

5   Bob Frost, "Tending the Earth, Mending the Soul," *Biography Magazine* (May 2003).

6   Ibid.

7   Growing Gardens, "Digging at the Root of Hunger" video, http://www.growing-gardens.org/our-programs.php.

8   Ibid.

9   Ibid.

10 Ibid.

11 Ecotrust, "Farm to School," http://www.ecotrust.org/farmtoschool/.

12 Growing Gardens, "Our Programs," http://www.growinggardens.org/our-programs.php.

13 Dave Cieslewicz, in Ben Block, "U.S. City Dwellers Flock to Raising Chickens," WorldWatch Institute, October 6, 2008, http://www.worldwatch.org/node/5900.

14 Dave Cieslewicz, Mayor Dave's Blog, "A Game of Chicken," July 23, 2009, http://www.cityofmadison.com/mayor/blog/index.cfm?Id=170.

15 Dave Cieslewicz, Mayor Dave's Blog, "Urban Agriculture," August 5, 2009, http://www.cityofmadison.com/mayor/blog/index.cfm?Id=177.

16 These points are adapted from Robin Ripley, "Eight Benefits of Raising Backyard Chickens," *Gardening Examiner,* January 18, 2010, http://www.examiner.com/gardening-in-national/eight-benefits-of-raising-backyard-chickens; Chickens In The City, "A Case for Backyard Chickens in Salem [Oregon]" (September 2010), http://www.chicken-revolution.com/Research_Packet_Sept_2010.pdf; and Mother Earth News, Chicken and Egg Page, http://www.motherearthnews.com/eggs.aspx#ixzz19QnwLVRF.

17 These arguments against backyard chickens are drawn from a variety of sources, but the most informative source is Chickens In The City, "A Case for Backyard Chickens in Salem [Oregon]" (September 2010), http://www.chicken-revolution.com/Research_Packet_Sept_2010.pdf.

18 Genetic Resources Action International (GRAIN), "Fowl Play: The Poultry Industry's Central Role in the Bird Flu Crisis" (February 2006), http://www.grain.org/briefings/?id=194.

19 Consumer Reports Health Blog, "Health Weekender: Egg Safety – In Defense of Home-Grown Eggs," July 24, 2009, http://blogs.consumerreports.org/health/2009/07/egg-safety-organic-eggs-health-benefits-of-eggs-salmonella-protect-yourself-from-foodborne-illness-.html.

20 Council President Richard Conlin, "One Small Step For Man, One Giant Leap For Goatkind," *Making It Work* (newsletter) 9, no. 8 (October 4, 2007), http://www.seattle.gov/Council/Conlin/miw/0708miw.htm#4.

21 Angela Galloway, "Seattle Homeowners May Keep Miniature Goats as Pets," *Seattle Post-Intelligencer*, September 26, 2007, www.seattlepi.com/local/333174_goat26.html.

22 These strategies are adapted from the website of the Goat Justice League at http://www.goatjusticeleague.org/Site/Legalizing_Goats.html.

23 Goat Justice League, "Urban Goats 101," http://www.goatjusticeleague.org/Site/Classes,_Urban_Goats_101.html.

24 Sharon Pian Chan, "'One Giant Step for Goatkind': Seattle Gives Them Pet Status," *Seattle Times,* September 25, 2007, http://seattletimes.nwsource.com/html/local-news/2003900621_minigoats25m.html.

25 Sustainable Food Denver, "Health Benefits," http://www.sustainablefooddenver.org/.

26 AskDrSears.com, "Feeding Infants & Toddlers: Got Goat's Milk?" http://askdrsears.com/html/3/t032400.asp.

27 Michael Graham Richard, "Rent-a-Goat in Action! Clearing Brush the Way Nature Intended It," TreeHugger, Travel & Nature, October 30, 2009, http://www.treehugger.com/files/2009/10/rent-a-ruminant-goats-clearing-brush-photos-before-and-after.php.

28 Living the Country Life, "Rent-A-Goat Program: Good Life for Goats," http://www.livingthecountrylife.com/animals/livestock/rent-a-goat-program/;jsessionid=REPNMQ0Y12RAECQCEASB5VQ?page=2.

29 Jon Gelbard, PhD, founder of Conservation Value, Inc., in a comment on "Rent-a-Goat in Action! Clearing Brush the Way Nature Intended It," by Michael Graham Richard, TreeHugger, Travel & Nature, October 30, 2009, http://www.treehugger.com/files/2009/10/rent-a-ruminant-goats-clearing-brush-photos-before-and-after.php.

30 Peter Hagar, "Meat Goats a Good Fit for Small Farms," *Plattsburg (NY) Post-Republican*, January 23, 2011, http://pressrepublican.com/coop_ext/x71343279/Meat-goats-a-good-fit-for-small-farms.

31 U.S. Department of Agriculture, Agricultural Research Service, USDA National Nutrient Database for Standard Reference, Release 23 (2010), Nutrient Data Laboratory Home Page, http://www.ars.usda.gov/ba/bhnrc/ndl.

32 Veronica Hinke, "Here & There: Beekeeping on a Chicago Rooftop Garden," *Herb Companion*, December/January 2007, http://www.herbcompanion.com/gardening/here-there-beekeeping-on-chicago-rooftop-garden.aspx.

33 Tiffany Crawford, "Bees Make Vancouver's City Hall a Hive of Activity," *Vancouver Sun*, March 19, 2010, http://www.vancouversun.com/.

34 Gerry Bellett, "'Bee Condos' Placed in City Parks to Boost Urban Colonies," *Vancouver Sun,* April 8, 2009.

35 Hugh Raffles, "Sweet Honey on the Block," *New York Times*, op-ed, July 6, 2010, http://www.nytimes.com/2010/07/07/opinion/07Raffles.html.

36 Will Allen, comments made during talks and a workshop at Lynchburg Grows, Lynchburg, Virginia, May 12–13, 2009.

37 Veronica Hinke, "Here & There: Beekeeping on a Chicago Rooftop Garden," *Herb Companion*, December/January 2007, http://www.herbcompanion.com/gardening/here-there-beekeeping-on-chicago-rooftop-garden.aspx.

38 Lance Sundberg, American Beekeeping Federation, "Honey Bee Facts," http://www.abfnet.org/display common.cfm?an=1&subarticlenbr=71

39 USDA Agricultural Research Service, "Colony Collapse Disorder," http://www.ars.usda.gov/News/docs.htm?docid=15572.

40 USDA, National Agricultural Statistics Service, "Honey" (February 2010), http://usda.mannlib.cornell.edu/usda/nass/Hone//2010s/2010/Hone-02-26-2010.pdf.

41 USDA Agricultural Research Service, "Questions and Answers: Colony Collapse Disorder," http://www.ars.usda.gov/News/docs.htm?docid=15572.

42 British Columbia Honey Producers Association, "CSI Apiary: Season Three – The Continued Unravelling of Colony Collapse Disorder," *BeesCene* 25, no 1, http://maarec.psu.edu/CCDPpt/CSIonCCDMar09.pdf.

43 Mid-Atlantic Apiculture Research and Extension Consortium, "Colony Collapse Disorder," FAQs, http://maarec.psu.edu/ColonyCollapseDisorder.html.

44 USDA Agricultural Research Service, "Questions and Answers: Colony Collapse Disorder," http://www.ars.usda.gov/News/docs.htm?docid=15572.

45 Marcia Caton Campbell and Danielle Salus, "Community and Conservation Land Trusts as Unlikely Partners? The Case of Troy Gardens, Madison, Wisconsin," *Land Use Policy* 20 (2003): 169–80.

46 Ibid.

47 Robert Putnam, *Bowling Alone* (New York: Simon and Schuster, 2000).

48 In response to concerns about differences in health status between and within countries – differences that many view as fully avoidable – the World Health Organization established a Commission on Social Determinants of Health (CSDH) in 2005 to provide advice on how to reduce them. The commission's final report was launched in August 2008. Of its three major recommendations, the second – "Tackle the inequitable distribution of power, money, and resources" – calls for "All groups in society to be empowered through fair representation in decision-making" and "Civil society to be enabled to organize and act in a manner that promotes and realizes the political and social rights affecting health equity." Both recommendations, along with others, inherently suggest the importance of organizations that are working toward these ends to build relationships and partnerships with each other and to work in concert. World Health Organization, "Social Determinants of Health," http://www.who.int/social_determinants/en/.

49 Alec Appelbaum, "Organic Farms as Subdivision Amenities," *New York Times*, July 1, 2009.

50 Seattle Department of Neighborhoods, "P-Patch Community Garden Program – Fact Sheet – 2010," www.seattle.gov/neighborhoods/ppatch/documents/FactSheet2010.pdf.

51 Seattle Department of Neighborhoods, "P-Patch Community Garden Program – Fact Sheet – 2009" (no longer available), and "P-Patch Community Garden Program – Fact Sheet – 2010," www.seattle.gov/neighborhoods/ppatch/documents/FactSheet2010.pdf.

52 Jim Diers, *Neighborhood Power: Building Community the Seattle Way* (Seattle: University of Washington Press, 2004), 102–3.

53 Seattle Department of Neighborhoods, "P-Patch Community Garden Program – Fact Sheet – 2009" (no longer available).

54 Neighborhood Gardens Association, "About NGA," http://www.ngalandtrust.org/history.html.

55 NeighborSpace. "About: NeighborSpace: Community Managed Open Space," http://neighbor-space.org/about.htm.

56 Paul M. Sherer, *The Benefits of Parks: Why American Needs More City Parks and Open Space* (San Francisco: Trust for Public Land, 2006).

57 Craig Springer, "Urban Land Trusts: Connecting Communities Through Shared Spaces," *Exchange* (newsletter of the Land Trust Alliance), Winter 2006, www.ngalandtrust.org/UrbanLandTrusts.pdf.

58 L.A. Neighborhood Land Trust, "Our Project," http://www.lanlt.org/.

59 Southside Community Land Trust, "City Farm," http://southsideclt.org/cityfarm.

60 Boston University School of Public Health, "Low Cost Compost Can Remedy Soil Contamination in Boston Community Gardens," *The Insider*, February 3, 2010, http://sph.bu.edu/insider/Recent-News/low-cost-compost-can-remedy-soil-contamination-in-boston-community-gardens.html.

61 Ibid.

62 Ibid.

63 Iris Zippora Ahronowitz, "Rooting the Community, Growing the Future: Two Massachusetts Urban Agriculture Organizations and Their Social Impacts" (undergraduate thesis, Harvard University, November 2003), http://thefoodproject.org/research, citing Laura Lawson, "Urban-Garden Programs in the United States" (PhD dissertation, University of California at Berkeley, 2000), and Joachim Wolshke-Bulmahn, "From the War Garden to the Victory Garden," *Landscape Journal* 11, no 1 (1992): 57.

64 Michael Pollan, "Farmer in Chief," *New York Times Magazine*, October 12, 2008, http://www.nytimes.com/ 2008/10/12/magazine/12policy-t.html?pagewanted=1&_r=1.

65 "Together We Are Growing Power," press kit, http://www.growingpower.org/assets/presskit.pdf.

66 Will Allen, comments made during talks and a workshop at Lynchburg Grows, Lynchburg, Virginia, May 12–13, 2009. Unless otherwise noted, comments from Allen and information about Growing Power derive from these presentations.

67 Will Allen, "The Food Fighter," *Outside*, March 2009.

68 Tram Nguyen, "Growing Power: Expanding Food Justice," *Colorlines*, January/February 2008.

69 Shannon Sloan-Spice, "Growing Hope, Growing Change, Growing Power," *Outpost Exchange* , July 1, 2009, 10–12.

70 Tracy McMillan, "Beets in the Hood," *Mother Jones*, March/April 2009.

71 Rhodes Yepsen, "Composting and Local Food Merge at Urban Garden," *BioCycle* 49, no. 11 (November 2008): 31–33.

72 Judy Jepson, "Urban Power Plants," *Exclusively Yours,* April 2008, 10–13.

73 Rhodes Yepsen, "Composting and Local Food Merge at Urban Garden," *BioCycle* 49, no. 11 (November 2008): 31–33.

74 Jennifer Yauck, "GLWI Delivers Perch to Growing Power," Great Lakes WATER Institute press release, April 30, 2008, http://www.glwi.uwm.edu/features/news/WATERNewsGrowingPowerperch04-30-08.php.

75 James Janega, "Perch Poised for Revival in Lake Michigan," *Holland Sentinel,* May 17, 2008, http://www.hollandsentinel.com/sports/x401616279/Perch-poised-for-revival-in-Lake-Michigan.

76 Jennifer Yauck, "GLWI Delivers Perch to Growing Power," Great Lakes WATER Institute press release, April 30, 2008, http://www.glwi.uwm.edu/features/news/WATERNewsGrowingPowerperch04-30-08.php.

77 Stefanie Ramp, "Growing Power Is in Full Bloom," *Shepherd Express,* September 22, 2005, http://www.shepherd-express.com/.

78 Growing Power, "Rainbow Farmers Cooperative," http://www.growingpower.org/rainbow_farmers_coop.htm.

79 Rhodes Yepsen, "Composting and Local Food Merge at Urban Garden," *BioCycle* 49, no. 11 (November 2008): 31–33.

80 Growing Power, "Compost," http://www.growingpower.org/compost.htm.

81 Alysha Schertz, "Growing Power: The Farm in the City," *BizTimes*, May 30, 2008, http://www.biztimes.com/news/2008/5/30/growing-power.

82 Growing Power, "Worms," http://www.growingpower.org/worms.htm.

83 Judy Jepson, "Urban Power Plants," *Exclusively Yours,* April 2008, 10–13.

84 Rhodes Yepsen, "Composting and Local Food Merge at Urban Garden," *BioCycle* 49, no. 11 (November 2008): 31–33.

85 Paul Kosidowski, "The Farmer Will Allen," *My Midwest*, March/April 2009, 92.

86 Judy Jepson, "Urban Power Plants," *Exclusively Yours,* April 2008, 10–13.

87 Greensgrow, "Lessons Learned," http://www.greensgrow.org/farm/modules/smartsection/item.php?itemid=56&keywords=lessons+learned.

88 Heather F. Clark, Daniel J. Brabander, and Rachel M. Erdil, "Sources, Sinks, and Exposure Pathways of Lead in Urban Garden Soil," *Journal of Environmental Quality* 35, no. 6 (2006): 2066–74, https://www.agronomy.org/publications/jeq/articles/35/6/2066.

89 The Food Project, "Soil Testing and Remediation," http://thefoodproject.org/soil-testing-and-remediation.

90 These steps are adapted from the Food Project, "Soil Testing and Remediation," http://thefoodproject.org/soil-testing-and-remediation.

91 Greensgrow, "Lessons Learned," http://www.greensgrow.org/farm/modules/smartsection/item.php?itemid=56&keywords=lessons+learned.

92 The Intervale Center, "Continuing the Mission," in *Explorer 2008: Your Free Guide to Intervale Farms, Local Food & Fun!* (annual newsletter).

93 Ibid.

94 Virginia A. Smith, "A Green Thumb, with Blisters on It," by Virginia A. Smith, *Philadelphia Inquirer*, April 17, 2009.

95 Greensgrow, "New LIFE Program Offers Healthy Options," posted June 18, 2010, http://www.greensgrow.org/farm/modules/news/article.php?storyid=195.

96 "LIFE, It's Better with Fresh Food," *Spirit Community Newspapers*, June 16, 2010, http://spiritnewspapers.com.

97 Larry Gabriel, "Life in the Desert," *Detroit Metro Times*, September 26, 2007.

98 Paul Mackun and Steven Wilson, "Population Distribution and Change: 2000 to 2010," U.S. Census Bureau, March 2011, http://www.census.gov/prod/cen2010/briefs/c2010br-01.pdf.

99 Ross A. Hammond and Ruth Levine, "The Economic Impact of Obesity in the United States," *Diabetes, Metabolic Syndrome and Obesity: Targets and Therapy*, no. 3 (August 2010): 285–95.

100 Greensgrow, "Lessons Learned," http://www.greensgrow.org/farm/modules/smartsection/item.php?itemid=56&keywords=lessons+learned.

101 Virginia Department of Historic Resources, PIF Resource Information Sheet (2008), http://www.dhr.virginia.gov/registers/Cities/Lynchburg/118-5294_Schenkel_Farm.pif.list.htm.

102 Hay Hardy, "Lynchburg Grows," *The Virginia Sportsman* (December 2007/January 2008).

103 Virginia Department of Historic Resources, PIF Resource Information Sheet (2008), http://www.dhr.virginia.gov/registers/Cities/Lynchburg/118-5294_Schenkel_Farm.pif.list.htm.

104 Hay Hardy, "Lynchburg Grows," *The Virginia Sportsman* (December 2007/January 2008).

105 Aaron Lee, "Jumpsuit Gardening," Lynchburg Grows blog, April 4, 2010, http://www.lynchburggrows.org/?p=594.

106 City-Data.com. "Holyoke, Massachusetts (MA) Poverty Rate Data," 2009, http://www.city-data.com/poverty/poverty-Holyoke-Massachusetts.html.

107 Area Connect, "Holyoke Massachusetts Population and Demographics Resources," http://holyoke.areaconnect.com/statistics.htm.

108 Neal Pierce, "Nourishing 'Nuestras Raíces' – 'Our Roots' in a Troubled Old City," *Nation's Cities Weekly*, May 15, 2006, http://www.nlc.org/ASSETS/95265E2EA9914D978DF2830172D12E1A/ncw051506.pdf.

109 Diabetes Initiative, "Advancing Diabetes Self Management," http://www.diabetesinitiative.org/programs/DIHHC.html.

110 Neal Pierce, "Nourishing 'Nuestras Raíces' – 'Our Roots' in a Troubled Old City," *Nation's Cities Weekly*, May 15, 2006, http://www.nlc.org/ASSETS/95265E2EA9914D978DF2830172D12E1A/ncw051506.pdf.

111 Corby Kummer, "A Papaya Grows in Holyoke," *The Atlantic,* April 2008, http://www.theatlantic.com/magazine/archive/2008/04/a-papaya-grows-in-holyoke/6702/.

112 Ibid.

113 Kay Oehler, Stephen Sheppard, and Blair Benjamin, "The Economic Impact of Nuestras Raíces on the City of Holyoke: Current and Future Projections," February 2007, Center for Creative Community Development, http://www.williams.edu/Economics/ArtsEcon/nuestras.html.

114 Ibid.

115 Town of Lincoln, "Open Space and Recreation Plan," March 2008, http://www.lincolntown.org/CLRP%20Open%20Space%20Committee/Open%20Space%20main.htm.

116 Iris Zippora Ahronowitz, "Rooting the Community, Growing the Future: Two Massachusetts Urban Agriculture Organizations and Their Social Impacts" (undergraduate thesis, Harvard University, November 2003), 68, http://thefoodproject.org/research.

117 The Food Project, "Our Farms," http://thefoodproject.org/our-farms.

118 The number often cited by Food Project staff is 250,000 pounds, but 200,000 pounds is the number reported in the *The Food Project: 2008–2009 Annual Report*, http://thefoodproject.org/sites/default/files/AnnualReport2009.pdf.

119 *The Food Project: 2008–2009 Annual Report*, http://thefoodproject.org/sites/default/files/AnnualReport2009.pdf.

120 Iris Zippora Ahronowitz, "Rooting the Community, Growing the Future: Two Massachusetts Urban Agriculture Organizations and Their Social Impacts" (undergraduate thesis, Harvard University, November 2003), 72, http://thefoodproject.org/research.

121 Roblyn Anderson Brigham and Jennifer Nahas, "The Food Project: A Follow-Up Study of Program Participants," March 2008, 12–13, http://thefoodproject.org/research.

122 Ibid., iv–v.

123 Iris Zippora Ahronowitz, "Rooting the Community, Growing the Future: Two Massachusetts Urban Agriculture Organizations and Their Social Impacts" (undergraduate thesis, Harvard University, November 2003), 64, http://thefoodproject.org/research.

124 *The Food Project: 2008–2009 Annual Report*, http://thefoodproject.org/sites/default/files/AnnualReport2009.pdf.

125 Ibid.

126 Ibid.

127 The Food Project, "The Blast Youth Initiative," http://www.thefoodproject.org/blast/Internal1.asp?id=358.

128 *The Food Project: 2008–2009 Annual Report*, http://thefoodproject.org/sites/default/files/AnnualReport2009.pdf.

129 Public Broadcasting Service (PBS), "The Civilian Conservation Corps," part of *The American Experience* series, http://www.pbs.org/wgbh/americanexperience/films/ccc/player/.

130 Will Allen, comments made during talks and a workshop at Lynchburg Grows, Lynchburg, Virginia, May 12–13, 2009.

131 Centers for Disease Control and Prevention, "Attention-Deficit/Hyperactivity Disorder: Data and Statistics," http://www.cdc.gov/ncbddd/adhd/data.html.

132 Will Allen, comments made during talks and a workshop at Lynchburg Grows, Lynchburg, Virginia, May 12–13, 2009.

133 Lianne Fisman, "Sowing a Sense of Place: An In-Depth Case Study of Changing Youths' Sense of Places"

(PhD dissertation, Massachusetts Institute of Technology, 2007), 313, http://dspace.mit.edu/handle/1721.1/39933.

134 U.S. Department of Health and Human Services, "Obesity Threatens to Cut U.S. Life Expectancy, New Analysis Suggests," March 16, 2005, http://www.nih.gov/news/pr/mar2005/nia-16.htm.

135 U.S. Department of Agriculture, "USDA Farm to School Initiatives Fact Sheet," http://www.fns.usda.gov/cnd/F2S/pdf/F2S_initiative_fact_sheet_040110.pdf

136 White House Task Force on Childhood Obesity, *Solving the Problem of Childhood Obesity within a Generation* (May 2010), http://www.letsmove.gov/pdf/TaskForce_on_Childhood_Obesity_May2010_FullReport.pdf.

137 The Farm to School programs described these examples are among those evaluated in *Bearing Fruit: Farm to School Program Evaluation Resources and Recommendations*, by Anupama Joshi and Andrea Misako Azuma (National Farm to School Network and the Center for Food & Justice Urban & Environmental Policy Institute at Occidental College, 2008).

138 National Farm to School Network, "About Us," http://www.farmtoschool.org/aboutus.php.

139 Emily Jackson, of the Appalachian Sustainable Agriculture Project, in a statement before the U.S. Senate Committee on Agriculture, Nutrition, and Forestry, regarding the *Farm Bill Reauthorization*, CQ Congressional Testimony, April 24, 2007.

140 Karrie Stevens Thomas, *Abernethy Scratch Kitchen Model 2005–2006 Baseline Assessment* (Portland, Ore.: Abernethy Elementary School, December 18, 2006), 7, http://www.ecotrust.org/farmtoschool/Abernethy_Kitchen_Assessment.pdf.

141 The Official Website for Birmingham, Alabama, "About Birmingham," http://www.birminghamal.gov/about-birmingham.aspx.

142 Rebecca Ruiz, "America's Most Obese Cities," Forbes.com, November 27, 2007, http://www.forbes.com/2007/11/14/health-obesity-cities-forbeslife-cx_rr_1114obese.html.

143 Centers for Disease Control and Prevention, "Overweight and Obesity: U.S. Obesity Trends – Trends by State 1985–2009," http://www.cdc.gov/obesity/data/trends.html#State.

144 Statehealthfacts.org, "Alabama: Percent of Adults Who Were Overweight in 2008," http://www.statehealthfacts.org/profileind.jsp?ind=89&cat=2&rgn=2 (updated yearly).

145 Bob Blalock, "Our View: Alabama's Obesity Problem Is Getting Worse and Needs To Be Tackled," Birmingham News Commentary, July 13, 2009, http://blog.al.com/birmingham-news-commentary/2009/07/our_view_alabamas_obesity_prob.html.

146 Associated Press, "State to Hit Workers with 'Fat Fee,'" MSN, August 26, 2008, http://articles.moneycentral.msn.com/Insurance/InsureYourHealth/AlabamaHitsObeseWorkersWithFee.aspx.

147 Jones Valley Urban Farm, "History," http://www.jvuf.org/about_history.php.

148 Jones Valley Urban Farm, "Education/ASAP," http://www.jvuf.org/ed_asap.php.

149 Jones Valley Urban Farm. Education/Weekend Workshops," http://www.jvuf.org/ed_secondsaturdays.php.

150 I heard this story from the university food supervisor at a presentation at a local food conference in Weyers Cave, Virginia, in the summer 2008. People involved with other universities have told me very similar stories.

151 University Leaders for a Sustainable Future, "Talloires Declaration," http://www.ulsf.org/programs_talloires.html.

152 International Sustainable Campus Network, "Charter and Guidelines," http://www.international-sustainable-campus-network.org/index.php?id=93.

153 The College Sustainability Report Card, "Key Findings," http://www.greenreportcard.org/report-card-2010/executive-summary/key-findings.

154 Berkeley College, "The Berkeley Dining Hall," http://www.yale.edu/berkeley/facilities_dining.html.

155 Andrew Martin, "Local Foods Flavor College Cafeterias," *Chicago Tribune*, April 24, 2005.

156 Ibid.

157 Wendy Y. Lawton, "Spreading the Gospel of Locally Grown," *George Street Journal*, September 24, 2004, http://www.brown.edu/Departments/Brown_Is_Green/documents/Spreadingthegospeloflocallygrown.pdf.

158 Charlotte Bruce Harvey, "The New Organic," *Brown Alumni Magazine*, July/August 2005, http://www.brownalumnimagazine.com/content/view/467/40/.

159 Andrew Martin, "Local Foods Flavor College Cafeterias," *Chicago Tribune*, April 24, 2005.

160 Brian Clark Howard, "Twelve of the Most Healthy and Sustainable College Cafeterias," The Daily Green, http://www.thedailygreen.com/environmental-news/latest/greenest-college-cafeterias-4608093.

161 Oberlin Online, "Dining: Oberlin College Sustainability Initiative Time Line," http://www.oberlin.edu/cds/social/sustainability.html.

162 Bill McKibben, *Deep Economy* (New York: Holt & Co, 2007), 63.

163 Guillaume P. Gruère and S. R. Rao, "A Review of International Labeling Policies of Genetically Modified Food to Evaluate India's Proposed Rule," *Journal of Agrobiotechnology Management and Economics* 10, no. 1 (2007): 51–64, http://www.agbioforum.org/v10n1/v10n1a06-gruere.htm.

164 Gary Nabhan, *Where Our Food Comes From* (Washington, DC: Island Press, 2009), 132.

165 Ibid.

166 U.N. Food and Agriculture Organization, *Protecting Animal Genetic Diversity for Food and Agriculture: Time for Action* (Rome: FAO, n.d.), http://www.fao.org/ag/magazine/pdf/angr.pdf.

167 *Crop Diversity at Risk: The Case for Sustaining Crop Collections* (Kent, UK: Dept. of Agricultural Sciences,

Imperial University Wye, n.d.), http://www.croptrust.org/documents/WebPDF/wyereport.pdf.

168 Millenium Ecosystem Assessment, "About," http://www.millenniumassessment.org/en/About.aspx#1 (no longer available).

169 Stuart H. M. Butchart, et.al. "Global Biodiversity: Indicators of Recent Declines," *Science*, April 29, 2010, doi:10.1126/science.1187512.

170 Cary Fowler, "One Seed At At Time: Protecting the Future of Food," TED Talks, July 2009, http://www.ted.com/talks/cary_fowler_one_seed_at_a_time_protecting_the_future_of_food.html.

171 Ibid.

172 Ibid.

173 U.N. Food and Agriculture Organization, "Harvesting Nature's Diversity," 1993, http://www.fao.org/docrep/004/V1430E/V1430E02.htm.

174 North Leupp Family Farm, "About Us," http://sites.google.com/site/leuppfarm/aboutnlff.

175 Ibid.

176 Woodruff Smith, "Volksgeist," in *New Dictionary of the History of Ideas* (2005), retrieved from Encyclopedia.com, http://www.encyclopedia.com/doc/1G2-3424300805.html.

177 Mvskoke Food Sovereignty Initiative, "Projects," http://www.mvskokefood.org/projects.html.

178 Cassandra Thompson, "Youth Running Like the Wind," *Mvskoke Food Sovereignty Initiative Newsletter*, June 2010, 4, http://www.mvskokefood.org/news/JuneNewsletter.pdf.

179 Mvskoke Food Sovereignty Initiative, "A Tribal Resolution of the Muscogee (Creek) Nation Establishing a Tribal Food and Fitness Policy Council," TR 10-079 (September 25, 2010), http://www.mvskokefood.org/flyers/Council.pdf.

180 Rhonda Beaver, in Mvskogee Food Sovereignty Initiative, "RWJ.m4v" (video about the Muscogee Nation Food and Fitness Policy Council," posted November 30, 2010, on YouTube, http://www.youtube.com/watch?v=uEESBOAxOAo&feature=player_embedded#at=14, and also posted on the MFSI Website's home page, http://www.mvskokefood.org/.

181 Save Wild Rice, "Don't Change the Wild Rice," http://savewildrice.org/.

182 Save Wild Rice, "Wild Rice: Maps, Genes and Patents," http://www.savewildrice.org/winona-article.

183 Ibid.

184 Ibid.

185 Winona LaDuke, "LaDuke: Minnesota's Manoomin Gets Legal Protection," *News From Indian Country*, n.d., http://indiancountrynews.net/index.php?option=com_content&task=view&id=853&Itemid=74.

186 Minnesota Department of Natural Resources, "Natural Wild Rice in Minnesota," February 15, 2008," http://files.dnr.state.mn.us/fish_wildlife/legislativereports/20080215_wildricestudy.pdf.

187 Native Harvest, "Mino-Miigim (Good Food) Program," http://nativeharvest.com/node/2.

188 Tim King, "Native Americans Take Back Land – White Earth, Minnesota," *The Progressive,* August 1994, as reprinted by BNET (CBS Business Network), http://findarticles.com/p/articles/mi_m1295/is_n8_v58/ai_15667738/.

189 NativeHarvest,"Sturgeon,"http://nativeharvest.com/node/5.

190 North Leupp Family Farm, "About Us," http://sites.google.com/site/leuppfarm/aboutnlff.

191 Michael Pollan, in a presentation to the Bioneers 2009 conference, October 16–18, 2009, San Rafael, California, posted on You-Tube, http://www.youtube.com/watch?v=ok-FkWr8LSo.

192 Union of Concerned Scientists, "Hidden Costs of Industrial Agriculture," http://www.ucsusa.org/food_and_agriculture/science_and_impacts/impacts_industrial_agriculture/costs-and-benefits-of.html.

193 Centers for Disease Control and Prevention. "Obesity: Halting the Epidemic by Making Health Easier: Report 2009," http://www.cdc.gov/nccdphp/publications/aag/dnpa.htm.

194 Michael Pollan, *The Omnivore's Dilemma* (New York: Penguin, 2007); see especially part 1, "Industrial Corn."

195 Bryan Walsh, "Getting Real About the High Price of Cheap Food," *Time*, August 21, 2009, http://www.time.com/time/health/article/0,8599,1917458,00.html. See also my own experiment with comparing the economy of junk calories versus healthy calories in Dave McNair, "Dutiful Foodie Conducts Shopping Experiment," *The Hook*, June 12, 2008, http://www.readthehook.com/dish/index.php/local-food-scene-dutiful-food-pyramidie-conducts-shopping-experiment/

196 Bryan Walsh, "Getting Real About the High Price of Cheap Food," *Time*, August 21, 2009. http://www.time.com/time/health/article/0,8599,1917458,00.html.

197 Food System Factoids, "Obesity Costs Greater than Iraq Costs, September 27, 2007, http://foodsystemfactoids.blogspot.com/2007/09/obesity-costs-greater-than-iraq-costs.html.

198 Robin Woods, "EPA Estimates Costs of Clean Water TMDL Program," U.S. Environmental Protection Agency, August 3, 2001.

199 Sir James Michael "Jimmy" Goldsmith, in an interview on *The Charlie Rose Show*, November 15, 1994, posted on YouTube, http://www.youtube.com/watch?v=yonUgZ2Y6Qs&feature=related (part 4 of the interview).

200 Darrell Frey, *Bioshelter Market Garden: A Permaculture Farm* (Gabriola Island, British Columbia: New Society Publishers, 2011), 28.

201 Michael Pollan, in a presentation to the Bioneers 2009 conference, October 16–18, 2009, San Rafael, California, posted on You-Tube, http://www.youtube.com/watch?v=ok-FkWr8LSo.

202 Michael Pollan, "Farmer in Chief," *New York Times Magazine,* October 12, 2008, http://www.nytimes.

com/2008/10/12/magazine/12policy-t.html?page wanted=1&_r=1.

203 These "green" seafood guidelines are adapted from the Northwest Atlantic Marine Alliance's website: "Green Seafood," http://namanet.org/factsheets/green-seafood.

204 Sunburst Trout Farm, "Our Story," http://www.sunburst trout.com/our-story.

205 Ibid.

206 Mackensy Lunsford, "Stream of Revenue," *Verve, Western North Carolina's Smartest Magazine for Women,* July–August 2009.

207 Ibid.

208 Ibid.

209 Worldwatch Institute, "Will Farmed Fish Feed the World?" http://www.worldwatch.org/node/5883.

210 Allan Nation, "Salatin-Designed Pastured Egg Production Lowers Labor Costs," *The Stockman Grass Farmer,* April 27, 2006, http://www.stockmangrassfarmer.net/cgi-bin/page.cgi?id=545.

211 These 2010 figures are extrapolated from 2006 figures citing $10,000 net if eggs are sold at $1.60 a dozen. See Allan Nation, "Salatin-Designed Pastured Egg Production Lowers Labor Costs," *The Stockman Grass Farmer,* April 27, 2006, http://www.stockmangrass farmer.net/cgi-bin/page.cgi?id=545.

212 Mark Neuzil, "Old-Time Innovator Runs Successful Poly-face Farm," MinnPost.com, http://www.minnpost.com/markneuzil/2008/10/22/3960/old-time_innovator_runs_successful_polyface_farm.

213 Allan Nation, "Virginia Grazier Joel Salatin Finds Pigs Can Profitably Create Pasture from Cut-Over Forest Lands," The Stockman Grass Farmer, June 2007, http://www.stockmangrassfarmer.net/cgi-bin/page.cgi?id=663.

214 Michael Pollan, *The Omnivore's Dilemma* (New York: Penguin, 2007), 224.

215 Polyface Farm, "Talk Topics," http://www.polyfacefarms.com/talk.aspx.

216 As stated on the Chipotle Mexican Grill website, www.chipotle.com.

217 Michael Pollan, "Behind the Organic-Industrial Complex," *New York Times Magazine*, May 13, 2001, http://www.nytimes.com/2001/05/13/magazine/13ORGANIC.html.

218 Michael Pollan, *The Omnivore's Dilemma* (New York: Penguin, 2007), 158.

219 Ibid., 184.

220 Kif Scheuer, "Wal-Mart's Organic Bomb," *Grist*, May 12, 2006, http://www.grist.org/article/wal-marts-organic-bomb.

221 Pallavi Gogoi, "Wal-Mart's Organic Offensive," *Bloomberg Businessweek*, March 29, 2006, http://www.businessweek.com/bwdaily/dnflash/mar2006/nf20060329_6971.htm.

222 William McKibben, *Deep Economy* (New York: Holt, 2007), 231. "The movement toward more local economies is the same direction we will have to travel to cope with the effects of these predicaments, not just to fend them off. The logic is fairly clear: in a world threatened by ever-higher energy prices and ever-scarcer fossil fuel, you're better off in a relatively self-sufficient county or state or region. In a world increasingly rocked by wild and threatening weather, durable economies will be more useful than dynamic ones."

223 For numerous charts showing the structure of the organic industry, see "Wal-Mart: Your First Choice for Organic Food???" Health Habits, April 30, 2008, http://www.healthhabits.ca/2008/04/30/walmart-your-first-choice-for-organic-food/.

224 Hickory Nut Gap Farm, "Farm History," http://www.hickorynutgapfarm.com/history.php.

225 Ibid.

226 John Todd, "The New Alchemists," http://www.ratical.org/co-globalize/DO_JohnTodd.html.

227 Rodale Institute, "About Us," http://www.rodaleinstitute.org/about_us.

228 A good description of life-cycle analysis and cost accounting is provided by the California State Parks Office of Historic Preservation ("Life Cycle Cost Accounting," http://www.parks.ca.gov/?page_id=25083):

Life Cycle Analysis or Assessment is the comprehensive examination of a product's environmental and economic aspects and potential impacts throughout its lifetime, including raw material extraction, transportation, manufacturing, use, and disposal. It is distinct from embodied energy, which is the energy consumed by all of the processes involved in extraction, manufacturing and delivery of the products used in the construction of a building, but does not include use or disposal.

Life Cycle cost accounting calculates the true cost of material components of products, their use and ultimate disposal or recycling throughout their lifetime. This allows for a comparison between similar products to determine which is the more efficient. The most efficient product is not always apparent at the start of an endeavor. Knowing the true cost of products can more accurately shape policies and designs.

229 I am not aware of anyone else using the term "whole-system natural efficiencies," though of course I cannot guarantee that it isn't in use elsewhere. (And if others *are* using this term, I'd be delighted to hear about it.) A potential alternative term is "value-based natural efficiencies."

230 The statistics in this box come from S. K. Duckett, J. P. S. Neel, J. P. Fontenot, and W. M. Clapham, "Effects of Winter Stocker Growth Rate and Finishing System on: III. Tissue Proximate, Fatty Acid, Vitamin and Cholesterol Content," *Journal of Animal Science* 87, no. 9 (2009): 2961–70, http://jas.fass.org/cgi/content/short/jas.2009-1850v1; and from Dr. Tilak Dhiman of Utah State University, as quoted in Allan Nation, "Grassfed Has a Far Better Story to Tell than Organics, CLA Research

Says," *Stockman Grass Farmer*, March 17, 2007, http://www.stockmangrassfarmer.net/cgi-bin/page.cgi?id=643.

231 Appalachian Sustainable Agriculture Project, "Farm to Hospital," http://www.asapconnections.org/farmto hospital.html.

232 U.S. Environmental Protection Agency, "Ag 101: Demographics," http://www.epa.gov/oecaagct/ag101/demographics.html.

233 "When we try to pick out anything by itself, we find it hitched to everything else in the Universe." – John Muir, *My First Summer in the Sierra* (1911).

234 Appalachian Sustainable Agriculture Project, "The Family Farm Tour," http://www.asapconnections.org/thefamilyfarmtour.html.

235 Iowa State University Extension, "What Producers Should Know About Selling to Local Foodservice Markets" (two-page brochure), March 2004, http://www.leopold.iastate.edu/pubs/other/files/PM2045.pdf.

236 PBS NOW, "Growing Local, Eating Local," a segment of "Enterprising Ideas: Social Entrepreneurs at Work" (2007), http://www.pbs.org/now/enterprisingideas/asd.html.

237 Ibid.

238 Dan Armstrong, "Why Beans and Grains," Southern Willamette Bean and Grain Project, http://www.mudcitypress.com/mudgrain.html.

239 Ibid.

240 Leah Koenig, "The Breadbasket of America: New England?" *The Atlantic*, March 22, 2010, http://www.theatlantic.com/food/archive/2010/03/the-breadbasket-of-america-new-england/37830/

241 North Carolina Organic Bread Flour Project, "Burgeoning Regional Wheat Initiative," May 25, 2009, http://ncobfp.blogspot.com/2009/05/burgeoning-regional-wheat-initiatives.html.

242 Central Virginia Grain Project, "Feed the People," http://grainproject.blogspot.com/.

243 Kootenay Grain Community Supported Agriculture, home page, http://www.kootenaygraincsa.ca/.

244 Wheatberry Farm and Baker, "The Pioneer Valley Heritage Grain CSA," http://www.localgrain.org/csa.html.

245 Huasna Valley Farm, "2010–2011 SLO County Grain CSA," www.huasnavalleyfarm.com.

246 Oklahoma Food Cooperative, "Producers and Products in the Grains and Flours Sections," http://www.oklahomafood.coop/category_list_products.php?category_id=3.

247 North Carolina Organic Bread Flour Project, "Burgeoning Regional Wheat Initiative," May 25, 2009, http://ncobfp.blogspot.com/2009/05/burgeoning-regional-wheat-initiatives.html.

248 Anita Crotty, "A Truly Local Loaf," The Daily Green, September 15, 2008, http://www.thedailygreen.com/healthy-eating/blogs/organic-sustainable-food/a-truly-local-loaf.

249 Morell's Bread, "About Us," http://web.me.com/eduardomorell/morellsbread/About.html (accessed January 1.

250 Cascade Baking Company, "About Us," http://www.cascadebaking.com/Cascade_about.htm.

251 Hungry Ghost Bread, "The Little Red Hen: Restoring Wheat In The Pioneer Valley," http://www.hungryghostbread.com/pages/the_little_red_hen_restoring_wheat_in_the_pioneer_valley.php.

252 *Diablo Canyon Rural Planning Area: Evolving Traditions in a New Economy* (2005), a collaboration of the Bar T Bar and Flying M Ranches, the Diablo Trust, and Coconino County; available from the Diablo Trust at info@diablotrust.org or as PDFs from the Diablo Trust website at www.diablotrust.org.

253 FoodProductionDaily.com, "Half of US Food Goes to Waste," November 25, 2004, http://www.foodproductiondaily.com/Supply-Chain/Half-of-US-food-goes-to-waste.

254 U.S. Environmental Protection Agency, "Basic Information about Food Waste," 2009, http://www.epa.gov/osw/conserve/materials/organics/food/index.htm.

255 J. Lundqvist, C. de Fraiture, and D. Molden, "Saving Water: From Field to Fork – Curbing Losses and Wastage in the Food Chain," policy brief for the Stockholm International Water Institute (SIWI), 2008.

256 Carol A. Breitinger, "How the Society of St. Andrew Got Started," Society of St. Andrew, 2009.

257 Society of St. Andrew, "History," http://www.endhunger.org/news_room/press/press_kit/History_Overview.pdf.

258 Society of St. Andrew, *2008 Annual Report: Envision A World Without Hunger*, 8.

259 Jane Black, "Notes from The Underground Market," *Washington Post*, June 8, 2010, http://voices.washingtonpost.com/all-we-can-eat/to-market-to-market/notes-from-the-underground-mar.html.

260 Helge Hellberg, as quoted in Rob Rogers, "Program Gives 'Gleaners' a New Perspective on Farming," *Marin Independent Journal*, July 5, 2009, http://www.marinij.com/westmarin/ci_12760030.

261 Ibid.

262 Faith Feeds, "How Do I Start Something Like This in My Community?" http://faithfeeds.wordpress.com/how/how-do-i-start-this/.

Acknowledgments

My thanks to those who facilitated and shaped the project, beginning with Will Allen, for doing what he does so well — inspiring others! With with one brief conversation he inspired me to propose this book, and I am forever indebted to him. Thanks also to Gwen Steege, for her enthusiasm and willingness to work collaboratively on the book concept, for being willing to take a chance on something outside the box, and most of all for her abiding faith in me; to Deborah Balmuth, for her confidence and trust in my judgment at critical moments, and for serving as a most gentle steward of the book; to Nancy Ringer, my editor, a real joy to work with, for tackling the brutal task of cutting more than one-third of the book with good humor and gusto, while protecting its heart and soul, to Jason Houston, for allowing his amazing photographer's vision to manifest the book's heart and soul; to Sally McMillan, for serving as my helpful and trusty agent for so many years, and for always being there whenever I needed her sound advice; and to Frank Dukes, director of the Institute for Environmental Negotiation at the University of Virginia, for supporting this project in every way conceivable.

My thanks to those who contributed to the book, beginning with Tim Beatley, friend and cohort in teaching food systems planning, for his belief in the importance of this project, for expanding the vision of the project, for his indefatigable desire to contribute, and for his genuine belief that I was up to the task — which kept me going through some rough trials. My thanks also go out to Debra Eschmeyer, outreach and communications director for the National Farm to School Network; Christine Gyovai, an associate at the Institute for Environmental Negotiation; Harvard undergraduate student Iris Ahronowitz, for permission to reprint a piece of her thesis concerning the Food Project; Mike Ellerbrock and Mike Foreman, for their willingness to contribute their time and expertise to the essay on the costs of the Green Revolution; UVA graduate student Jessie Ray, project manager

extraordinaire; UVA graduate student Lisa Hardy, dauntless transcriber of massive quantities of in-person interviews; and UVA graduate student Megan Bucknum, UVA graduate student Regine Kennedy, UVA graduate student Robin Proebsting, UVA undergraduate student Ben Chrisinger, and Columbia University undergraduate student Coogan Brennan, all terrific researchers and contributing writers.

My thanks to those who were generous with their time in helping me identify successful projects that could or should serve as models for others: Jeanette Abi-Nader, Martin Bailkey, Betsy Johnson, Jerry Kaufman, Wayne Roberts, Mark Winne, and the Virginia Food System Council.

The best behind-the-scenes cheerleaders were my family. My especial thanks to Cecil, my friend and life partner, who with never-ending good humor sacrificed two years of weekends — all in support of this project. Above all, I am grateful for his daily support, encouragement, and reminders to "focus." Thanks also to Sergei Kucherov, my attentive and supportive brother, who helped me navigate three of my first site visits, serving as "audio man" and carefully instructing me on how to back up absolutely everything. And thanks to his friend, Barbara, who served as our trusty driver and photographer, and who urged me to just keep at it, keep gathering data, and trust that I would find a way to connect all the threads.

My thanks to Jayne Riew and Jonathan Haidt, who made a special trip out to the country to give me in-person feedback on my first draft of the introduction, raising the bar and setting me on my course as only supportive friends can do.

Most importantly, my thanks to all the food projects that generously allowed us to pester them with visits, calls, and e-mails! If I have by any chance omitted someone, please forgive my oversight. It is not intentional, and the fault is entirely mine.

Resources

Farming

Agriculture Manuals
THE FOOD PROJECT
http://thefoodproject.org/manuals

The Boston-based Food Project (page 136) offers the *Rural Grower's Manual* (by Don Zasada, published in 2008) and the *Urban Grower's Manual* (by Amanda Cather, published in 2008) as free PDFs on its website. Both cover the mechanics of running a sustainable production farm, working with the community and young people.

American Grassfed Association
877-774-7277
www.americangrassfed.org

This advocacy group for grass-fed meat is working with the USDA to establish a legal definition for grass-fed and to establish a certification standard and process for farmers.

"Best Practices for Accepting EBT at the Farmers Market"
APPALACHIAN SUSTAINABLE AGRICULTURE PROJECT
www.asapconnections.org

A free downloadable publication from the ASAP website. This pamphlet offers advice for farmers' market on how to reach out to EBT populations and how to prepare the market to process EBT transactions.

Brownfields and Land Revitalization
U.S. ENVIRONMENTAL PROTECTION AGENCY
www.epa.gov/brownfields/urbanag

The "Urban Agriculture" page of the EPA's Brownfields Program site provides information for people interested in pursuing agriculture projects as a part of brownfield redevelopment and reuse.

"Farm Promotion and Support"
APPALACHIAN SUSTAINABLE AGRICULTURE PROJECT
www.asapconnections.org

A free downloadable publication from the ASAP website. Targeted toward county-level economic development and tourism development authorities, it's loaded with practical, helpful information.

"A Farmer's Guide to Hosting Farm Visits for Children"
UNIVERSITY OF CALIFORNIA, SUSTAINABLE AGRICULTURE RESEARCH AND EDUCATION PROGRAM
www.sarep.ucdavis.edu/Grants/Reports/Kraus/97-36FarmersGuide.htm

Written by Market Cooking for Kids, a project of the nonprofit organization, Center for Urban Education about Sustainable Agriculture (CUESA). Based in San Francisco, this guide provides farmers with advice on planning farm visits, a range of activities for different ages, and management and safety considerations.

Farmers' Market Manual
THE FOOD PROJECT
http://thefoodproject.org/manuals

Written by Kristin Brennan and published in 2003. This free PDF manual provides guidance on setting up the market, selecting produce, training workers and young people, marketing, and keeping business records.

Holistic Management International
505-842-5252
www.holisticmanagement.org

Founded in 1984 by Allan Savory (now with Savory Institute; see below), HMI is based in New Mexico and provides training, education, and consultation worldwide to organizations that wish to promote land stewardship and restore grasslands, rangelands, and savannas. Holistic management seeks to make economic, environmental, and social benefits the "triple bottom line" for restoring and managing healthy land.

National Institute of Food and Agriculture
FORMERLY THE COOPERATIVE STATE RESEARCH, EDUCATION, AND EXTENSION SERVICE
U.S. DEPARTMENT OF AGRICULTURE
www.crees.usda.gov

NIFA offers a host of tools and resources for farmers through its regional offices, including soil testing services.

Practical Farmers of Iowa

515-232-5661

www.practicalfarmers.org

This nonprofit, educational organization was founded in 1985 to research, develop, and promote profitable, ecologically sound, community-enhancing approaches to agriculture. PFI assists farmers with a wide range of production and marketing programs.

Savory Institute

www.savoryinstitute.com

Founded by Allan Savory and based in New Mexico, the Savory Institute provides diverse workshops, training, and services in Holistic Management.

Soil Foodweb, Inc.

www.soilfoodweb.com

An excellent resource for soil and compost management.

Urban Agriculture and Community Food Security in the United States

COMMUNITY FOOD SECURITY COALITION

503-954-2970

http://foodsecurity.org

Though published in 2003, this report on urban agriculture, prepared by the Community Food Security Coalition, is still current, as the issues identified are still alive and well, and the potential solutions still viable. It's available as a free downloadable PDF from the coalition's website.

Weston A. Price Foundation

www.westonaprice.org

Founded in 1999, the nonprofit Weston A. Price Foundation is "dedicated to restoring nutrient-dense foods to the human diet through education, research and activism." Local chapters, which can be found through the website, are likely customers for local foods grown using sustainable methods.

Training for New Farmers

Beginningfarmers.org

http://beginningfarmers.org

Designed for new farmers, this website, developed by Taylor Reid and Jim Bingen at Michigan State University, offers an amazing array of guidance on topics from financing and finding land to production and marketing.

Center for Agroecology & Sustainable Food Systems

831-459-3240

http://casfs.ucsc.edu

Based at the University of California at Santa Cruz, this program offers an apprenticeship in ecological horticulture, which provides training for organic gardening and small-scale farming.

Land Stewardship Project

617-722-6377

www.landstewardshipproject.org

This project runs the Farm Beginnings program, a training and support program for prospective farmers, offering mentorship from and networking with established farmers, educational seminars and resources, and technical assistance in establishing a business plan.

New England Small Farm Institute

413-323-4531

www.smallfarm.org

The institute's "Exploring the Small Farm Dream" course, offered in seven different states and online, is designed to help aspiring farmers learn what it takes to start and manage a commercial agricultural business.

Biodiversity and Food Heritage

American Livestock Breeds Conservancy

919-545-0022

http://albc-usa.org

This nonprofit organization works to protect and preserve endangered livestock breeds. Their website includes information about rare and endangered breeds along with numerous links to additional resources for those interested in raising them.

Center for Biodiversity and Conservation

212-769-5742

http://cbc.amnh.org

Created by the American Museum of Natural History, the center integrates scientific research, education, and outreach to encourage conservation.

Genetic Engineering Policy Alliance

www.gepolicyalliance.org

This network is a source of archival information about all things relating to genetically engineered crops, including health risks, environmental impacts, initiatives, and liability and legal protection. Though the network of organizations and individuals working in California may not be active or current, it still offers information that is useful everywhere.

Heritage Breeds Conservancy, Inc.

nehbcinfo@nehbc.org

www.nehbc.org

This organization works with farmers in New England to establish heritage breeds in production herds, as part of an effort to preserve America's livestock legacy.

Heritage Foods USA

718-389-0985

www.heritagefoodsusa.com

Heritage Foods USA, an arm of Slow Food USA, markets meat from antibiotic-free and pastured rare and heritage animal breeds and other heritage foods. By marketing these foods, it hopes to support their continued production by farmers.

Northeast Organic Wheat

growseed@yahoo.com

http://growseed.org/now.html

Northeast Organic Wheat is a consortium of teams in Vermont, New York, Massachusetts, and Maine that are restoring rare, heritage wheats and hosting field days at demonstration farms in each state to introduce farmers to heritage varieties and teach growers on-farm breeding techniques.

Seed Savers Exchange

563-382-5990

www.seedsavers.org

This nonprofit organization is dedicated to preserving and sharing heirloom seed varieties. Their site includes information about heirloom plant varieties and an extensive catalog of heirloom seeds.

Southern Exposure Seed Exchange

540-894-9480

www.southernexposure.com

This seed company seeks to preserve heritage seeds by cultivating, selling, and storing them. It maintains a seed bank at its facility in central Virginia, where it conducts germination testing.

Renewing America's Food Traditions (RAFT)

www.environment.nau.edu/raft

RAFT is a coalition of nonprofits dedicated to "rescuing America's diverse foods and food traditions," as the group's website notes. From conservationists to historians, agriculturists, and chefs, RAFT's supporters promote the cultivation and consumption of uniquely American foods that are at risk of extinction.

Farm to School

California School Garden Network

www.csgn.org

This network offers more than a hundred garden-based lessons to create, expand, and sustain garden-based learning experiences.

Chez Panisse Foundation

510-843-3811

www.chezpanissefoundation.org

Founded by Alice Waters. Aims to reform school lunches nationwide by supporting a model program in Berkeley, California, that provides, and educates children about, healthy eating opportunities.

Cornell Farm to School Program

CORNELL UNIVERSITY

http://farmtoschool.cce.cornell.edu

This program supports Farm to School programs throughout New York through education, research, and program development.

Farm to College

COMMUNITY FOOD SECURITY COALITION

717-240-1361

www.farmtocollege.org

A program offering a host of resources to help colleges navigate the logistics of setting up a program to bring local food to their campuses.

FoodCorps

www.food-corps.org

A new branch of the AmeriCorps program, FoodCorps offers opportunities for young adults to work for a year in food-oriented public service, to help bring fresh foods into America's schools through gardens and nutrition- and agriculture-based education.

French Fries and the Food System
THE FOOD PROJECT

http://thefoodproject.org/books-manuals

Written by Sara Coblyn for the Food Project in Boston (page 136), this book provides curricula for building a school gardening program. It's available for purchase from the Food Project or booksellers nationwide.

KidsGardening.org
NATIONAL GARDENING ASSOCIATION

800-538-7476

www.kidsgardening.org

This website, a project of the National Gardening Association, has a wealth of information on grants, activities, resources, and advice for school gardens. Learn about other school projects here. Also, you can find free curricula and individual project materials.

Life Lab Science Program
831-459-2001

www.lifelab.org

Based in Santa Cruz, and focused on assisting California schools, this program has developed a Cadillac website with information on all aspects of school gardening programs – field trips, summer camps, classroom curricula, professional development, and school gardens.

National Farm to School Network

www.farmtoschool.org

This advocacy group helps schools improve student access to and knowledge about fresh, healthy food by connecting them with local and regional farms and food-related curricula.

School Garden Wizard

schoolgardenwizard@chicagobotanic.org

www.schoolgardenwizard.org

Created for the K–12 school community through a partnership between the United States Botanic Garden and Chicago Botanic Garden, this site provides step-by-step information and resources for starting and sustaining a school garden.

Community Gardens

American Community Gardening Association

877-275-2242

www.communitygarden.org

The ACGA website contains a wealth of information and resources for community gardens, including sample forms for garden-plot registration, garden contract, garden rules, land-use agreement, and release of claims for volunteers; start-up guides; how-to guides for urban gardening; and a database of existing gardens that new projects may look to for mentorship.

Community Food Security Coalition

http://foodsecurity.org

This site offers a wealth of information on community gardening, including guidebooks and reports, funding opportunities, training and technical information, and conferences.

Municipal Research and Services Center of Washington

www.mrsc.org

The nonprofit MRSC of Seattle, Washington, offers a website with extensive links to information on all aspects of community gardening, including community food assessment tools, successful community gardening programs, zoning and policy, school gardens, and organizations.

P-Patch Leadership Handbook 2009
P-PATCH COMMUNITY GARDENING PROGRAM
DEPARTMENT OF NEIGHBORHOODS

www.seattle.gov/neighborhoods/ppatch/
gardeningresources.htm

A detailed guide that outlines the various roles and responsibilities for making a community garden program successful.

Working with Volunteers

AmeriCorps
800-833-3722

www.americorps.gov

Many food projects in this book have benefited from hosting AmeriCorps volunteers. The AmeriCorps program places people in jobs that address critical community needs in education, public safety, health, and the environment. The AmeriCorps Vista program places people in jobs that fight poverty. The AmeriCorps National Civilian Community Corps (NCCC) program places people in projects that address critical needs related to natural and other disasters, infrastructure improvement, environmental stewardship and conservation, energy conservation, and urban and rural development. The AmeriCorps FoodCorps program places people in food-oriented community service jobs, focusing on Farm to School programs.

Volunteer Program Manual
THE FOOD PROJECT
http://thefoodproject.org/manuals
An invaluable guide to recruiting and working with volunteers written by Greg Gale for the Food Project in Boston (page 136) and published in 2000. Available free as a downloadable PDF.

World Wide Opportunities on Organic Farms
www.wwoof.org
This organization connects organic farms that seek volunteer assistance with people who wish to learn more about organic farming. It is not unusual for people to "wwoof" for a year, traveling across several continents as they work at different organic farms.

Fish Farming and Seafood

Friendly Aquaponics
www.friendlyaquaponics.com
A tremendous resource for aquaponic systems of all sizes.

Marine Stewardship Council
www.msc.org
Works worldwide to set standards for sustainable fishing and seafood traceability, reward fisheries that practice good management, and assist retailers in increasing the sustainable seafood they offer.

Northwest Atlantic Marine Alliance
978-281-6934
http://namanet.org
An organization that works to transform the market for local, small-scale fishing communities.

Seafood Watch
MONTEREY BAY AQUARIUM
877-229-9990
www.montereybayaquarium.org/cr/seafoodwatch.aspx
Tracks ocean issues, suggests how to take action, provides recipes, and makes seafood recommendations by region.

Backyard Chickens

BackYardChickens.com
www.backyardchickens.com
This website calls itself the number one learning site for raising backyard chickens. It includes a learning center, a forum for discussions, a place to buy and sell equipment, and more.

Chicken Revolution
www.salemchickens.com
Barbara Palermo, a backyard poultry advocate, created this site as part of an effort to legalize chickens in Salem. It offers advice and information for backyard chicken enthusiasts of all kinds and in all places.

Urban Chickens
http://urbanchickens.org
This website offers probably the most comprehensive listing of zoning ordinances in each state, although many states still haven't provided information. Also, you'll find a wonderful map that allows you to locate urban chicken owners near you.

Gleaning and Food Recovery

A Citizen's Guide to Food Recovery
U.S. DEPARTMENT OF AGRICULTURE
www.usda.gov/news/pubs/gleaning/content.htm
Published in 1999 by the USDA, this web-based guide is very informative and useful for food recovery projects. One chapter offers ideas for how corporations, nonprofits, and individual citizens can help recover food. Another chapter offers lessons learned from an AmeriCorps summer of gleaning case study, including tips for managing volunteers. Two interesting ideas here: the recipients of the food may be the most satisfied and effective gleaning volunteers, and minimum-security and alternative prison inmates can be productive gleaners.

Gather It! How to Organize an Urban Fruit Harvest
SOLID GROUND
www.solid-ground.org/Programs/Nutrition/FruitTree/Documents
Written by Gail Savina and published in 2009 by Solid Ground, a nonprofit community development organization based in Seattle, this is the bible for foraging and gleaning. It's available as a free downloadable PDF from the group's website at www.solid-ground.org (click on "About Us" and then "Publications").

Neighborhood Fruit
http://neighborhoodfruit.com
This online mapping and swapping system has registered more than 10,000 public fruit trees around the country. Fruit tree owners on both public and private lands can register their trees, and anyone looking for fruit can log in and find out where those trees are.

Organizations Profiled in This Book

My thanks to the representatives from each of these groups and organizations who so graciously allowed us to pester them with visits, calls, and e-mails. If I have omitted anyone here, please forgive my oversight. It is not intentional, and the fault is my own.

Abernethy Elementary School
2421 SE ORANGE
PORTLAND, OR 97214
503-916-6190
www.pps.k12.or.us/schools/abernethy
Representatives: Sarah Sullivan, Wendy Willis

The Appalachian Center for Economic Networks
94 COLUMBUS ROAD
ATHENS, OH 45701
740-592-3854
www.acenetworks.org
Representative: Leslie Schaller

Appalachian Sustainable Agriculture Project
306 WEST HAYWOOD STREET
ASHEVILLE, NC 28801
828-236-1282
www.asapconnections.org
Representatives: Charlie Jackson, Emily Jackson, Peter Marks

Appalachian Sustainable Development
PO BOX 791
ABINGDON, VA 24212
276-623-1121
www.asdevelop.org
Representative: Anthony Flaccovento

Australis Aquaculture
1 AUSTRALIA WAY
TURNERS FALLS, MA 01376
888-602-2772
www.thebetterfish.com
Representative: Carol Devine, Josh Goldman

Beardsley Community Farm
1719 REYNOLDS STREET
KNOXVILLE, TN 37921
865-546-8446
http://beardsleyfarm.org
Representatives: Ben Epperson, John Harris

Burgh Bees
info@burghbees.com
www.burghbees.com

Chicago Honey Co-op
www.chicagohoneycoop.com

Community GroundWorks
3601 MEMORIAL DRIVE, SUITE 4
MADISON, WI 53704
608-240-0409
www.troygardens.org
Representatives: Marcia Caton Campbell, Nathan Larson, Christie Ralston, Greg Rosenberg, Claire Strader

Community Roots
30 SOUTH 31ST STREET
BOULDER, CO 80305
303-499-0866
www.communityrootsboulder.com
Representative: Kip Nash

Country Natural Beef
2277 BISHOP ROAD
VALE, OR 97918
541-473-3355
www.countrynaturalbeef.com
Representatives: Mary and Lowell Forman, Doc and Connie Hatfield, John Hyde and family

DanRose Farms
C/O DEVELOPING INNOVATIONS IN NAVAJO
EDUCATION INC. (DINÉ INC.)
9975 CHESTNUT ROAD
FLAGSTAFF, AZ 86004
928-606-4998
Representatives: Rose Marie and Dan Williams

D.C. Central Kitchen
425 2ND STREET NW
WASHINGTON, DC 20001
202-234-0707
www.dccentralkitchen.org
Representatives: Joelle Johnson, Brian MacNair,
Allison Sosna, Jasmine Touton

Diablo Burger
120 NORTH LEROUX STREET
FLAGSTAFF, AZ 86001
928-774-3274
www.diabloburger.com
Representative: Derrick Widmarck

Environmental Youth Alliance
517-119 WEST PENDER STREET
VANCOUVER, BC V6B 1S5
604-689-4446
www.eya.ca

Farm to Table D.C.
http://farm-table.com
Representative: Oren Molovinsky

The Food Project
10 LEWIS STREET
LINCOLN, MA 01773
781-259-8621
http://thefoodproject.org
Representative: Jen James

Forage Oakland
www.forageoakland.blogspot.com
Representative: Asiya Wadud

ForageSF
INFO@FORAGESF.COM
http://foragesf.com
Representatives: Kevin Feinstein, Iso Rabins

Gateway Greening
2211 WASHINGTON AVENUE
ST. LOUIS, MO 63103
314-588-9600
www.gatewaygreening.org
Representatives: Hannah Butz, Travis Hall, Gwenne Hayes-
Stewart, Mara Higdon, Lauren Maul, Annie Mayrose, Ariel
Roads-Buback

Gladheart Farms
29 LORA LANE
ASHEVILLE, NC 28803
828-280-7595
www.gladheartfarms.com
Representative: Michael Porterfield

The Goat Justice League
www.goatjusticeleague.org
Representative: Jennie Grant

Greensgrow Farm
2501 EAST CUMBERLAND STREET
PHILADELPHIA, PA 19125
www.greensgrow.org
Representatives: Cody Richter, Mary Seton-Corboy

Growing Gardens
2003 NE 42ND AVENUE, #3
PORTLAND, OR 97213
503-284-8420
www.growing-gardens.org
Representatives: Rodney Bender, Caitlin Blethen, Debra
Lippoldt, Nell Tessman

Growing Minds
C/O APPALACHIAN SUSTAINABLE AGRICULTURE
PROJECT
306 WEST HAYWOOD STREET
ASHEVILLE, NC 28801
828-236-1282
www.growing-minds.org
Representative: Emily Jackson

Growing Power, Inc.
5500 WEST SILVER SPRING DRIVE
MILWAUKEE, WI 53218
414-527-1546
www.growingpower.org
Representatives: Will Allen, Martin Bailkey, Jerry Kaufman,
Karen Parker

Hickory Nut Gap Farm

57 SUGAR HOLLOW ROAD

FAIRVIEW, NC 28730

828-628-1027

www.hickorynutgapfarm.com

Representatives: Jamie and Amy Ager

The Home Gardening Project Foundation

8060 UPPER APPLEGATE ROAD

JACKSONVILLE, OR 97530

541-899-1114

www.jeffnet.org/~hgpf

Representative: Dan Barker

Honeyestewa Farm

C/O DEVELOPING INNOVATIONS IN NAVAJO
EDUCATION INC. (DINÉ INC.)

9975 CHESTNUT ROAD

FLAGSTAFF, AZ 86004

928-606-4998

Representatives: Carl Honeyestewa and his parents, Marie
and Luther Honeyestewa

Indian Line Farm

57 JUG END ROAD

GREAT BARRINGTON, MA 01230

413-528-8301

www.indianlinefarm.com

Representative: Elizabeth Keen, Al Thorpe

Innisfree Village Farm

C/O INNISFREE VILLAGE

5505 WALNUT LEVEL ROAD

CROZET, VA 22932

434-823-5400

www.innisfreevillage.org

Representative: Peter Traverse

Intervale Center

180 INTERVALE ROAD

BURLINGTON, VT 05401

802-660-0440

www.intervale.org

Intervale Community Farm

128 INTERVALE ROAD

BURLINGTON, VT 05401

802-658-2919

http://intervalecommunityfarm.com

Representatives: Adam Housmann, Glenn McRae,
Will Raap

Janus Youth Programs

707 NE COUCH STREET

PORTLAND, OR 97232

503-233-6090

www.jyp.org

Representatives: Amber Baker, Michelle Hanna,
Rosalie Karp, Dennis Morrow, Agnes Sola, Etabo Wasongola,
Tera Wick

Jones Valley Urban Farm

701 25TH STREET NORTH

BIRMINGHAM, AL 35203

205-439-7213

www.jvuf.org

Representatives: Bryding Adams, Edwin Marty, Rachel
Reinhart

Kelly Elementary School

9030 SE COOPER

PORTLAND, OR 97266

503-916-6350

www.pps.k12.or.us/schools/kelly

Lynchburg Grows

1339 ENGLEWOOD STREET

LYNCHBURG, VA 24506

434-846-5665

www.lynchburggrows.org

Representatives: Meade Anderson, Dereck Cunningham, Scott
Lowman, John Matheson, Michael Van Ness

Marin Organic

PO BOX 962

POINT REYES STATION, CA 94956

415-663-9667

www.marinorganic.org

Representative: Scott Davidson

Mvskoke Food Sovereignty Initiative

100 EAST 7TH, SUITE 101
OKMULGEE, OK 74447
918-756-5915
www.mvskokefood.org
Representative: Vicky Karhu

Navajo Nation Traditional Agricultural Outreach

C/O DEVELOPING INNOVATIONS IN NAVAJO
EDUCATION INC. (DINÉ INC.)
9975 CHESTNUT ROAD
FLAGSTAFF, AZ 86004
928-606-4998
www.navajofarms.org
Representatives: Kyril Calsoyas, the Nez family, Jamescita
Peshlakai, Thomas Walker, Jr.

North Leupp Family Farm

PO BOX 5427
LEUPP, DINÉ NATION, AZ 86035
http://northleuppfarm.org
Representatives: Stacey Jensen, Tyrone Thompson, Dennis
Walker

Nuestras Raíces

329 MAIN STREET
HOLYOKE, MA 01040
413-535-1789
www.nuestras-raices.org
Representatives: Kevin Andaluz, Jesus Espinosa, Gerard
Ramos, Julia Rivera, Dan Ross, Tito Santana

People's Grocery

909 7TH STREET
OAKLAND, CA 94607
510-652-7607
www.peoplesgrocery.org
Representatives: Jumoke Hinton Hodge, Nikki Henderson,
Max Kurtz-Cadji

Polyface Farm

43 PURE MEADOWS LANE
SWOOPE, VA 24479
www.polyfacefarms.com
Representatives: Joel Salatin, Sheri Salatin, Phil Petrilli (of
Chipotle, Inc.)

P-Patch Community Gardening Program

DEPARTMENT OF NEIGHBORHOODS
700 5TH AVENUE, SUITE 1700
P.O. BOX 94649
SEATTLE, WA 98124
206-684-0264
www.seattle.gov/neighborhoods/ppatch
Representative: Rich MacDonald

Radical Roots Farm

3083 FLOOK LANE
KEEZLETOWN, VA 22832
540-269-2228
www.radicalrootsfarm.com
Representatives: Dave and Lee O'Neill

Seattle Urban Farm Company

11550 NORTH PARK AVENUE NORTH
SEATTLE, WA 98133
206-816-9740
www.seattleurbanfarmco.com
Representatives: Brad Halm, Colin McCrate

SEEDS, INC.

706 GILBERT STREET
DURHAM, NC 27701
919-683-1197
www.seedsnc.org
Representatives: Brenda Brodie, Lucy Harris, Kavanah
Ramsier

Seeds of Solidarity

165 CHESTNUT HILL ROAD
ORANGE, MA 01364
978-544-9023
www.seedsofsolidarity.org
Representative: Deborah Habib

Society of St. Andrew

3383 SWEET HOLLOW ROAD
BIG ISLAND, VA 24526
800-333-4597
www.endhunger.org
Representative: Steven Waldmann

Solid Ground Community Fruit Tree Harvest

1501 NORTH 45TH STREET
SEATTLE, WA 98103-6708
Representatives: Sadie Beauregard, Mike Buchman

South End/Lower Roxbury Open
Space Land Trust
PO BOX 180923
BOSTON, MA 02118
617-437-0999
www.landtrustgardens.org
Representative: Betsy Johnson

South Village
130 ALLEN ROAD EAST
SOUTH BURLINGTON, VT 05403
802-861-7600
www.southvillage.com
Representative: Will Raap

Southside Community Land Trust
109 SOMERSET STREET
PROVIDENCE, RI 02907
401-273-9419
www.southsideclt.org
Representative: Katherine Brown

Sunburst Trout Farm
128 RACEWAY PLACE
CANTON, NC 28716
800-673-3051
www.sunbursttrout.com
Representatives: Lila Eason, Sally Eason

Three Sisters Farm
134 ORBITZ ROAD
SAND LAKE, PA 16145
724-376-2797
www.bioshelter.com
Representative: Darrell Frey

The Urban Farm
www.urbanfarm.org
Representative: Greg Peterson

Walking Fish
PO BOX 2357
BEAUFORT, NC 28516
www.walking-fish.org
Representatives: Debbie Callaway, Lyn Chestnut,
Henry Coppola, Mark Hooper, Chuck Just, Bill Rice, Paul
Russell, Josh Stoll, Amy Tornquist

Warren Wilson College Farm
CPO 6255
PO BOX 9000
ASHEVILLE, NC 28815
828-771-3014
www.warren-wilson.edu/~farm
Representatives: Chase Hubbard, Karen Joslyn,
Ian Robertson

White Earth Land Recovery Project
607 MAIN AVENUE
CALLAWAY, MN 56521
888-274-8318
http://nativeharvest.com
Representative: Winona LaDuke

Yazzie Farm
C/O DEVELOPING INNOVATIONS IN NAVAJO
EDUCATION INC. (DINÉ INC.)
9975 CHESTNUT ROAD
FLAGSTAFF, AZ 86004
928-606-4998
Representative: Jonathan Yazzie

Contributors

Tanya Denckla Cobb is a writer, professional environmental mediator, and teacher of food-system planning at the University of Virginia. She is passionate about bringing people together to discover common ground and create solutions for mutual gain. She has worked at the grassroots, cofounding a community forestry nonprofit and mediating for community mediation centers. At the state level, she facilitated the birth of the Virginia Natural Resources Leadership Institute and the Virginia Food System Council, and served as Executive Director of the Virginia Urban Forest Council. At her home in Virginia, she enjoys the restorative energy of gardening and cooking from the garden.

Photographer Jason Houston explores the intersection between social and environmental issues in the United States and around the world. His work has been published widely and exhibited in galleries and public spaces across the country; he has served as a juror, curator, and reviewer for many different photography festivals and organizations; and he regularly presents on using the narrative tools of photography combined with the power of art to influence cultural change. He lives in western Massachusetts, where he also works as picture editor for *Orion* magazine.

Timothy Beatley is the Teresa Heinz Professor of Sustainable Communities at the School of Architecture at the University of Virginia. Much of Beatley's work focuses on the subject of sustainable communities and the creative strategies by which cities and towns can fundamentally reduce their ecological footprints, while at the same time becoming more livable and equitable places. He has written more than fifteen books, including *Biophilic Cities: Integrating Nature into Urban Design*. Beatley holds a PhD in city and regional planning from the University of North Carolina at Chapel Hill.

Coogan Brennan contributed while taking a hiatus from studies at Columbia University.

Megan Bucknum is the Farm to School program associate at Fair Food Philadelphia. An environmental and food systems planner, she worked for the Wallace Center at Winrock International, where she helped coordinate the Value Chain Research Collaboration and conducted research for a joint project with FamilyFarmed.org to analyze the feasibility of a local food packhouse in Virginia. Bucknum holds a master's in urban and environmental planning from the University of Virginia.

Michael J. Ellerbrock is a professor of agricultural and applied economics, a Cooperative Extension specialist, and director of the Center for Economic Education at Virginia Tech. His interests include environmental and macro economics, agricultural and economic ethics, natural resources conflict resolution, educational philosophy, science and religion, and K–12 social studies curricula. He has received numerous teaching awards, including Virginia Tech's William E. Wine award for exceptional teaching talent. Ellerbrock holds a PhD in applied economics from Clemson University.

Debra Eshmeyer is the communications and outreach director of the National Farm to School Network and the program director of FoodCorps, as well as a Food and Society Fellow. A leading voice for school food reform, she presented at the White House Childhood Obesity Taskforce Forum. She manages the Centers for Disease Control's national media initiative on school gardens, farmers' markets, and healthy corner stores and served as an editor for *Food Justice*, as a consultant for Jamie Oliver's *Food Revolution*, and as a contributor to the documentary *Lunch Line*.

Christine Muehlman Gyovai is principal of Dialogue and Design Associates, a plant lover, and an avid permaculture designer. She is an environmental planner and permaculture educator with over twelve years of experience in facilitation and training with a focus on building sustainable and resilient communities. Gyovai holds an M.P. in urban and environmental planning from the University of Virginia and is certified in mediation and permaculture design.

Regine Kennedy is interpretive planner and designer at 106 Group in St. Paul, Minnesota. Kennedy served in the Peace Corps in the Ukraine, where she worked with organizations promoting civil society initiatives and with elementary schools teaching English. She happily resides in St. Paul, where she enjoys hiking and biking and is a sustainability and local food system advocate. Kennedy holds a master's in urban and environmental planning from the University of Virginia.

Researchers **Ben Chrisinger**, **Jessica Ray**, and **Robin Proebsting** contributed while completing their studies. All now hold a master's in urban and environmental planning from the University of Virginia.

Index

Page references in bold indicate main entries; references in italic indicate photos.